Contemporary Calculus V

For the

students

A free, color PDF version is available online at
http://scidiv.bellevuecollege.edu/dh/Calculus_all/Calculus_all.html

Dale Hoffman
Bellevue College
Author web page: http://scidiv.bellevuecollege.edu/dh/

Version: May 2016

CONTEMPORARY CALCULUS V: Contents

Note:
Each section contains Practice Problems throughout the section. The solutions to these Practice Problems are at the end of that section, after the Problem Set for the section.

Each section also contains a Problem Set. The solutions to the odd problems of each Problem Set are at the end of each chapter.

These materials are also available free and in color on the web at:
http://scidiv.bellevuecollege.edu/dh/Calculus_all/Calculus_all.html

Chapter 14: Multiple Integrals

Chapter 15: Vector Calculus

14.0 INTRODUCTION TO DOUBLE INTEGRALS AND THEIR APPLICATIONS

Chapters 4 and 5 introduced integrals of functions of a single variable, y = f(x), as well as several of their applications. Chapter 14 introduces integrals for functions of more than one variable, shows how to calculate their values, and starts to examine some of their applications. Later chapters will use these double integrals extensively.

The applications will include calculating

* Volumes in 3D
* Total masses of solids whose densities are not constant
* Moments of solids around each axis
* Centers of mass of some solids
* Moments of inertia
* Surface areas

Moving from 2D Areas to 3D Volumes

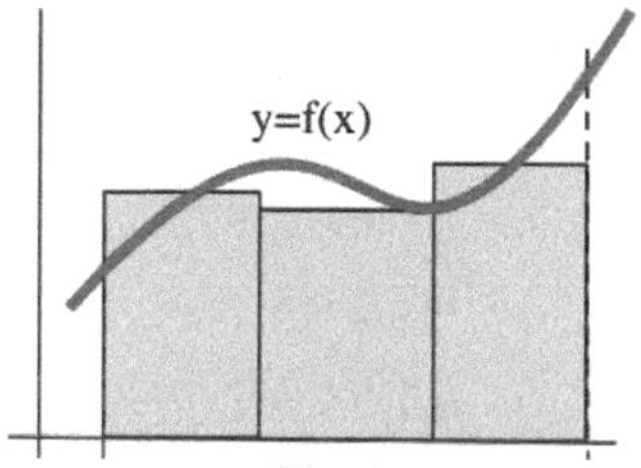

Fig. 1

Many of the ideas about integrals of a single variable are also important for these new integrals but things do get more complicated. In Chapter 4 we began by approximating the area under a curve by partitioning the x-axis domain, calculating the area of the rectangles over each segment of the partition, and then adding those areas together to get an approximation of the total area (Fig. 1). We begin by doing something similar to approximate the volume under a surface by portioning the 2D domain in the xy-plane into rectangles, calculating the volume of boxes over each rectangular piece, and then adding those volumes together to approximate the total volume (Fig. 2).

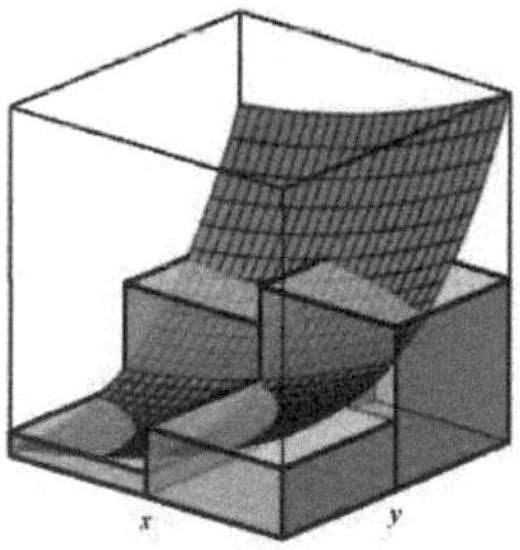

Fig. 2

The approximation of the 2D area became simply a matter of multiplying the length of the partition segment length by the height of the function and then adding those values together. The approximation of the 3D volume will be similar: multiply the area of the partition base by the height of the function, and then adding those values together.

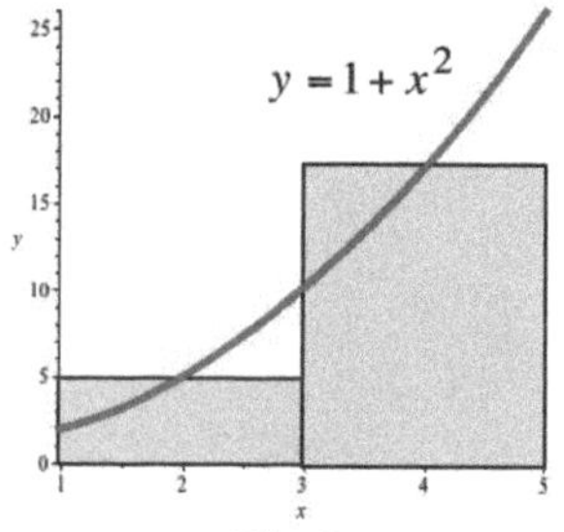

Fig. 3

Example 1: Approximate the area between the function $f(x) = 1 + x^2$ using the partition {1, 3, 5} and evaluate the function at the midpoint of each subinterval (Fig. 3).

Solution: The partition consists of the intervals [1, 3] and [3, 5] with midpoints x=2 and x=4 so the

approximation is $\sum_{i=1}^{2} f(x_i)\cdot \Delta x_i = f(2)\cdot 2 + f(4)\cdot 2 = 10 + 34 = 44.$

(Note: We know from our earlier work that the exact value of

this area is $\int_1^5 1 + x^2 \; dx = x + \frac{1}{3}x^3 \Big|_1^5 = 45\frac{1}{3}$)

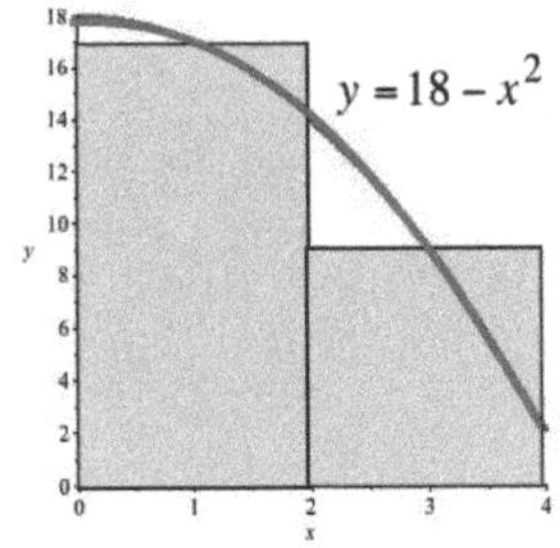

Fig. 4

Practice 1: Approximate the area between the function $g(x) = 18 - x^2$ using the partition {0, 2, 4} and evaluate the function at the midpoint of each subinterval (Fig. 4).

We can do something similar to approximate a volume.

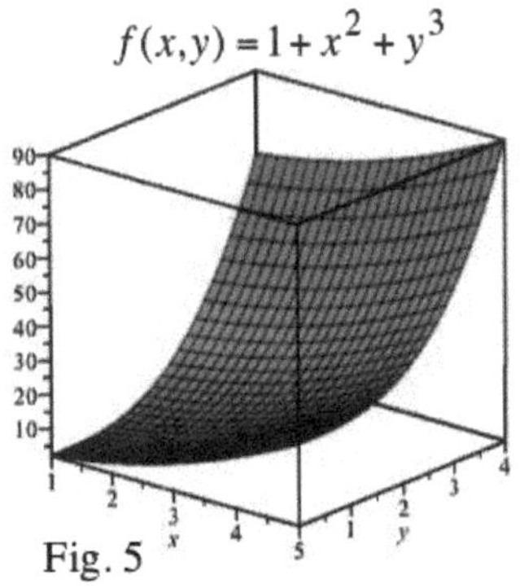

Fig. 5

Example 2: Approximate the volume of between the surface $f(x,y) = 1 + x^2 + y^3$ (Fig. 5) over the rectangle $1 \le x \le 5$ and $0 \le y \le 4$ by partitioning the x interval into subintervals [1,3] and [3,5] and the y interval into subintervals [0,2] and [2,4] and then evaluate the function at the midpoint of each sub-rectangle. The dots in Fig. 6 are at the midpoints of each sub-rectangle and the numbers by each dot is the value of the function at that midpoint.

Solution: $\sum_{i=1}^{2} \sum_{j=1}^{2} f(x_i, y_j)\cdot \Delta x_i \cdot \Delta y_j = f(2,1)\cdot 2\cdot 2 + f(4,1)\cdot 2\cdot 2 + f(2,3)\cdot 2\cdot 2 + f(4,3)\cdot 2\cdot 2$

$= 6\cdot 4 + 18\cdot 4 + 32\cdot 4 + 44\cdot 4 = 400$

(In the next section we will be able to calculate that the exact volume is $437\frac{1}{3}$ so our approximation was very crude.)

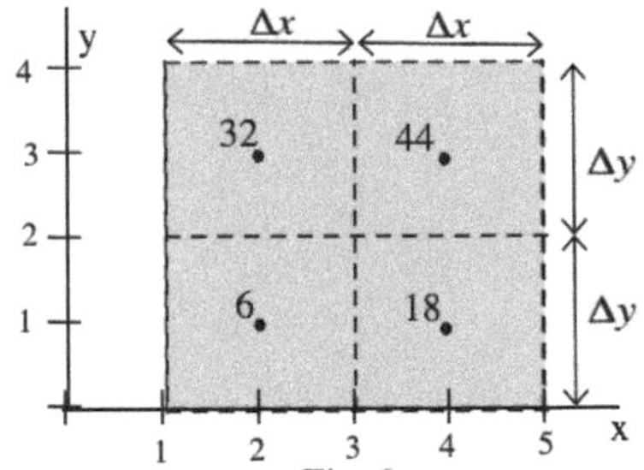

Fig. 6

Practice 2: Fig. 7 shows the depths (meters) at various locations in a backyard swimming pool. Approximate the total volume of the pool.

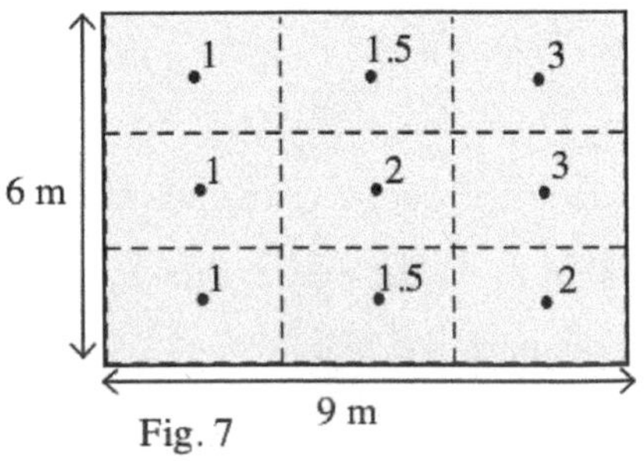

Fig. 7

If we have partitioned the domain into rectangles with dimensions Δx and Δy , then the area of each sub-rectangle is $\Delta A = \Delta x \cdot \Delta y$ and the volume of each box is $\Delta V = f(x_i, y_i) \cdot \Delta A$. But sometimes the natural partition is not rectangles.

Example 3: A gardener wants to estimate how much water a sprinkler puts out in an hour. She places several small cans at various distances from the sprinkler and measures the depth of the water in each can after one hour. Her data is given in Fig. 8. Estimate the hourly output of the sprinkler.

distance of can from sprinkler	depth of water in can
1 ft	0.3 ft
3 ft	0.4 ft
5 ft	0.3 ft
7 ft	0.2 ft

Fig. 8

Solution: In this case it is more useful to partition the circular pattern into concentric rings, and then calculate the volume of water for each ring. The area of the innermost ring (a circle) is $\Delta A_1 = \pi \cdot 2^2$, and the others are

$\Delta A_2 = \pi \cdot (4^2 - 2^2)$, $\Delta A_3 = \pi \cdot (6^2 - 4^2)$, and $\Delta A_4 = \pi \cdot (8^2 - 6^2)$.

The depths D_i are given in the table, and the total volume is

$\sum D_i \cdot \Delta A_i = 4\pi(0.3) + 12\pi(0.4) + 20\pi(0.3) + 28\pi(0.2) \approx 55.3 \ ft^3$.

Since each cubic foot of water is 7.5 US gallons, the sprinkler is putting out approximately 415 US gallons per hour.

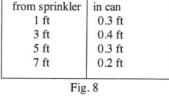

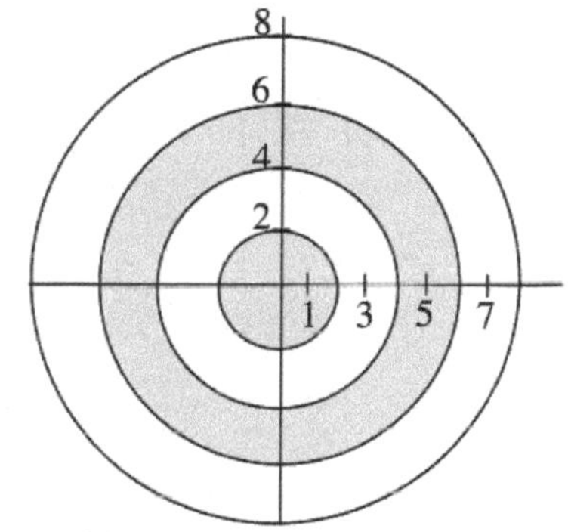

Fig. 9: Sprinkler pattern

(Note: Obviously this is a very crude estimate. She would get a much better estimate if there were more cans and the distances between the cans was smaller.)

Using Averages to Estimate Volumes

Sometimes in practice we just need an approximate value for a volume. We may need an estimate of the volume of a small pond in order to know how much of a chemical to put in the pond in order to stop an algae bloom. Or we may need an estimate of the volume of a hole on a construction site in order to know how much gravel we need to order to fill the hole. In these cases a good estimate of the volume is all that we need, and an estimate of the average depth of the pond or hole can enable us to estimate the volume.

In Section 4.7 we used integrals to calculate the average value of a function f≥0 on an interval domain as

$$\{\text{ average value of f(x) on interval I}\} = \frac{1}{\text{length of I}} \cdot \{\text{area between f and I}\} = \frac{1}{\text{length of I}} \cdot \int_I f(x)\, dx \,.$$

A similar approach can let us estimate volume based on the average value of a function f≥0 on a domain R:

$$\{\text{average value of f(x,y) on region R}\} = \frac{1}{\text{area of R}} \cdot \{\text{volume between f and R}\}$$

so $\{\text{volume between f and R}\} = \{\text{area of R}\} \cdot \{\text{average value of f(x,y) on region R}\}$.

If we have approximate values for the area of R and for the mean value of f on R, then we simply need to multiply those values to get an approximation of the volume. It seems too easy, but it is useful.

Approximating a Volume

$$\left\{\begin{array}{c}\text{approximate volume} \\ \text{of f over R}\end{array}\right\} = \left\{\begin{array}{c}\text{approximate} \\ \text{area of R}\end{array}\right\} \cdot \left\{\begin{array}{c}\text{approximate average value} \\ \text{of f(x,y) on region R}\end{array}\right\}$$

Example 4: The numbers by each dot in Fig. 6 (from Example 2) is the height of the function at that location. Use that information to approximate the volume between f and the xy-plane.

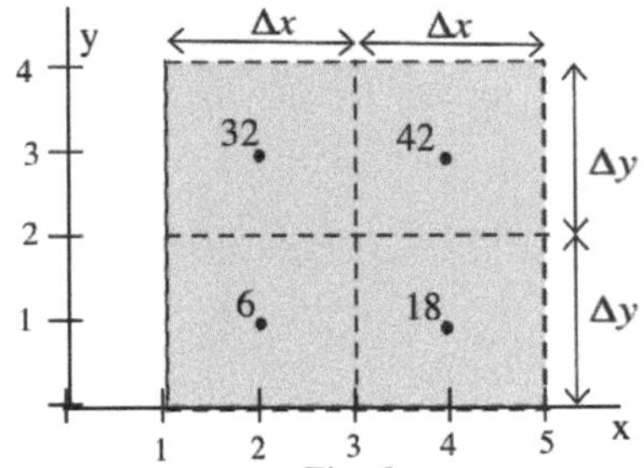

Fig. 6

Solution: The average of the four height values is (32+44+18+6)/4=25 and the area of the domain is (4)(4)=16 so our volume estimate is (25)(16)=300.

Practice 3: Use the information in Practice 2 and the average depth to estimate the volume of the pool.

Example 5: The numbers by each dot in Fig. 10 is the depth (feet) of the pond at each location. The area of the pond is approximately 12,000 square feet. Estimate the volume of the pond.

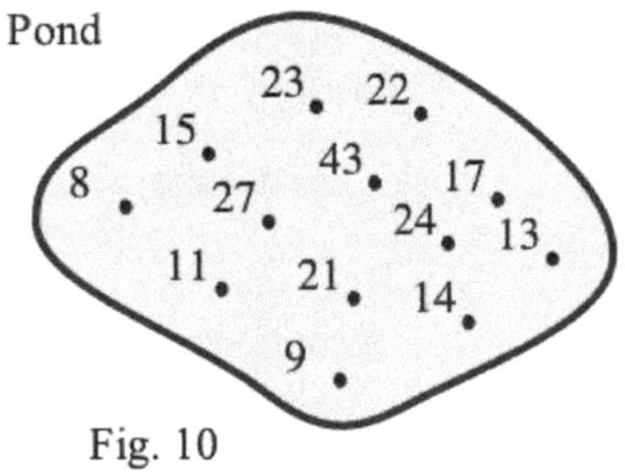

Fig. 10

Solution: The average depth is (247)/13=19 feet and the area is approximately 12,000 ft^2 so the approximate volume is (19 ft)(12,000 ft^2) = 228,000 ft^3.

In the following sections we will develop the calculus to determine volumes exactly (just as we did earlier to use integrals to calculus areas exactly), but the main ideas state as finite accumulations and lead to double integrals.

Problems

1. Fig. 11 shows the domain R of the function $z = f(x,y) = 1 + 3x + y^2$ and a partition of R into rectangles. Use the value of f at the **lower left** (x,y) point in each rectangle to approximate the volume between the graph z=f(x,y) and the xy-plane over R.

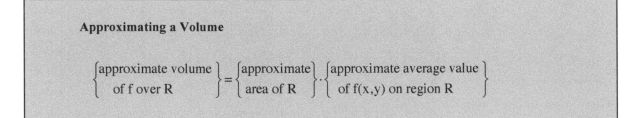

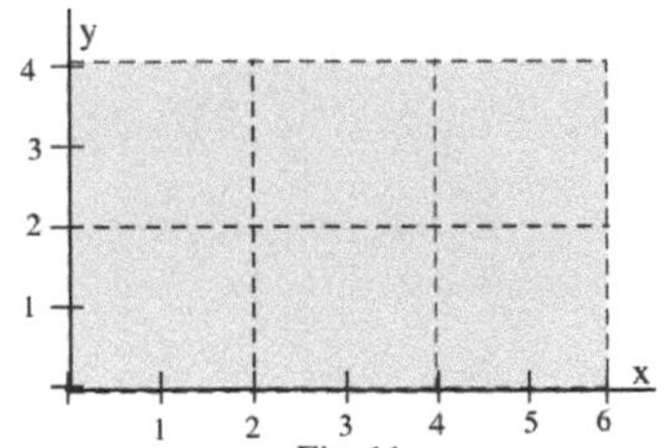

Fig. 11

2. (a) Use the same function and rectangles as in Problem 1, but evaluate the function at the **upper right** (x,y) point in each rectangle to approximate the volume between the graph z=f(x,y) and the xy-plane over R. (b) Evaluate the function at the **midpoint** (x,y) in each rectangle to approximate the volume.

3. Fig. 12 shows the domain R of the function $f(x,y) = x^2 + y^2$ and a partition of R into rectangles.
(a) Use the value of f at the **lower left** (x,y) point in each rectangle to approximate the volume between the graph z=f(x,y) and the xy-plane over R.
(b) Use the value of f at the **midpoint** (x,y) point in each rectangle to approximate the volume between the graph z=f(x,y) and the xy-plane over R.

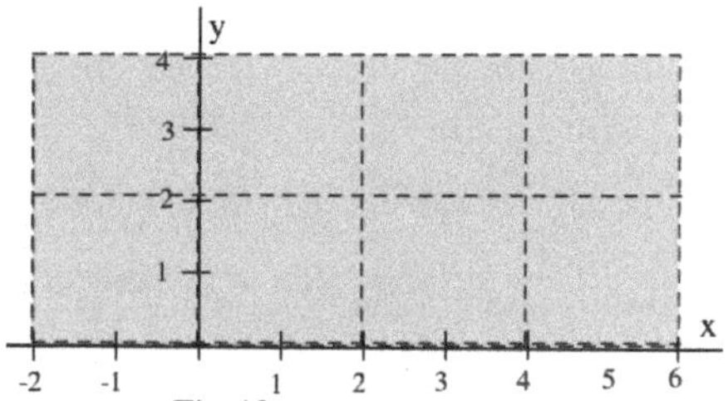

Fig. 12

4. Use the same function and rectangles as in Problem 3, but evaluate the function at the **upper right** (x,y) point in each rectangle to approximate the volume between the graph z=f(x,y) and the xy-plane over R.

5. A gardener places several small cans at a variety of locations (Fig. 13) and measures the depth of the water in each can after one hour. Estimate the hourly water output of the sprinkler.

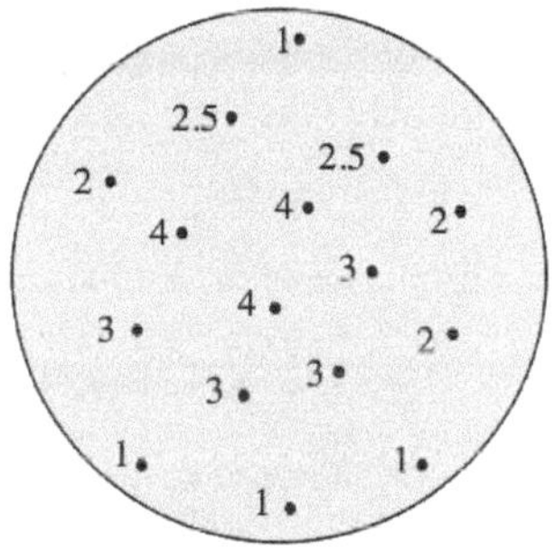

Fig. 13: water depths (inches)

6. A back-and-forth sprinkler has a rectangular distribution pattern. Fig. 14 shows the water depth (inches) at several locations after 15 minutes. Assume that the distribution is symmetric and estimate the hourly water output of the sprinkler.

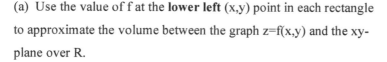

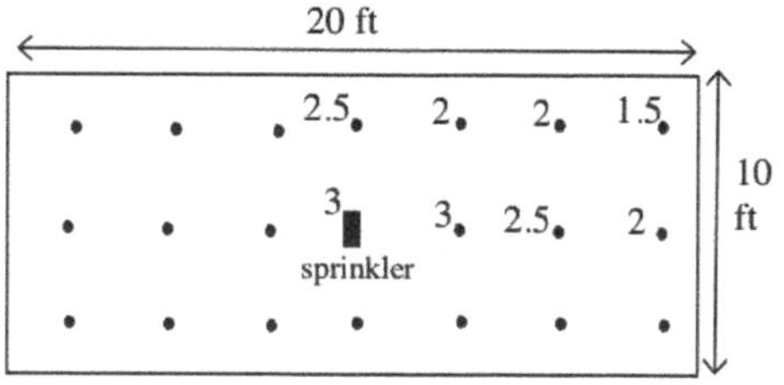

Fig. 14

7. A hole in the ground has a 4 ft. by 5 ft. rectangular opening, and the depths at the four corners are 3 ft., 4 ft. 6 ft. and 3 ft. Estimate the volume of the hole.

8. A hole in the ground has a 3 ft. by 5 ft. rectangular opening, and the depths at the four corners are 6 ft., 4 ft. 7 ft. and 6 ft. Estimate the volume of the hole.

9. A circular hole has a radius of 2 feet, and several depth measurements around the edge of the hole are 3 ft., 4 ft., 5 ft., 3 ft., and 4 ft. Estimate the volume of the hole.

10. A circular hole has a radius of 2 feet, and several depth measurements around the edge of the hole are 2 ft., 4 ft., 3 ft., 2 ft., and 3 ft. Estimate the volume of the hole.

11. Fig. 15 shows the depths (meters) at several locations in a small pond that has a surface area of 90 m^2. Approximate the volume of the pond.

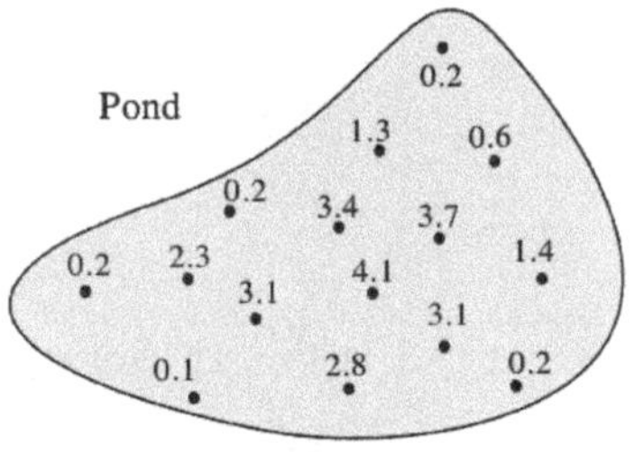

Fig. 15

12. A spacecraft has landed on a strange flat object and has measured its density ($10^3 kg/m^3$) at several locations (Fig. 16). From photographs during the approach to the object scientists have estimated the thickness of the object to be about 1.4 m thick the area to be about 3500 m^2 . Approximate the mass of the object.

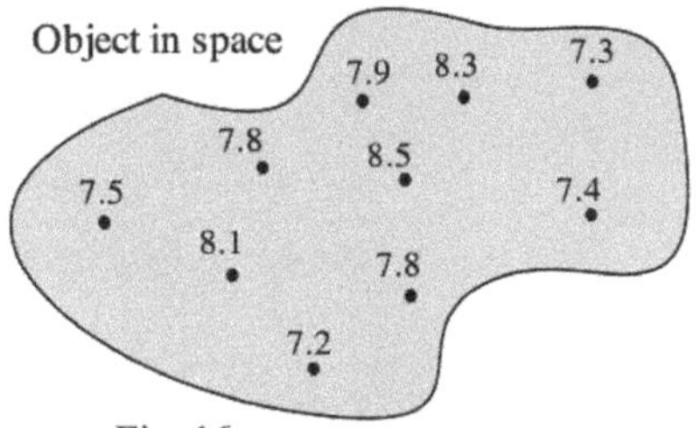

Fig. 16

13. Fig. 17 shows level elevation curves (m) for a small hill that needs to be removed for a construction project. Pick the elevations at several locations and use them to get a reasonable estimate of the volume of material to be removed. This also requires as estimate of the area of the bottom of the hill which is roughly elliptical. (ellipse area = $\frac{1}{4}\pi$(width)(length))

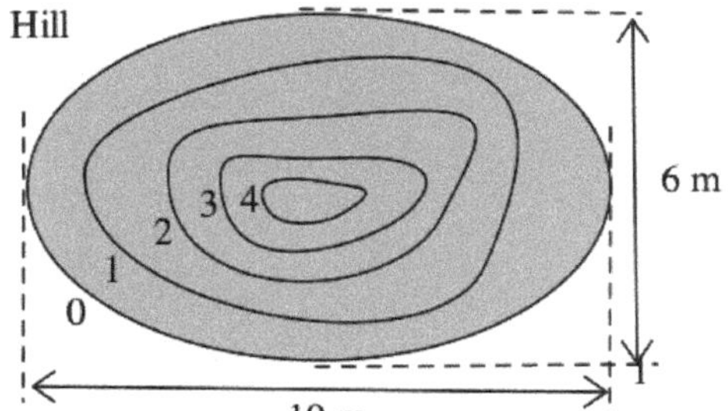

Fig. 17

14. In Example 3 (Fig. 9) we estimated the volume of the water from the sprinkler I one hour to be 55.3 ft^3 . But if we use the (average value)(area) we get {average value} = (0.3+0.4+0.3+0.2)/4 =0.3 ft and the area is 64 ft^2 so our estimate for the volume is (0.3 ft)(64 ft^2)$\approx$60.3 ft^3 .
Which estimate do you think is better and why?

15. Fig. 18 shows the level curves for z = f(x,y) on a region R. Use the midpoint sample points to estimate the value of $\iint_R f(x,y)\ dA$.

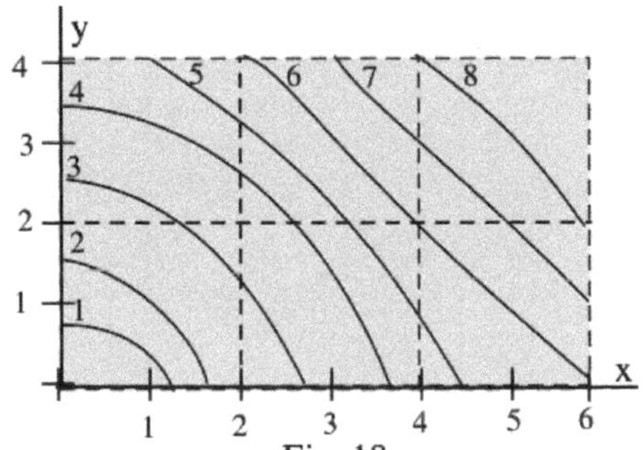

Fig. 18

16. Fig. 19 shows the level curves for z = f(x,y) on a region R. Use the midpoint sample points to estimate the value of $\iint_R f(x,y)\ dA$.

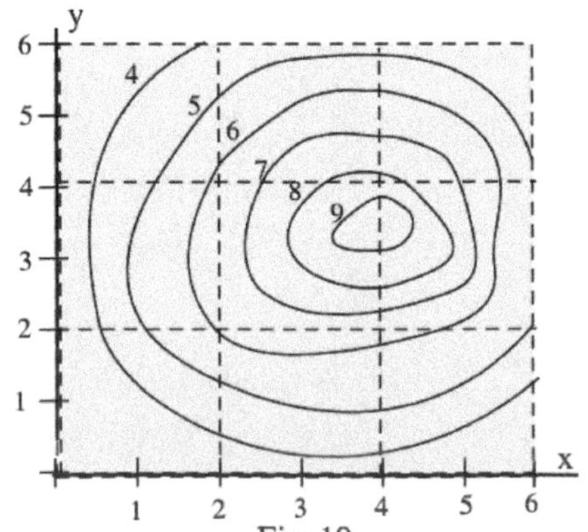

Fig. 19

Practice Answers

Practice 1: The intervals [0,2] and [2,4] have lengths 2 and 2 and midpoints x=1 and x=3. Then

$$\sum_{i=1}^{2} f(x_i)\cdot \Delta x_i = f(1)\cdot 2 + f(3)\cdot 2 = 34+18 = 52.$$ (The exact area is $50\frac{2}{3}$.)

Practice 2: $\sum_{i=1}^{3}\sum_{j=1}^{3} f(x_i,y_j)\cdot \Delta x_i \cdot \Delta y_j = \sum_{i=1}^{3}\sum_{j=1}^{3} f(x_i,y_j)\cdot 2\cdot 3 = (1+1+1+1.5+2+1.5+3+3+2)\cdot 6 = 96\ m^3.$

Practice 3: The average depth is $(1+1+1+1.5+2+1.5+3+3+2)/9 = \frac{16}{9}$ m and the area of the pool is

(6)(9)= 54 m^2 so the approximate volume is $\left(\frac{16}{9}\ m\right)\left(54\ m^2\right) = 96\ m^3$.

14.1 DOUBLE INTEGRALS OVER RECTANGLES

Volume of a Solid Region in 3D

To calculate the area between the curve y = f(x) and an interval on the x-axis (Fig. 1) we partitioned the interval [a,b] (Fig. 2), created an approximation of the area using a Riemann sum, and then took the limit of the Riemann sum as the widths of the subintervals approached 0 to get the area as a definite integral:

$$\text{Area} = \int_a^b f(x)\ dx$$

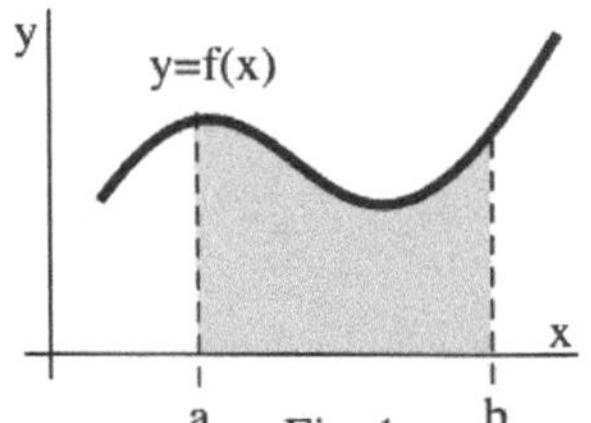

Fig. 1

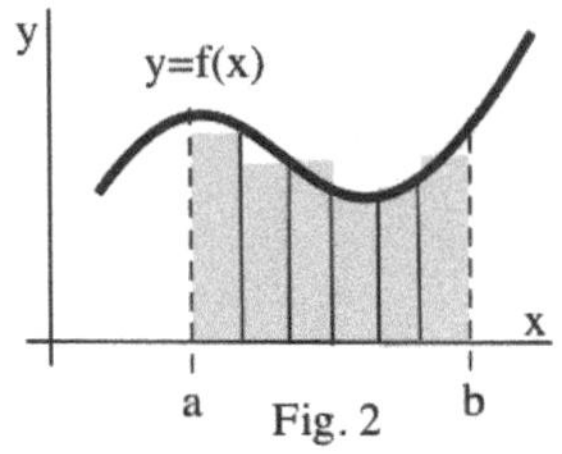

Fig. 2

In order to calculate the volume (Fig. 3) between a surface z = f(x,y) ≥ 0 and a rectangular $R = \{(x,y):\ a \le x \le b\ \ and\ \ c \le y \le d\}$ of the xy-plane, we will do something similar, but now the domain of integration is a region in the xy-plane, and our result will be a double integral:

$$\text{Volume} = \iint_R f(x,y)\ dA = \int_a^b \int_c^d f(x,y)\ dy\ dx\ ,$$

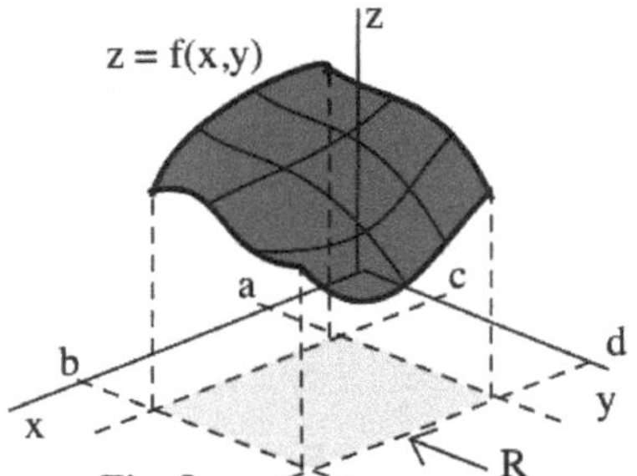

Fig. 3

For the rectangular region R we begin by partitioning along both the x-axis and along the y-axis (Fig. 4) to create small rectangles with areas $\Delta A_{ij} = \Delta x_i \cdot \Delta y_j$ and we select any point $(x_i^*,\ y_j^*)$ in this rectangle. Then the volume of the box above this ij-rectangle with height $f(x_i^*,\ y_j^*)$ is $\text{Volume}_{ij} = f(x_i^*,\ y_j^*)\cdot \Delta x_i \cdot \Delta y_j = f(x_i^*,\ y_j^*)\cdot \Delta A_{ij}$.

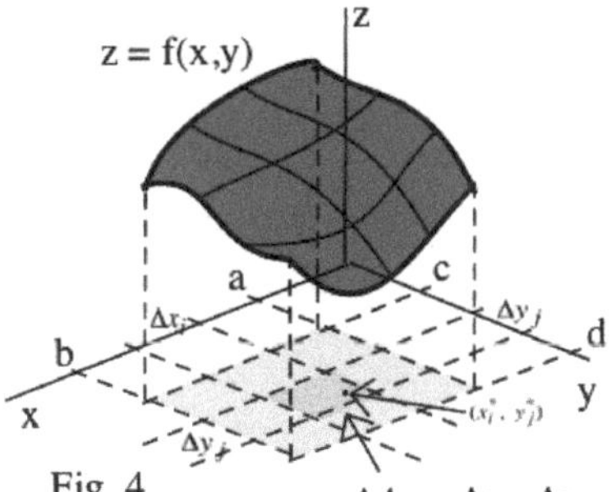

Fig. 4

$\Delta A_{ij} = \Delta x_i \cdot \Delta y_j$

Then the total approximate volume the sum of all of these little volumes:

$$\text{Approximate total volume} = \sum_j \sum_i f(x_i^*,\ y_j^*)\cdot \Delta x_i \cdot \Delta y_j\ .$$

Taking the limit as all of the Δx_i and Δy_j approach 0,

$$\text{Exact total volume} = \int_a^b \int_c^d f(x,y)\ dy\ dx = \iint_R f(x,y)\ dA\ .$$

Calculating the value of a double integral is no more difficult (nor any easier) than calculating the value of a single integral except we need to integrate twice.

Example 1: For $f(x,y) = 20 - x^2 y$ and $R = \{(x,y):\ 0 \le x \le 3,\ 1 \le y \le 2\}$ evaluate $\iint_R f(x,y)\ dA$.

Solution: In this example (Fig. 5) $\iint_R f(x,y)\ dA = \int_0^3 \int_1^2 20 - x^2 y\ dy\ dx$.

First we evaluate the inside integral $\int_1^2 20 - x^2 y\ dy$ **treating x as a constant**:

(this is just the inverse of partial differentiation)

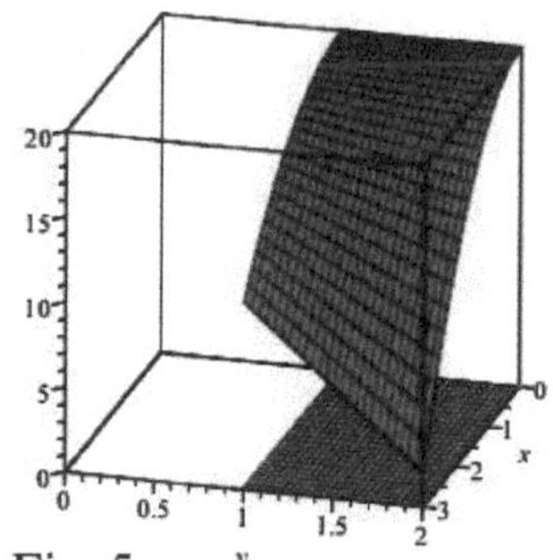

Fig. 5

$$\int_1^2 20 - x^2 y\ dy = 20y - \frac{1}{2} x^2 y^2 \Big|_{y=1}^{2}$$

$$= \frac{1}{2} x^2 (2)^2 - \frac{1}{2} x^2 (1)^2 = 20 - \frac{3}{2} x^2 .$$

Then $\int_0^3 \left\{ \int_1^2 20 - x^2 y\ dy \right\} dx$

$$= \int_0^3 \left\{ 20 - \frac{3}{2} x^2 \right\} dx = 20x - \frac{1}{2} x^3 \Big|_{x=0}^{3} = \frac{93}{2}.$$

Practice 1: For the same function and region R, evaluate $\int_1^2 \int_0^3 20 - x^2 y\ dx\ dy$. (This is the same solid as Example 1, but now we start by evaluating the inside integral $\int_0^3 x^2 y\ dx$, treating y as a constan.)

> Theorem:
>
> If $f(x,y) \geq 0$ and f is integrable over the rectangle R,
>
> then the volume V of the solid that lies above R and under the surface $z = f(x,y)$ is
>
> $$V = \iint_R f(x,y)\ dA .$$

Note: If $f(x,y) \geq 0$ on R then the double integral gives volume. If f is sometimes negative, then the double integral gives a "signed volume" in a manner similar to how a single integral gives "signed area."

These properties of double integrals follow from the properties of summations:

(1) $\iint_R f(x,y) + g(x,y)\ dA = \iint_R f(x,y)\ dA + \iint_R g(x,y)\ dA$

(2) $\iint_R K\, f(x,y)\, dA = K \iint_R f(x,y)\, dA$.

(3) If $f(x,y) \geq g(x,y)$ for all (x,y) in R, then $\iint_R f(x,y)\, dA \geq \iint_R g(x,y)\, dA$.

A few important points:

* Always work from the inside out: first evaluate the inside integral.
* For $\int f(x,y)\, dx$ integrate with respect to x and **treat y as a constant**.
* For $\int f(x,y)\, dy$ integrate with respect to y and **treat x as a constant**.

It was not an accident that the answers to Example 1 and Practice 1 were the same since both versions represented the volume of the same solid.

> Fubini's Theorem:
>
> If f is integrable over the rectangle $R = \{ (x,y) : a \leq x \leq b \text{ and } c \leq y \leq d \} = [a, b] \times [c, d]$
>
> then $\iint_R f(x,y)\, dA = \int_a^b \int_c^d f(x,y)\, dy\, dx = \int_c^d \int_a^b f(x,y)\, dx\, dy$

Fubini's Theorem says that we can integrate in either order and still get the same result — sometimes one order of integration is much easier than the other order.

Example 2: Evaluate $\iint_R 3 + y \cdot \sin(xy)\, dA$ where $R = [1,2] \times [0,\pi]$.

Fig. 6

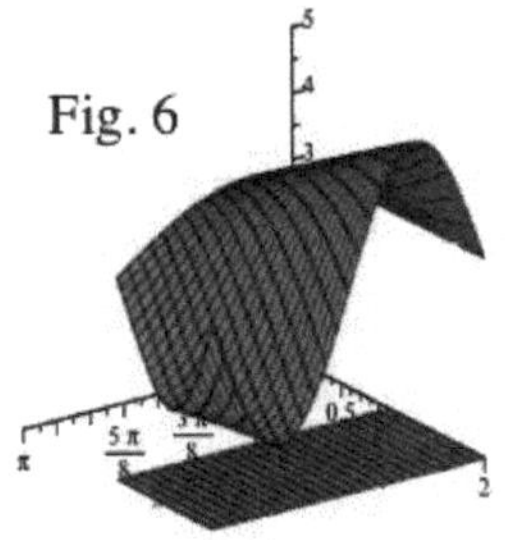

Solution: The notation $R = [1,2] \times [0,\pi]$ means the rectangle $1 \leq x \leq 2$ and $0 \leq y \leq \pi$ (Fig. 6). By Fubini's Theorem we have a choice of evaluating

(a) $\int_0^\pi \int_1^2 3 + y \cdot \sin(xy)\, dx\, dy$ or (b) $\int_1^2 \int_0^\pi 3 + y \cdot \sin(xy)\, dy\, dx$.

(a) $\int_0^\pi \int_1^2 3 + y \cdot \sin(xy)\, dx\, dy = \int_0^\pi \left\{ \int_1^2 3 + y \cdot \sin(xy)\, dx \right\} dy$

$= \int_0^\pi \left\{ 3x - \cos(xy) \Big|_{x=1}^{2} \right\} dy = \int_0^\pi \left\{ 6 - \cos(2y) - 3 + \cos(1y) \right\} dy$

$= 3y - \frac{1}{2}\sin(2y) + \sin(y) \Big|_{y=0}^{\pi} = 3\pi$

(b) $\int_1^2 \int_0^{\pi} 3 + y \cdot \sin(xy)\, dy\, dx = \int_1^2 \left\{ \int_0^{\pi} 3 + y \cdot \sin(xy)\, dy \right\} dx$ so first we need to evaluate $\int_0^{\pi} 3 + y \cdot \sin(xy)\, dy$ and that requires Integration by Parts, a more difficult situation than the method in part (a).

Example 3: Find the volume of the solid S that is bounded by the elliptic paraboloid $x^2 + 2y^2 + z = 16$, the planes $x = 2$ and $y = 2$, and the three coordinate planes (xy, xz, and yz–planes. (Fig. 7)

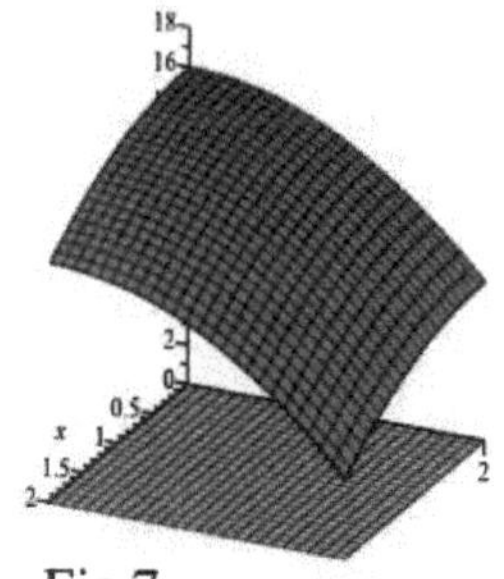

Fig.7

Solution: S lies under the surface $f(x,y) = z = 16 - x^2 - 2y^2$ and above the square $0 \le x \le 2,\ 0 \le y \le 2$. Then

$$V = \int_0^2 \int_0^2 16 - x^2 - 2y^2\, dx\, dy = \int_0^2 \left\{ \int_0^2 16 - x^2 - 2y^2\, dx \right\} dy$$

$$= \int_0^2 \left\{ 16x - \tfrac{1}{3}x^3 - 2xy^2 \Big|_{x=0}^{2} \right\} dy = \int_0^2 \left\{ \tfrac{88}{3} - 4y^2 \right\} dy = \tfrac{88}{3}y - \tfrac{4}{3}y^3 \Big|_{y=0}^{2} = 48.$$

A Simple Application: Average Value of f(x,y) on R

In section 4.7 we saw that the average value of an integrable function on the interval [a,b] is $\frac{1}{b-a}\int_a^b f(x)\,dx$, that is, the average value is the integral of the function divided by the length of the domain. We have a similar result for f(x,y) on the rectangular domain D:

> The average value H of an integrable function f(x,y) on domain D is $H = \frac{1}{\text{area of D}} \cdot \iint_D f(x,y)\, dA$.

The "box" (Fig. 8a) with height H={average value of f on D} has volume V ={area of D}{average height H} (Fig. 8b). But we know the volume is $V = \iint_D f(x,y)\, dA$ so

(area of D){average height H}= $\iint_D f(x,y)\, dA$ and then

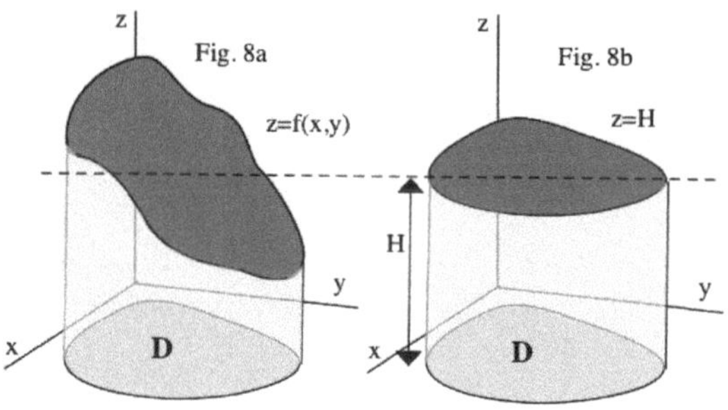

$$H = \{\text{average value of f on D}\} = \frac{1}{\text{area of D}} \cdot \iint_D f(x,y)\ dA \cdot$$

Example 4: Determine the average value of the paraboloid $z = 16 - x^2 - 2y^2$ on the domain D = [0,2]x[0,2] .

Solution: From Example 3 we know $V = \int_0^2 \int_0^2 (16 - x^2 - 2y^2)\ dx\ dy = 48$ and {area of D} = 4 so

$$H = \left(\frac{1}{4}\right)(48) = 12$$

Practice 2: Determine the average value of f(x,y) = $3 + y \cdot \sin(xy)$ on the domain $D = [1,2]\text{x}[0,\pi]$. (Fig. 6)

Areas, Volumes and Double Integrals

Practice 3: Evaluate $\int_a^b \int_c^d 1\ dx\ dy$ (a<b, c<d) and interpret the meaning of the result.

So far our discussion has involved double integrals of positive functions to get volumes, but sometimes it is useful to calculate the double integral of the very simple function f(x,y) = 1 to get an area:

$$\{\text{area of D}\} = \iint_D 1\ dA$$

This is valid even when D is not a rectangle and is especially useful in those situations.

We began the study of single integrals by trying to find the area between a positive function y=f(x) and the x-axis, but then extended that idea to more general functions that were not always positive. In the more general case, $\int_a^b f(x)\ dx$ represented the "signed area" between f(x) and the x-axis. If the areas above and below the x-axis were equal, then $\int_a^b f(x)\ dx = 0$. In a similar manner, if z=f(x,y) is sometimes negative on D, then $\iint_D f(x,y)\ dA$ will represent the "signed volume" between f(x,y) and D in the xy-plane.

Example 5: Evaluate the double integral of f(x,y) = 2x-y over the rectangle D={(x,y): -1≤x≤1 and 0≤y≤2}.

Solution: $\int_0^2 \int_{-1}^1 (2x - y)\ dx\ dy = \int_0^2 (x^2 - xy)\Big|_{x=-1}^{x=1}\ dy = \int_0^2 -2y\ dy = -4$

so more of the volume between f(x,y) and D lies below the xy-plane than lies above the xy-plane. (Fig. 9)

Fig. 9: f(x,y)=2x-y

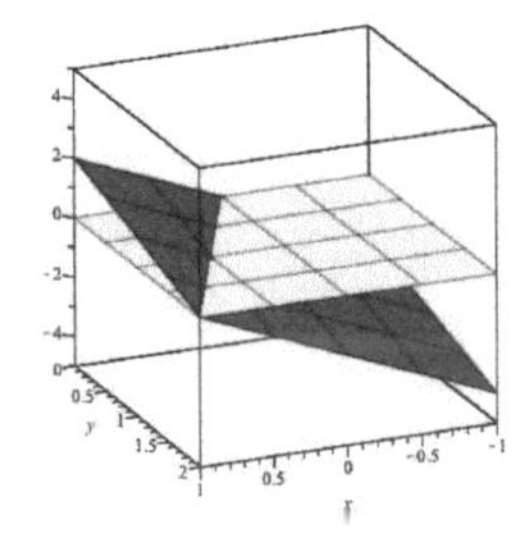

Practice 4: Suppose f(x,y) = Ax+By is a plane through the origin and $D = [-a,a]\text{x}[-b,b]$ is a rectangle that is symmetric about the origin. Show that the volume between the plane and D that is above the xy=plane is the same of as the volume that is below the xy-plane.

This section has focused on "what is a double integral" and "how to calculate the value of a double integral over a rectangular domain." In Section 14,2 we will extend these ideas to domains that are not rectangles and in Section 14.3 we will then use these double integral calculations in several applied settings.

PROBLEMS

For problems 1 – 4, calculate $\int_0^2 f(x,y)\,dx$ and $\int_0^1 f(x,y)\,dy$

1. $f(x,y) = x^2y^3$
2. $f(x,y) = 2xy - 3x^2$
3. $f(x,y) = x \cdot e^{x+y}$
4. $f(x,y) = \dfrac{x}{y^2+1}$

In problems 5 – 9, evaluate the double integrals.

5. $\displaystyle\int_{x=0}^{x=2}\int_{y=0}^{y=3} 3+4x+2y\ dy\ dx$

6. $\displaystyle\int_{x=0}^{x=2}\int_{y=1}^{y=2} 7+3x-6y\ dy\ dx$

7. $\displaystyle\int_{x=0}^{x=4}\int_{y=0}^{y=1} 1+e^x+\cos(y)\ dy\ dx$

8. $\displaystyle\int_1^3\int_2^5 x\cdot\cos(y)\ dx\ dy$

9. $\displaystyle\int_0^2\int_0^3 x^2+y^2\ dx\ dy$

10. $\displaystyle\int_0^{\pi}\int_0^{\pi} \cos(x+y)\ dx\ dy$

11. $\displaystyle\int_0^4\int_0^2 x\sqrt{y}\ dx\,dy$

12. $\displaystyle\int_{-1}^1\int_0^1 (x^3y^3+3xy^2)\ dy\,dx$

13. $\displaystyle\int_0^3\int_0^1 \sqrt{x+y}\ dx\,dy$

14. $\displaystyle\int_0^{\pi/4}\int_0^3 \sin(x)\ dy\,dx$

15. $\displaystyle\int_0^{\ln(2)}\int_0^{\ln(5)} e^{2x-y}\ dx\,dy$

In problems 16-24 decide which order of integration is easier, $\int f(x,y)\ dx$ or $\int f(x,y)\ dy$, and then calculate the easier antiderivative.

16. $f(x,y) = x\cdot e^{xy}$
17. $f(x,y) = y\cdot\sqrt{x^2+y^2}$
18. $f(x,y) = \dfrac{4x}{x^2+y^2}$
19. $f(x,y) = \sin(x)\cdot e^{x+y}$
20. $f(x,y) = \dfrac{\cos(x)}{x+y}$
21. $f(x,y) = \sqrt{x+y^2}$

22. $f(x,y) = x \cdot \ln(y+3)$ 23. $f(x,y) = e^{(y^2)}$ 24. $f(x,y) = \cos(x^2)$

In problems 25-28, evaluate the double integrals. One order of integration may be easier than the other.

25. $\iint_R (2y^2 - 3xy^3)\ dA$ where $R = \{ (x,y) : 1 \le x \le 2, 0 \le y \le 3 \}$.

26. $\iint_R x \sin(y)\ dA$ where $R = \{ (x,y) : 1 \le x \le 4, 0 \le y \le \pi/6 \}$.

27. $\iint_R x \sin(x+y)\ dA$ where $R = [0, \pi/6] \times [0, \pi/6]$.

28. $\iint_R \frac{1}{x+y}\ dA$ where $R = [1,2] \times [0,1]$.

In problems 29-34 determine the average value of f(x,y) on the given domain.

29. $f(x,y) = 3+4x+2y$ on $R = [0,2]X[0,3]$ 30. $f(x,y) = 7+3x-6y$ on $R = [0,2]X[1,2]$

31. $f(x,y) = x^2+y^2$ on $R = [0,2]X[0,3]$ 32. $f(x,y) = \cos(x+y)$ on $R = [0,\pi]X[0,\pi]$

In problems 33 and 34 the depths (in meters) of a small rectangular pond are shown. Give a good estimate of the volume of water in the pond. You should be able to justify why your estimate is reasonable.

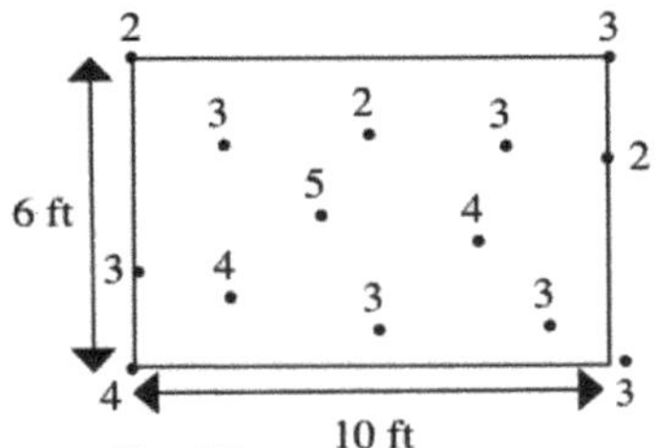

Fig. 10

33. Length = 4 m. Width = 3 m. Depths at the four corners are 2.4, 3.3, 2.1 and 3.2 m.

34. Depths (ft) are shown at various locations (Fig. 10).

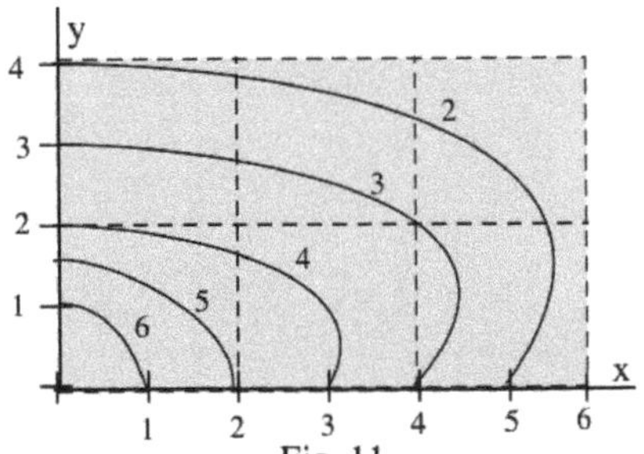

Fig. 11

In problems 35 and 36 some height contours are shown for a small hill. Give a good estimate of the volume of the hill. You should be able to justify why your estimate is reasonable.

35. See Fig. 11. All of the measurements are in meters.

36. See Fig. 12. All of the measurements are in meters.

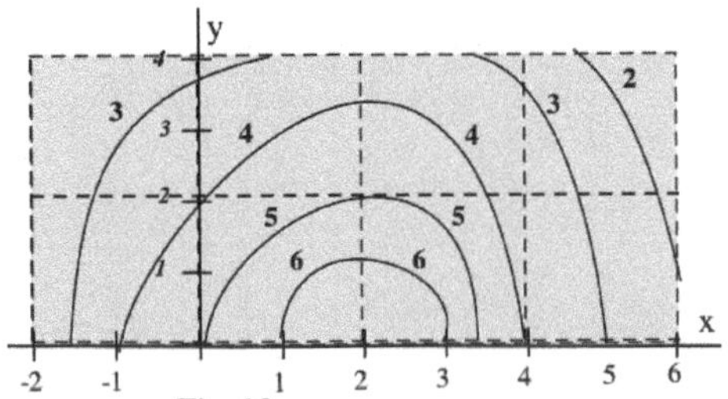

Fig. 12

Practice Answers

Practice 1: $\int_1^2\int_0^3 20 - x^2y \, dx \, dy$ means $\int_1^2 \left\{ \int_0^3 20 - x^2y \, dx \right\} dy$ so first we evaluate the inside integral $\int_0^3 x^2y \, dx$ **treating y as a constant**:

$$\int_0^3 20 - x^2y \, dx = 20x - \frac{1}{3}x^3y \Big|_{x=0}^{3} = 20(3) - \frac{1}{3}(3)^3y = 60 - 9y .$$

Then $\int_1^2 \left\{ \int_0^3 x^2y \, dx \right\} dy = \int_1^2 \left\{ 60 - 9y \right\} dy$

$$= 60y - \frac{9}{2}y^2 \Big|_{y=1}^{2} = \left(60(2) - \frac{9}{2}(2)^2\right) - \left(60(1) - \frac{9}{2}(1)^2\right) = (102) - \frac{111}{2} = \frac{93}{2}.$$

Practice 2: In Example 2 we evaluated $\int_0^\pi\int_1^2 3 + y\cdot\sin(xy) \, dx \, dy = 3\pi$.

The area of D is $(2-1)\cdot(\pi - 0) = \pi$ so the average value is $(3\pi)/\pi = 3$.

Practice 3: $\int_a^b\int_c^d 1 \, dx \, dy = \int_a^b x\Big|_{x=c}^{x=d} dy = \int_a^b (d - c) \, dy = (d - c)\cdot y\Big|_{y=a}^{y=b} = (d-c)\cdot(b-a)$

This is the area of the rectangular domain.

Practice 4: $\int_{-a}^{a}\int_{-b}^{b} Ax + By \; dy \; dx = \int_{-a}^{a} Axy + \frac{B}{2}y^2\Big|_{y=-b}^{y=b} dx = \int_{-a}^{a} 2abx \; dx = abx^2\Big|_{x=-a}^{x=a} = 0$

So the volume between the plane f(x,y)=Ax+By and D that is above the xy=plane is the same as the volume that is below the xy-plane.

14.2 DOUBLE INTEGRALS OVER GENERAL REGIONS

The double integrals over rectangular regions in the section 14.1 are relatively straightforward, but many applied situations have domains that are not rectangular, and this section considers those. In section 14.1 our focus was on finding antiderivatives (twice) and evaluating the double integrals. In this section the main difficulties involve setting up the endpoints of the integrals. This is very important and is vital for later sections.

The starting ideas here are the same as in 14.1:

* partition the general domain D into small Δx_i by Δy_j rectangles, (Fig. 1)
* create a double Riemann sum of the areas of the boxes above each rectangle,

  $$\sum_j \sum_i f(x_i, y_j) \cdot \Delta x_i \cdot \Delta y_j = \sum_j \sum_i f(x_i, y_j) \cdot \Delta A_{ij}$$

* take the limits as all of the Δx_i and Δy_j values approach 0

  so they fill the domain, and
* create a double integral $\iint_D f(x,y)\, dA$.

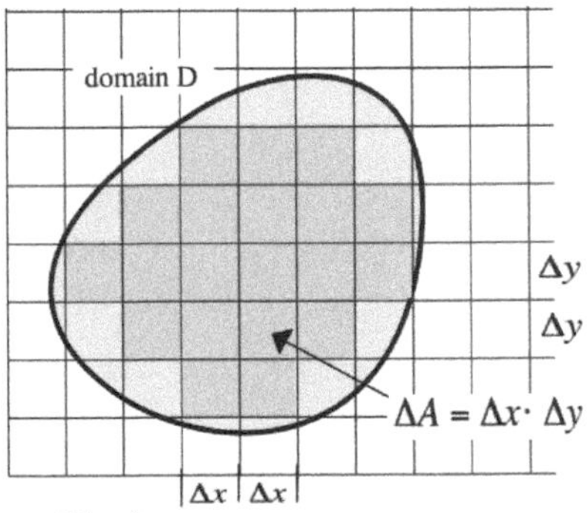

Fig. 1

The notation $\iint_D f(x,y)\, dA$ is neat and compact and is commonly used, but in order to compute the value of the double integral we need to explicitly rewrite it as $\iint_D f(x,y)\, dx\, dy$ or $\iint_D f(x,y)\, dy\, dx$ with the appropriate endpoints of integration.

Note: All of our work with double integrals assumes that the domain D is "bounded" (fits inside a finite rectangle) and that the boundary curve of D is "nice" (consists of a finite number of smooth curves with finite length). In most applications, these are not major restrictions.

Two Easiest Cases

Case V: For each x* value, D contains a single vertical segment of y values (Fig. 2).

Case H: For each y* value, D contains a single horizontal segment of x values .

Other: It is possible for the domain D to be both Case V and Case H or neither.

Fig. 2

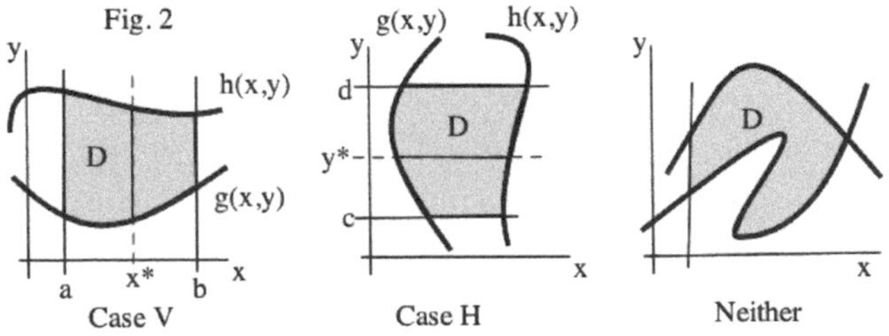

Case V: For each value of x between a and b, D consists of those y with $g(x) \le y \le h(x)$. For each value of x, we only care about f(x,y) for y values between g(x) and h(x).

$$\text{Then } \iint_D f(x,y)\, dA = \int_{x=a}^{x=b} \int_{y=g(x)}^{y=h(x)} f(x,y)\ dy\ dx \ \ \text{(or simply } \int_a^b \int_{g(x)}^{h(x)} f(x,y)\ dy\ dx\text{).}$$

Example 1: f(x,y) = 1 + 4x + 2y and D is the region in Fig. 3. Calculate the volume of the solid between D and the surface z=f(x,y).

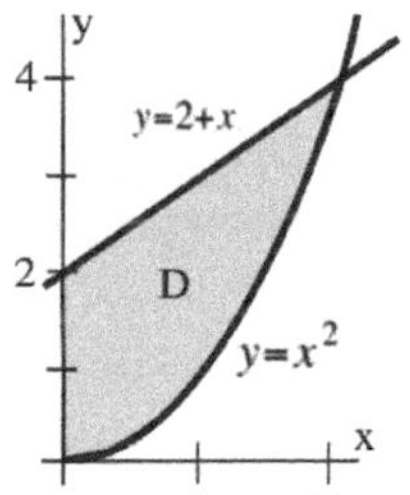

Fig. 3

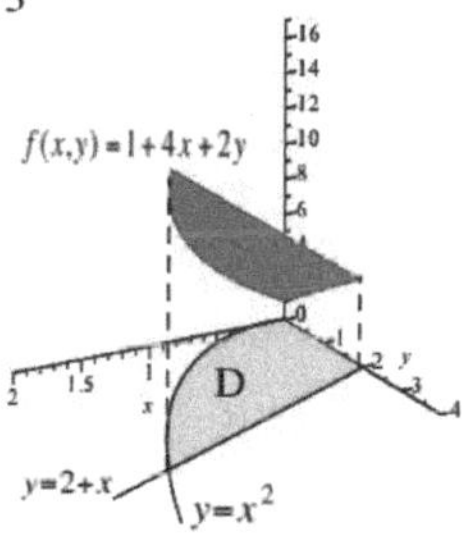

Solution: For each x between 0 and 2, the y values are between $2+x$ and x^2 so

$$\text{Volume} = \iint_D f(x,y)\, dA = \int_{x=0}^{x=2} \int_{y=x^2}^{y=2+x} 1+4x+2y\ dy\ dx .$$

We deal with this double integral in the same way we did in section 14.1 by starting with the inside integral.

$$\int_{y=x^2}^{y=2+x} 1+4x+2y\ dy = y+4xy+y^2 \Big|_{y=x^2}^{y=2+x}$$

$$= \{(2+x)+4x(2+x)+(2+x)^2\} - \{(x^2)+4x(x^2)+(x^2)^2\} = -x^4-4x^3+4x^2+13x+6 .$$

$$\text{Then } \int_{x=0}^{x=2} \{-x^4-4x^3+4x^2+13x+6\}\ dx = \frac{394}{15} \approx 26.27$$

If the units of x, y and z=f(x,y) are centimeters, then the double integral units are cm^3 .

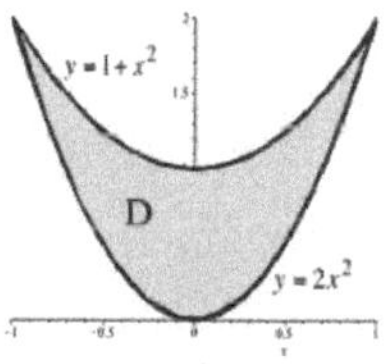

Fig. 4

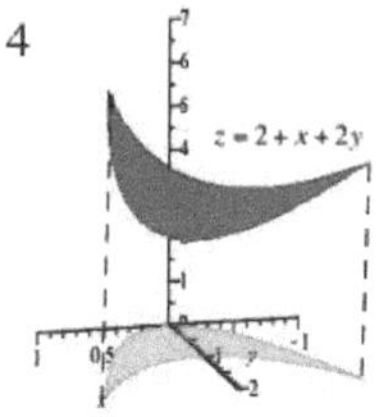

Practice 1: f(x,y) = 2 + x + 2y and D is the smiley region in Fig. 4. Calculate $\iint_D f(x,y)\, dA$ = volume of the solid between D and the surface z=f(x,y).

Example 2: Determine the integration endpoints for $\iint_D f(x,y)\ dy\ dx$ on the domain D shown in Fig. 5.

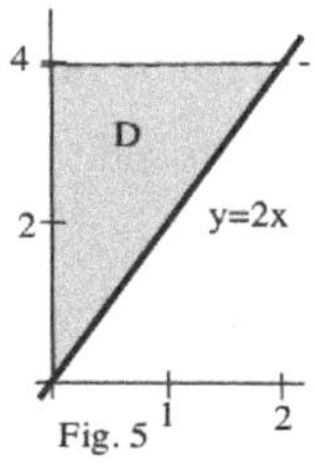

Fig. 5

Solution: Since the dx is on the outside, the endpoints of the outside integral will be from x=0 to x=2, But for each x value between 0 and 2 (Fig. 6), a vertical slice at x enters the domain D when y=2x (at the bottom) and exits when y=4 (at the top) so the endpoints of the inside integral are y=2x and y=4:

$$\int_{x=0}^{x=2} \int_{y=2x}^{y=4} f(x,y)\ dy\ dx \ \text{ or simply } \int_0^2 \int_{2x}^4 f(x,y)\ dy\ dx .$$

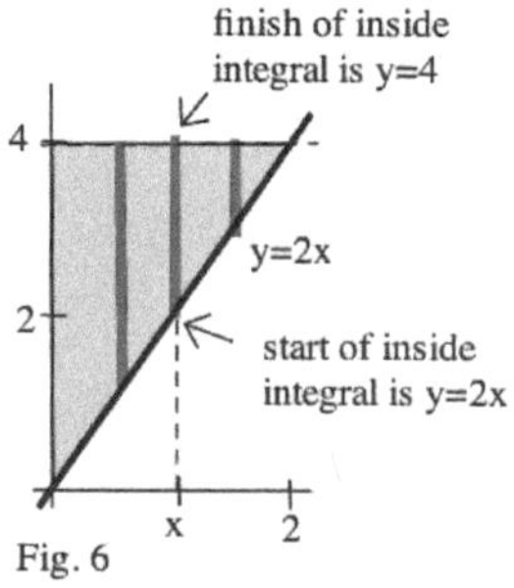

Fig. 6

Practice 2: Determine the integration endpoints for $\iint_D f(x,y)\,dy\,dx$ on the domain D shown in Fig. 7.

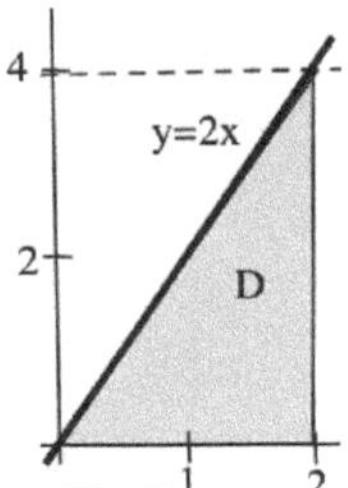

Fig. 7

Case H: For each value of y between c and d, the domain D consists of those x values with $g(y) \le x \le h(y)$. For each value of y, we only care about f(x,y) for x values between g(y) and h(y).

Then $\iint_D f(x,y)\,dA = \int_{y=c}^{y=d} \int_{x=g(y)}^{x=h(y)} f(x,y)\,dx\,dy$ (or simply $\int_c^d \int_{g(y)}^{h(y)} f(x,y)\,dx\,dy$).

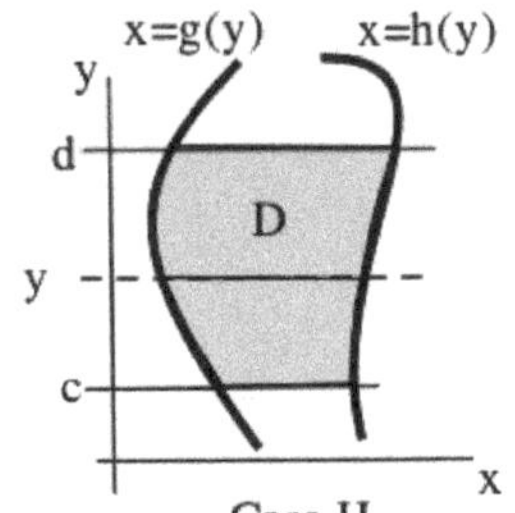

Fig. 8 Case H

Note: If the boundary function has the form y=f(x) then we need to solve for x as a function of y. For example, if the boundary is y = f(x) = 1+2x, then we need to solve for x to get x = g(y) = (y-1)/2.

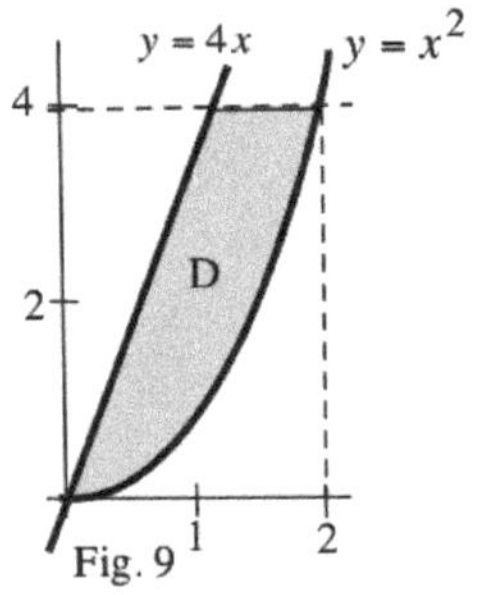

Fig. 9

Example 3: Determine the integration endpoints for $\iint_D f(x,y)\,dx\,dy$ on the domain D shown in Fig. 9.

Solution: Clearly the horizontal slices go from y = 0 to y = 4 so those are the endpoints of the outside integral. Then we need to convert the y=4x and $y = x^2$ from functions of x to functions of y: x = y/4 and $x = \sqrt{y}$. For each y value between 0 and 4 we need to see when a horizontal slice at y enters and exits the domain (Fig. 10).

The result is $\iint_D f(x,y)\,dx\,dy = \int_{y=0}^{y=4} \int_{x=y/4}^{x=\sqrt{y}} f(x,y)\,dx\,dy$.

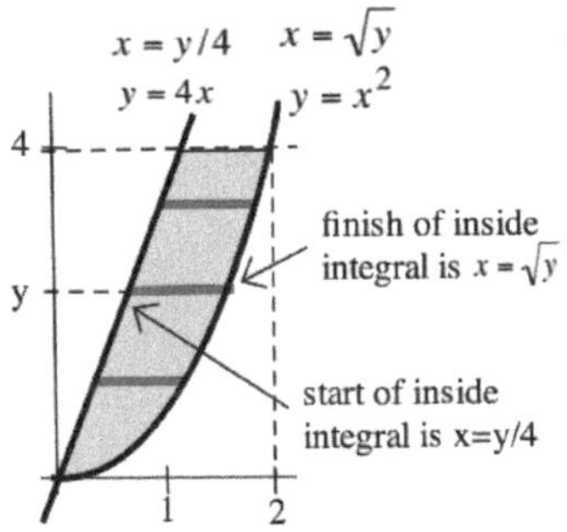

Fig. 10

Practice 3: Determine the integration endpoints for $\iint_D f(x,y)\,dx\,dy$ on the domain D shown in Fig. 11.

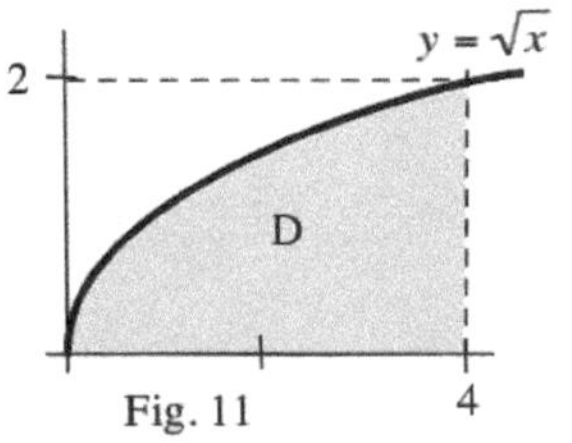

Fig. 11

Splitting the Domain & Switching the Order of Integration

Sometimes the domain requires that we do the double integral in two (or more) pieces.

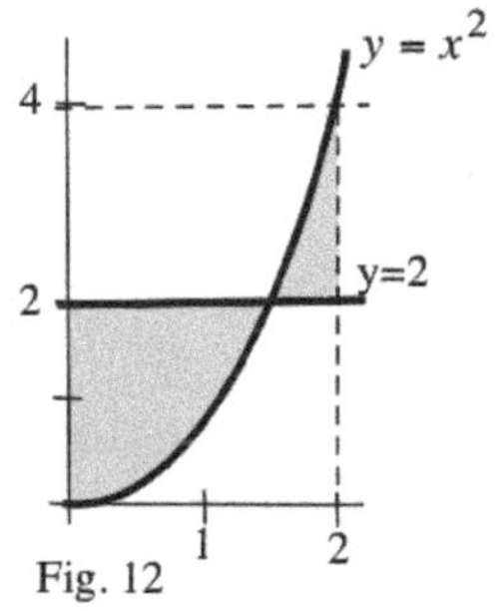

Fig. 12

Example 4: Determine the integration endpoints for $\iint_D f(x,y)\,dy\,dx$ on the shaded domain D shown in Fig. 12.

Solution: For $0 \le x \le 1$, the bottom boundary is $y = x^2$ and the top boundary is y=2. For $0 < x \le 2$ the top and bottom boundary functions are switched sp we need to use two double integrals:

$$\iint_D f(x,y)\,dy\,dx = \int_{x=0}^{x=1} \int_{y=x^2}^{y=2} f(x,y)\,dy\,dx + \int_{x=1}^{x=2} \int_{y=2}^{y=x^2} f(x,y)\,dy\,dx$$

Practice 4: Determine the integration endpoints for $\iint_D f(x,y)\,dx\,dy$ on the shaded domain D shown in Fig. 12. Note the switch to dx dy in the double integral -- this is also an example of switching the order of integration.

Example 5: Switch the order of integration of $\int_{x=-2}^{x=2} \int_{y=x^2}^{y=4} f(x,y)\,dy\,dx$.

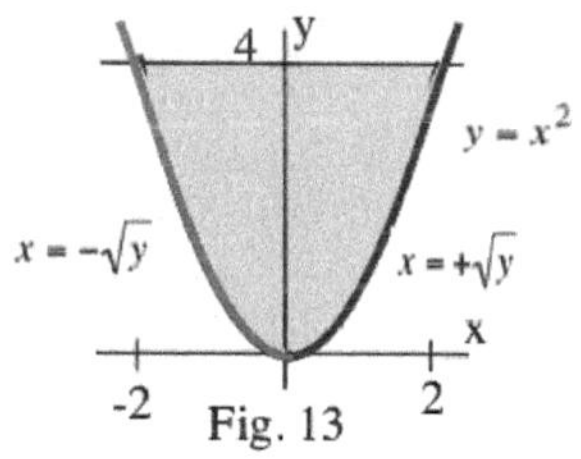

Fig. 13

Solution: From the given double integral we can create a picture of the domain (Fig. 13) and then write the double integral with the order dx dy:

$$\int_{x=-2}^{x=2} \int_{y=x^2}^{y=4} f(x,y)\,dy\,dx = \int_{y=0}^{y=4} \int_{x=-\sqrt{y}}^{x=+\sqrt{y}} f(x,y)\,dx\,dy .$$

Why are we doing this?

Sometimes one order for a double integral is much easier to evaluate than the other order, and then it is worthwhile to be able to convert to the easier way.

Example 6: Evaluate $\iint_R e^{(x^2)}\,dA$ for R = {region for y between 0 and 1 and x between 3y and 3}.

Solution: It is natural to set up the integrals as $\int_{y=0}^{1}\int_{x=3y}^{3} e^{(x^2)}\,dx\,dy$ since that is the description of the region R, but as soon as we try the inside integral $\int_{x=3y}^{3} e^{(x^2)}\,dx$ we are stuck because $e^{(x^2)}$ does not have an elementary antiderivative. Fig. 14 shows the region R

When we switch the order the integrals become $\int_{x=0}^{3}\int_{y=0}^{x/3} e^{(x^2)}\,dy\,dx$.

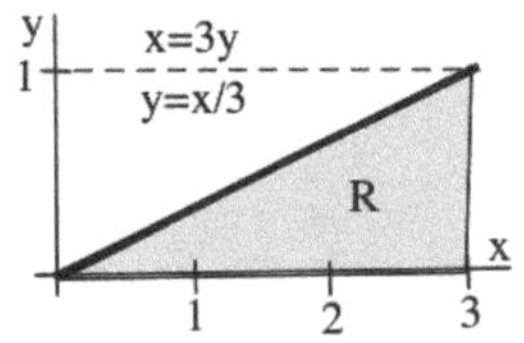

Fig. 14

Now things are easier: $e^{(x^2)}$ does not depend on y so

$\int_{y=0}^{x/3} e^{(x^2)}\,dy = y\cdot e^{(x^2)}\Big|_{y=0}^{y=x/3} = \frac{1}{3}y\cdot e^{(x^2)}$. Then the outside integral is $\int_{x=0}^{3} \frac{1}{3}x\cdot e^{(x^2)}\,dx$.

With the simple change of variable $u = x^2$ and $du = 2x\,dx$ we get

$$\int \frac{1}{6}e^{(u)}\,du = \frac{1}{6}e^{(u)} = \frac{1}{6}e^{(x^2)}\Big|_{x=0}^{x=3} = \frac{1}{6}e^{9} - \frac{1}{6}.$$

Practice 5: Which is easier? (a) $\int_{x=0}^{1}\int_{y=x}^{1} \cos(y^2)\,dy\,dx$ or (b) $\int_{y=0}^{1}\int_{x=0}^{y} \cos(y^2)\,dx\,dy$

Properties of Double Integrals

These properties all are similar to the familiar properties of single integrals and follow from the properties of summations. We will not prove them here and have already used them in the examples. You should understand what each of them means geometrically.

(1) $\iint_D c\cdot f(x,y)\,dA = c\cdot\iint_D f(x,y)\,dA$

(2) $\iint_D \{f(x,y)+g(x,y)\}\,dA = \iint_D f(x,y)\,dA + \iint_D g(x,y)\,dA$

(3) If $g(x,y) \le f(x,y)$ for all (x,y) in D, then $\iint_D g(x,y)\,dA \le \iint_D f(x,y)\,dA$

(4) $\iint_D 1\,dA = \text{area}(D)$

(5) If $m \le f(x,y) \le M$ for all (x,y) in D, then $m\cdot\text{area}(D) \le \iint_D f(x,y)\,dA \le M\cdot\text{area}(D)$.

(6) If D1 and D2 are disjoint domains and $D = D1 \cup D2$ (= all (x,y) in $D1$ or $D2$)

then $\iint_D f(x,y)\, dA = \iint_{D1} f(x,y)\, dA + \iint_{D2} f(x,y)\, dA$.

In this section the focus has been on setting up the endpoints of integration for non-rectangular domains. Some computer programs can evaluate double integrals, but only after the user has determined the endpoints. Even with technology you need to be able to determine those endpoints. And since it is sometimes much easier to integrate in one order than the other, it is important that you be able to write the double integral in either order.

Using Technology -- MAPLE

Some of the double integrals that appear in later sections and many of them in applications are very difficult to evaluate "by hand." Fortunately technology can be very helpful. The computer program MAPLE is very powerful and relatively simple to use.

In Example 1 we were able to evaluate the double integral of $f(x,y) = 1+4x+2y$ on the region $R = \{(x,y):\ 0 \le x \le 2,\ x^2 \le y \le 2+x\}$ by hand, but the MAPLE command

int(1+4*x+2*y, y=x^2 .. 2+x, x=0 ..1);

gives the same result, $\frac{379}{30}$, much quicker. And MAPLE can evaluate much more difficult integrals such as $\int_{x=0}^{1} \int_{y=0}^{x+2} xy + \cos(xy)\ dy\ dx$. The MAPLE command int(x*y+cos*x*y), y=0 .. x+2, x=0 .. 1.0) : quickly gives 3.387845686 .

MAPLE syntax: int(formula for f(x,y) , y= lower endpoint .. upper endpoint, x= lower x .. upper x) ;

The program Mathematica and the online WolframAlpha can also evaluate double integrals.

PROBLEMS

For problems 1 – 6 sketch the domain of integration and evaluate the integrals.

1. $\int_{x=0}^{x=3} \int_{y=x}^{y=2x} x\ dy\ dx$

2. $\int_{x=0}^{x=2} \int_{y=0}^{y=2x} 3+4x+2y\ dy\ dx$

3. $\int_{y=1}^{y=4} \int_{x=y}^{x=2y} 7+3x-4y\ dx\ dy$

4. $\int_{x=0}^{x=1} \int_{y=x^2}^{y=x} x^2+\sqrt{y}\ dy\ dx$

5. $\int_{x=0}^{x=1} \int_{y=0}^{y=x} \sin(x^2)\ dy\ dx$

6. $\int_{y=0}^{y=2} \int_{x=\sqrt{y}}^{x=3} x+y^2\ dx\ dy$

In problems 7-12 a shaded domain D is shown. Determine the endpoints for both $\iint_D f\,dy\,dx$ and $\iint_D f\,dx\,dy$.

7. Fig. 14

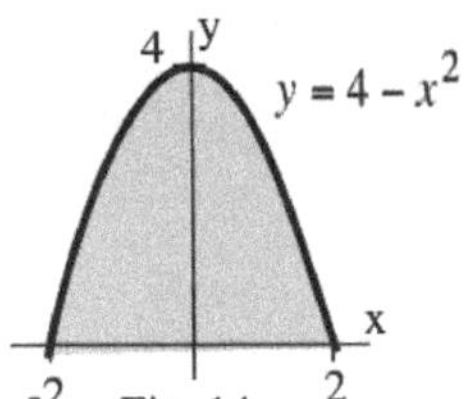

Fig. 14

8. Fig. 15

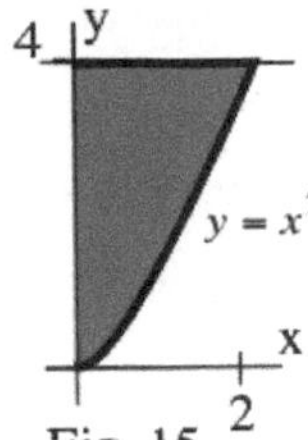

Fig. 15

9. Fig. 16

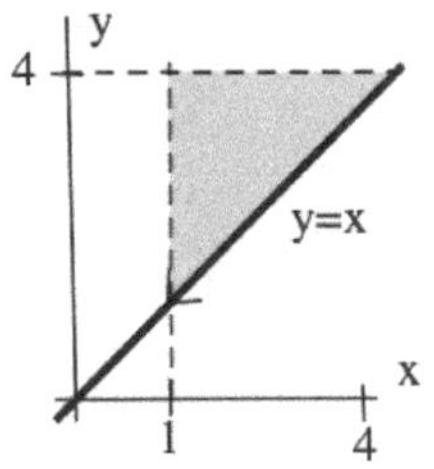

Fig. 16

10. Fig. 17

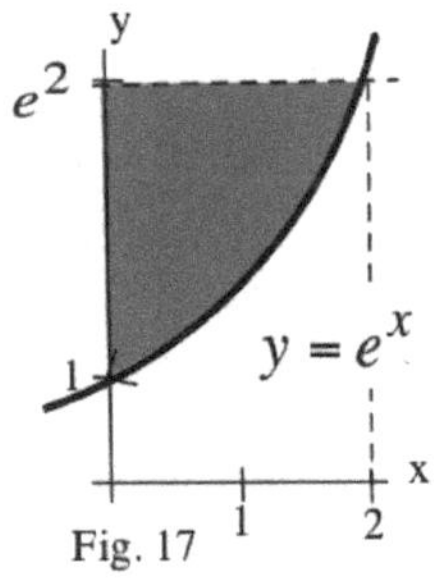

Fig. 17

11. Fig. 18

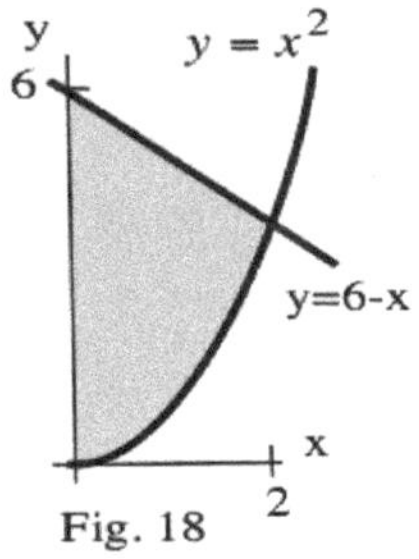

Fig. 18

12.

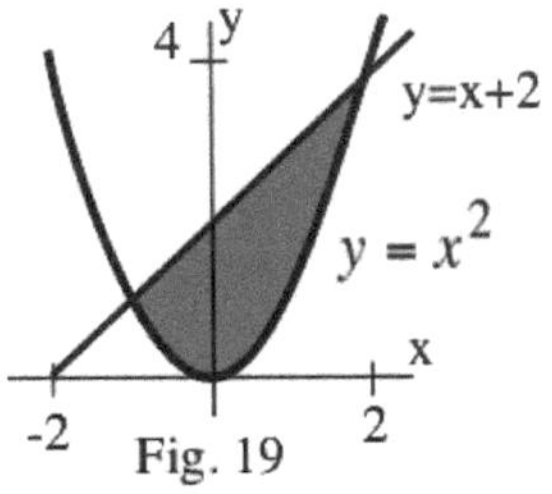

Fig. 19

In problems 13 - 20 sketch the domain of integration, set up the appropriate double integral and evaluate it.

13.. $\iint_D xy\,dA$ where $D = \{ (x,y) : 0 \le x \le 1, x^2 \le y \le \sqrt{x} \}$.

14. $\iint_D (x + y)\,dA$ where $D = \{ (x,y) : \pi/6 \le x \le \pi/4, \sin(x) \le y \le \cos(x) \}$.

15. $\iint_D \left(y - xy^2\right) dA$ where $D = \{ (x,y) : -y \le x \le 1+y, 0 \le y \le 1 \}$.

16. $\iint_D x^2 y\,dA$ where $D = \{ (x,y) : x^2 \le y \le 2+x, -1 \le x \le 2 \}$.

17. $\iint_D 1 + x \cdot \cos(y)\,dA$ where D is the bounded region between $y = 0$ and $y = 4 - x^2$.

18. $\iint_D \left(x^2 + y\right) dA$ where D is the bounded region between $y = x^2$ and $y = 8 - x^2$.

19. $\iint_D 4y^3\,dA$ where D is the bounded region between $y = 2 - x$ and $y^2 = x$.

20. $\iint_D 6x + y\,dA$ where D is the bounded region between $y = x^2$ and $y = 4$.

In problems 21-24 determine the average value of the function f(x,y) on the given domain.

21. $f(x,y) = 3 + 4x + 2y$ where $D = \{(x,y): 0 \le x \le 2 \text{ and } 0 \le y \le 2x\}$

22. $f(x,y) = 1 + 2x + 3y$ where $D = \{(x,y): -2 \le x \le 2 \text{ and } 0 \le y \le 4 - x^2\}$

23. $f(x,y) = x + y^2$ where $D = \{(x,y): 0 \le x \le 3 \text{ and } 0 \le y \le 3 - x\}$

24. $f(x,y) = 2 + xy$ where $D = \{(x,y): 0 \le x \le 2 \text{ and } x^2 \le y \le 6 - x\}$

In problems 25 – 36, change the order of integration. (It helps to sketch the region.)

25. $\int_0^1 \int_0^x f(x,y)\, dy\, dx$

26. $\int_{-2}^2 \int_{x^2}^4 f(x,y)\, dy\, dx$

27. $\int_0^6 \int_0^{3-y/2} f(x,y)\, dx\, dy$

28. $\int_0^4 \int_0^{\sqrt{x}} f(x,y)\, dy\, dx$

29. $\int_1^2 \int_0^{\ln(x)} f(x,y)\, dy\, dx$

30. $\int_0^4 \int_{y/2}^2 f(x,y)\, dx\, dy$

31. $\int_0^2 \int_0^{4-2x} f\, dy\, dx$

32. $\int_0^3 \int_0^{9-x^2} f\, dy\, dx$

33. $\int_0^4 \int_{\sqrt{x}}^2 f\, dy\, dx$

34. $\int_1^3 \int_2^{2y} f\, dx\, dy$

35. $\int_0^1 \int_1^{e^x} f\, dy\, dx$

36. $\int_0^4 \int_{-\sqrt{y}}^{\sqrt{y}} f\, dx\, dy$

In problems 37 and 38 the depths (in meters) of a small pond are shown. Give a good estimate of the volume of water in the pond. You should be able to justify why your estimate is reasonable.

37. Fig. 20

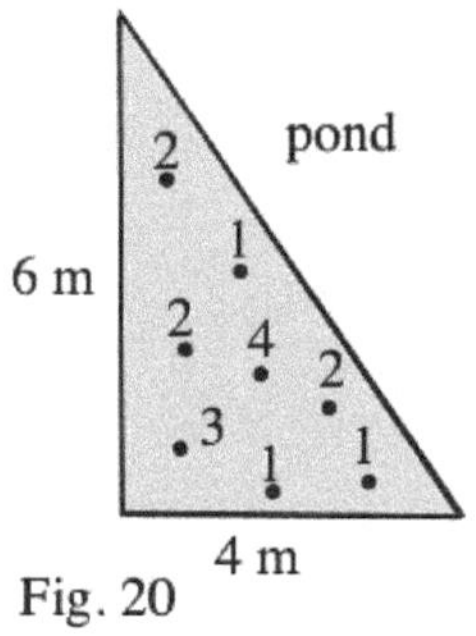

Fig. 20

38. Fig. 21

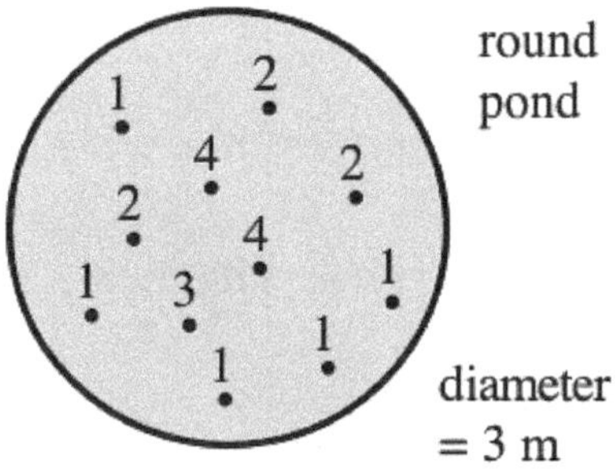

Fig. 21

PRACTICE ANSWERS

Practice 1: $\iint_D (x+2y)\, dA = \int_{-1}^{1} \int_{2x^2}^{1+x^2} (x+2y)\, dy\, dx$

Starting with the inside integral $\int_{2x^2}^{1+x^2} 2+x+2y\ dy = 2y+xy+y^2 \Big|_{2x^2}^{1+x^2}$

$= \{2(1+x^2)+x(1+x^2)+(1+x^2)^2\} - \{2(2x^2)+x(2x^2)+(2x^2)^2\} = -3x^4 - x^3 + x + 3$

Then $\int_{-1}^{1} -3x^4 - x^3 + x + 3\ dx = = -\frac{3}{5}x^5 - \frac{1}{4}x^4 + \frac{1}{2}x + 3x \Big|_{-1}^{1} = \left(\frac{53}{20}\right) - \left(\frac{-43}{20}\right) = \frac{24}{5}$

The Maple command int(f(x,y),y=2*x^2..1+x^2,x=-1..1); gives 24/5.

Practice 2: See Fig. 22. The outside integral (corresponding to the dx) goes from x=0 to x=2. The inside integral begins at y=0 and ends at y=2x.

$\iint_D f(x,y)\, dy\, dx = \int_{x=0}^{x=2} \int_{y=0}^{y=2x} f(x,y)\ dy\ dx$

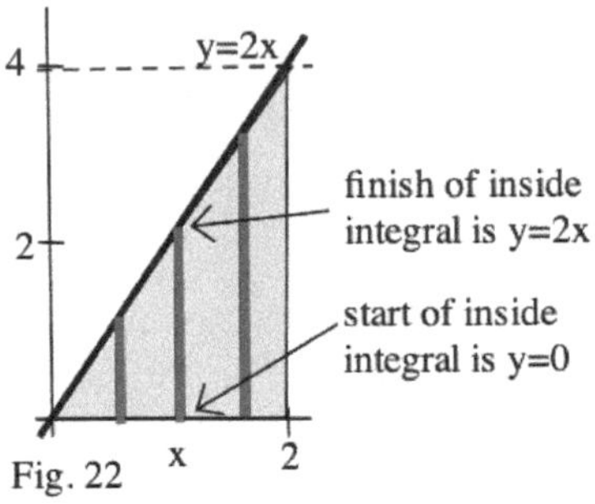

Fig. 22

Practice 3: See Fig. 23. The outside integral goes from y=0 to y=2. $y=\sqrt{x}$ becomes $x=y^2$. Then for each y value between 0 and 2, the horizontal slice enters the domain at $x=y^2$ and exits at x=4.

$\iint_D f(x,y)\, dx\, dy = \int_{y=0}^{y=2} \int_{x=y^2}^{x=4} f(x,y)\ dx\ dy$

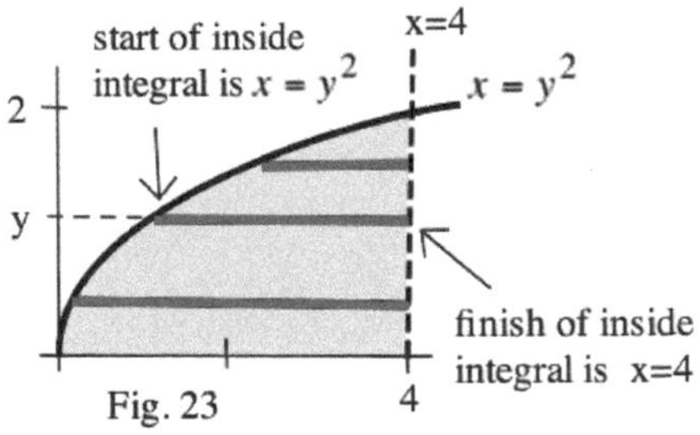

Fig. 23

Practice 4: Since the dy is on the outside, we need to partition the 0≤y≤4 interval. Then

$\iint_D f(x,y)\, dx\, dy = \int_{y=0}^{y=2} \int_{x=0}^{x=\sqrt{y}} f(x,y)\ dx\ dy + \int_{y=2}^{y=4} \int_{x=\sqrt{y}}^{x=2} f(x,y)\ dx\ dy$

Practice 5: Version (b) is easier sine we can evaluate $\int_{x=0}^{y} \cos(y^2)\ dx = x \cdot \cos(y^2) \Big|_{x=0}^{x=y} = y \cdot \cos(y^2)$

and we cannot evaluate $\int_{y=x}^{1} \cos(y^2)\ dy$.

14.3 DOUBLE INTEGRALS IN POLAR COORDINATES

As you learned in earlier classes some shapes (Fig. 1) are much simpler to describe using polar coordinates, and sometimes we need to calculate volumes of regions whose domains are those shapes. The process is straightforward, but first we need a way to partition a polar coordinate region. Since the polar variables are r and θ, it is natural to partition the domain into "polar rectangles" (Fig. 2) and then proceed as we did in Section 14.1 to build Riemann Sums and, by taking limits, get to double integrals.

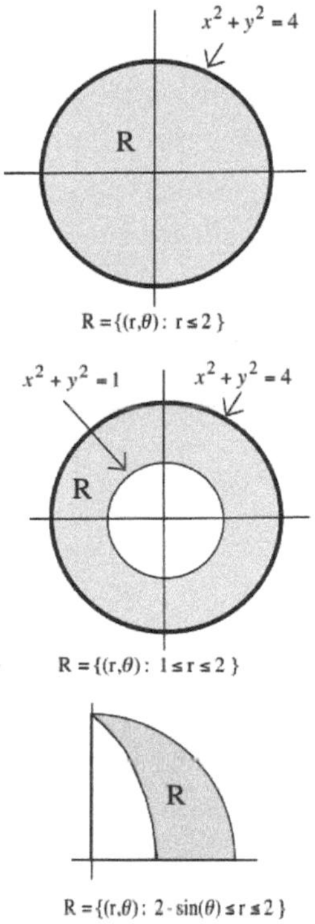

Fig. 1

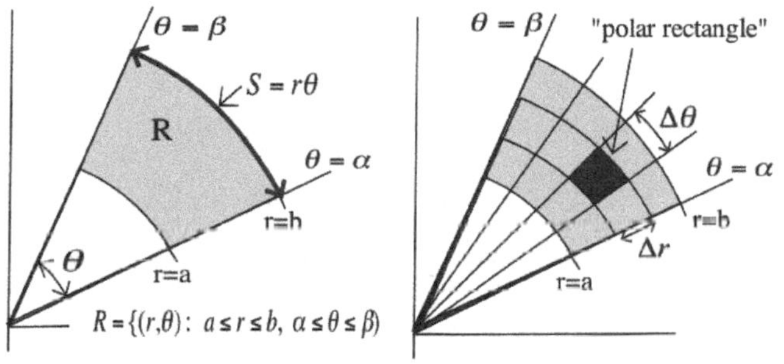

Fig. 2: Polar Rectangle

The area of each sub-rectangle is approximately $\Delta A \approx r\cdot \Delta r\cdot \Delta\theta$ (Fig. 3) so the volume above each sub-rectangle will be $\Delta V = f(x,y)\cdot \Delta A$. But we are in polar coordinates, so we rneed to eplace x and y with their polar coordinate values $x = r\cdot \cos(\theta)$ and $y = r\cdot \sin(\theta)$.

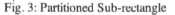

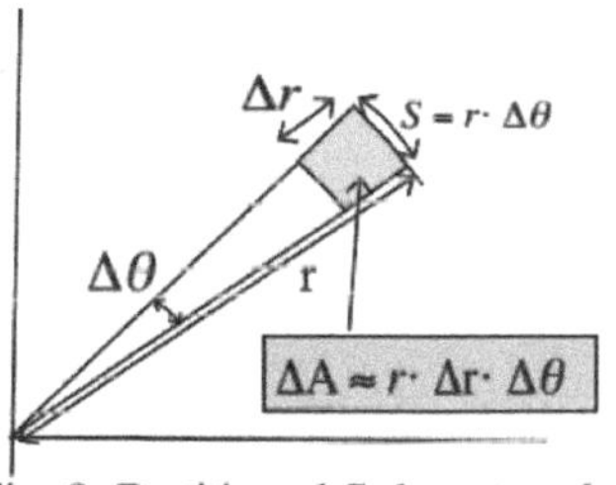

Fig. 3: Partitioned Sub-rectangle

Then the total volume is approximately the value of the double Riemann sum

$$\text{volume} \approx \sum \sum f(x,y)\cdot \Delta A = \sum_{r} \sum_{\theta} f(r\cdot \cos(\theta), r\cdot \sin(\theta))\cdot \Delta A\cdot$$

Taking limits as Δr and $\Delta\theta$ both approach 0, the exact volume is a double integral.

Volume in Polar Coordinates

If $f(x,y) \ge 0$ is continuous on the polar rectangle $R=\{ (r,\theta) : a \le r \le b, \alpha \le \theta \le \beta)\}$, then

The volume between the domain R and the surface $z = f(x,y)$ is

$$\text{Volume} = \iint_R f(x,y)\cdot dA = \int_{\theta=\alpha}^{\theta=\beta} \int_{r=a}^{r=b} f(r\cdot \cos(\theta),\ r\cdot \sin(\theta))\cdot r\cdot dr\cdot d\theta$$

This may seem complicated but it is simply volume = height $\cdot$ (base area) with height=function (in polar coordinates) and base area $= r \cdot dr \cdot d\theta$. (Note: Don't forget the r in the base area.)

Example 1: Find the volume between the plane f(x,y)=5+x+2y and the circular region $R = \{(x,y): x^2 + y^2 \le 4\}$. (Fig. 4)

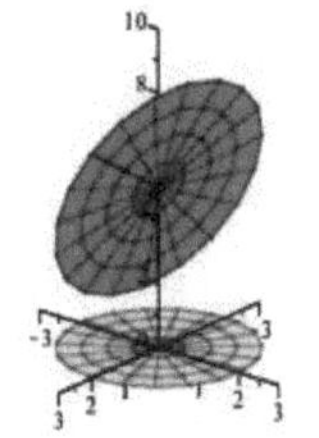

Fig. 4

Solution: Translating the problem into polar coordinates we have

$f(x,y) = f(r\cdot\cos(\theta), r\cdot\sin(\theta)) = 5 + r\cdot\cos(\theta) + 2\cdot r\cdot\sin(\theta)$

and $R = \{(r,\theta):\ 0 \le r \le 2,\ 0 \le \theta \le 2\pi\}$ so

$$\text{volume} = \iint_R f(x,y)\cdot dA = \int_{\theta=0}^{\theta=2\pi}\int_{r=0}^{r=2} \{5 + r\cdot\cos(\theta) + 2\cdot r\cdot\sin(\theta)\}\cdot r\cdot dr\cdot d\theta$$

which can be evaluated in the usual way starting with the inside integral.

$$\int_{r=0}^{r=2} \{5 + r\cdot\cos(\theta) + 2\cdot r\cdot\sin(\theta)\}\cdot r\cdot dr = \int_{r=0}^{r=2} \{5r + r^2\cdot\cos(\theta) + 2\cdot r^2\cdot\sin(\theta)\}\ dr$$

$$= \frac{5}{2}r^2 + \frac{1}{3}r^3\cdot\cos(\theta) + \frac{2}{3}r^3\cdot\sin(\theta)\Big|_{r=0}^{r=2} = \frac{5}{2}(4) + \frac{1}{3}(8)\cdot\cos(\theta) + 2\cdot\frac{1}{3}(8)\cdot\sin(\theta)$$

Then
$$\int_{\theta=0}^{\theta=2\pi} 10 + \frac{8}{3}\cdot\cos(\theta) + \frac{16}{3}\cdot\sin(\theta)\ d\theta = 10\theta + \frac{8}{3}\cdot\sin(\theta) - \frac{16}{3}\cdot\cos(\theta)\Big|_{\theta=0}^{\theta=2\pi}$$

$$= \{20\pi + 0 - \frac{16}{3}\} - \{0 + 0 - \frac{16}{3}\} = 20\pi \approx 62.83 \ .$$

The volume is $20\pi \approx 62.83$.

Practice 1: If we restrict the domain to $R = \{(x,y): x^2 + y^2 \le 4,\ 0 \le y\}$ then the polar coordinate form of the domain is $R = \{(r,\theta):\ 0 \le r \le 2,\ 0 \le \theta \le \pi\}$. Verify that the volume for the same plane f(x,y)=5+x+2y over this new domain is $\frac{32}{3} + 10\pi \approx 42.08$. Since the domain R in Practice 1 is half of the domain R in Example 1, why isn't the new volume half of the previous volume?

Example 2: Find the volume between the surface $f(x,y) = e^{-(x^2+y^2)}$ and the circular domain $R = \{(x,y): x^2 + y^2 \le 4\}$. (Fig. 5)

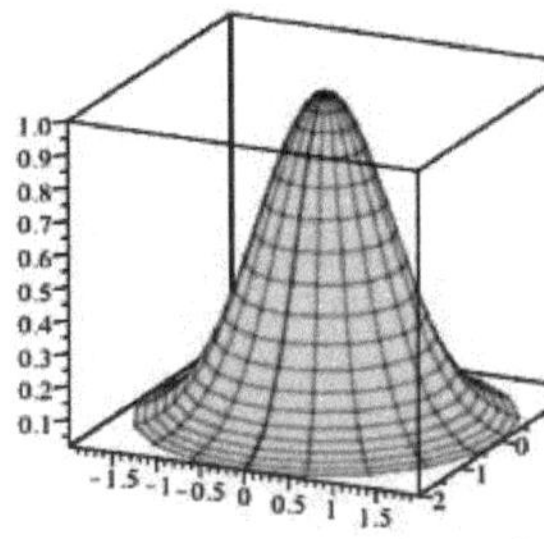

Fig. 5: $f(x,y) = e^{-(x^2+y^2)}$

Solution: This function looks complicated, but when we translate into polar coordinates, things get much easier.

$x = r\cdot\cos(\theta)$ and $y = r\cdot\sin(\theta)$ so $x^2 + y^2 = r^2$. And R becomes $R = \{(r,\theta):\ 0 \le r \le 2,\ 0 \le \theta \le 2\pi\}$. Then

$$\text{Volume} = \iint_R f \cdot dA = \int_{\theta=0}^{\theta=2\pi} \int_{r=0}^{r=2} e^{-r^2} \cdot r \cdot dr \cdot d\theta \cdot$$

The inside integral $\int_{r=0}^{r=2} e^{-r^2} \cdot r \cdot dr = -\frac{1}{2} e^{-r^2} \Big|_{r=0}^{r=2} = \frac{1}{2}\left(1 - e^{-4}\right)$ (using the substitution $u = -r^2$)

$$\text{Volume} = \iint_R f \cdot dA = \int_{\theta=0}^{\theta=2\pi} \frac{1}{2}\left(1 - e^{-4}\right) \cdot d\theta = \frac{1}{2}\left(1 - e^{-4}\right) \cdot \theta \Big|_{\theta=0}^{\theta=2\pi}$$

$$= \frac{1}{2}\left(1 - e^{-4}\right) \cdot \theta \Big|_{\theta=0}^{\theta=2\pi} = \left(1 - e^{-4}\right)\pi \approx 0.98\pi \cdot$$

Note: The $x^2 + y^2$ in the rectangular function is often a signal that the polar version may be easier since $x^2 + y^2 = r^2$. Then the substitution $u = r^2$ means $du = 2 \cdot r \cdot dr$ so $dA = r \cdot dr \cdot d\theta = \frac{1}{2} d\theta$.

Practice 2: (a) Find the volume between the surface $z = f(x,y) = e^{-(x^2+y^2)}$ and the circular domain $R = \{(x,y) : x^2 + y^2 \le A^2\}$, a circle of radius A centered at the origin.

(b) Show that as $A \to \infty$ then Volume $\to \pi$. This means the volume between the surface $z = f(x,y) = e^{-(x^2+y^2)}$ and the entire xy-plane is π.

Areas and Average Values with Double Integrals in Polar Coordinates

If the z=f(x,y) is always equal to 1, then $\text{Volume} = \iint_R 1 \cdot dA = \{\text{base area}\}\{\text{height} = 1\} = \text{base area} \cdot$

> **Area in Polar Coordinates**
>
> If R is a closed and bounded region in the polar coordinate plane, then
>
> $\text{Area of R} = \iint_R 1 \cdot dA = \iint_R 1\, r \cdot dr \cdot d\theta$ In polar coordinates $dA = r \cdot dr \cdot d\theta$

The average value of a continuous function z=f(x,y) over a polar coordinate region is the same as we have used for rectangular coordinate regions: Average Value $= \frac{1}{\text{area}} \cdot$ volume.

> **Average Value of a Function in Polar Coordinates**
>
> If R is a closed and bounded region in the polar coordinate plane, and fx,y) is continuous on R, then
>
> $$\text{Average Value of f on R} = \frac{1}{\text{area of R}} \cdot \{\text{volume between f and R}\} = \frac{\iint_R f\, dA}{\iint_R 1\, dA}$$

Example 3: Let $z=f(x,y)=e^{-(x^2+y^2)}$ be the height of a solid ice sculpture over the circular base $R=\{(x,y): x^2+y^2 \le 4\}$. (This is Example 2.) If this sculpture is in a water-tight cylinder and then melts, how high will the resulting water be in the cylinder? (Fig. 6)

Solution: We have already done the calculus for this volume in Example 2: volume = $0.98\,\pi$ so we just need to divide this volume by the area of the circular base, $4\,\pi$. The average height is approximately $0.98/4 = 0.245$.

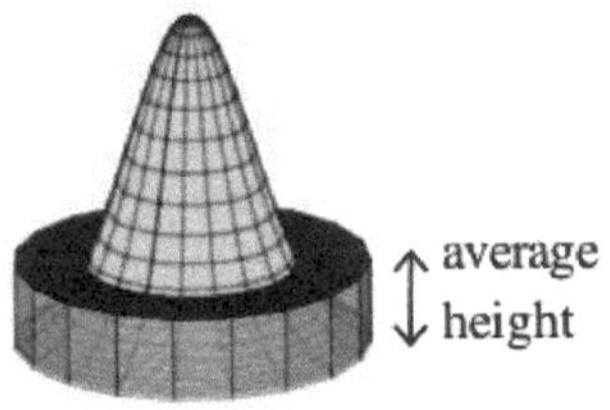

Fig. 6

Practice 3: Find the average value of $z=f(x,y)=5+x+2y$ on the domain $R=\{(x,y): x^2+y^2 \le 4\}$. (This is Example 1.)

Problems

For each given region R, decide whether to use polar or rectangular coordinates to evaluate a double integral with domain R.

1.

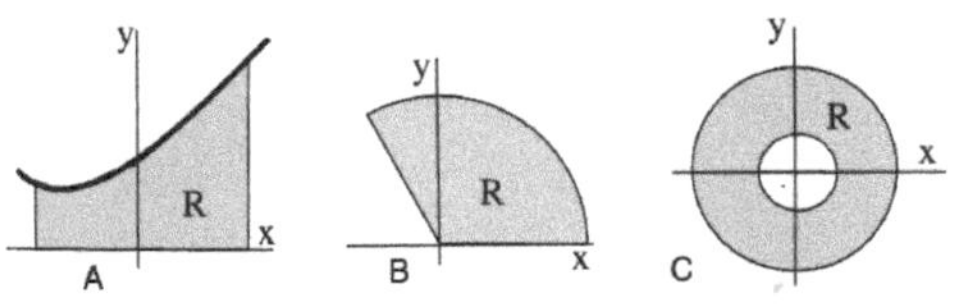

2.

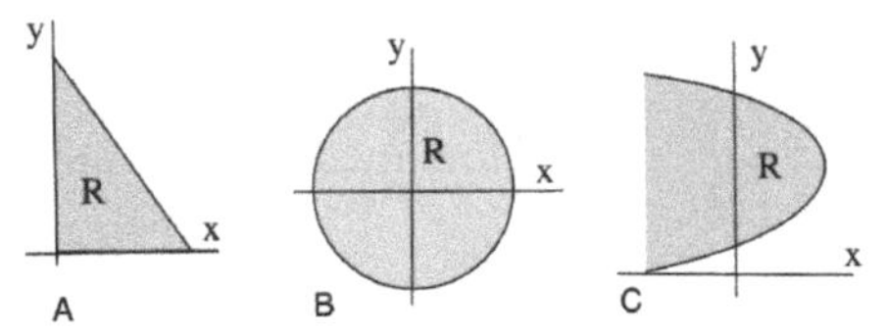

3.

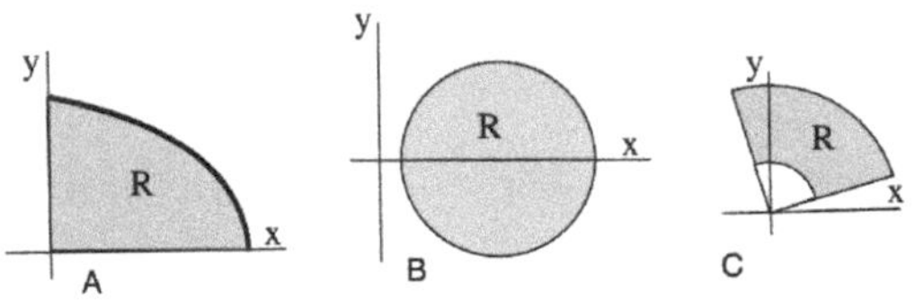

4.

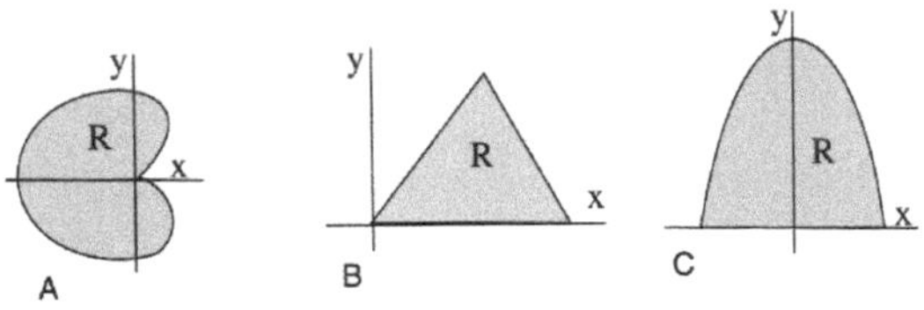

For Problems 5 to 8 find the volume between the surface $z=f(x,y)$ and the given region R.

5. $f(x,y)=7+3x+2y$ with $R=\left\{(x,y):\ x^2+y^2 \le 9\right\}$.

6. $f(x,y)=9-2x+4y$ with $R=\left\{(x,y):\ x^2+y^2 \le 5\right\}$.

7. $f(x,y)=A+Bx+Cy$ with $R=\left\{(x,y):\ x^2+y^2 \le D^2\right\}$. (A, B, C, D are positive constants.)

8. $f(x,y)=A+Bx+Cy$ with $R=\left\{(x,y):\ E^2 \le x^2+y^2 \le D^2\right\}$. (A, B, C, D, E are positive constants.)

For Problems 9 to 12 sketch the domain of integration, use polar coordinates to evaluate each double integral.

9. Evaluate $\iint_R \sqrt{9-x^2-y^2}\ dA$ where $R=\{(x,y):\ x^2+y^2\le 9\}$.

10. Evaluate $\iint_R \sin(x^2+y^2)\ dA$ where $R=\{(x,y):\ 0\le x,\ 0\le y,\ x^2+y^2\le 9\}$.

11. Evaluate $\iint_R 2+xy\ dA$ where R is the region in inside the circle $x^2+y^2=4$ and above the x-axis..

12. Evaluate $\iint_R e^{x^2+y^2}\ dA$ where R is the set of (x,y) more than 1 unit and less than 3 units from the origin.

For Problems 13 to 18 use polar coordinates to find the volume of each solid.

13. Under the plane $z=5+2x+3y$ and above the disk $x^2+y^2\le 16$.

14. Under the paraboloid $z=x^2+y^2$ and above the disk $x^2+y^2\le 9$ for (x,y) in the first quadrant.

15. Between the surfaces $z=1+x+y$ and $z=8+2x+3y$ for $x^2+y^2\le 1$ and $0\le y$.

16. Above the paraboloid $z=6-x^2-y^2$ and under the plane $z=9$ for $x^2+y^2\le 4$.

17. Between the surface $z=f(x,y)=\dfrac{1}{1+x^2+y^2}$ and the xy-plane for (x,y) in the first quadrant and $1\le x^2+y^2\le 4$.

18. (a) Between $z=\dfrac{1}{(1+x^2+y^2)^3}$ and the disk $x^2+y^2\le C^2$.

(b) Between $z=\dfrac{1}{(1+x^2+y^2)^3}$ and the entire xy-plane.

In problems 19 to 24, change the rectangular coordinate integral into an equivalent polar integral and evaluate the polar integral. It is usually helpful to sketch the domain of the integral.

19. $\displaystyle\int_{-1}^{1}\int_{0}^{\sqrt{1-x^2}} dy\ dx$

20. $\displaystyle\int_{-1}^{1}\int_{0}^{\sqrt{1-x^2}} (x^2+y^2)\ dy\ dx$

21. $\displaystyle\int_{0}^{1}\int_{0}^{\sqrt{1-x^2}} y\ dy\ dx$

22. $\displaystyle\int_{0}^{1/\sqrt{2}}\int_{y}^{\sqrt{1-y^2}} (x^2+y^2)\ dx\ dy$

23. $\displaystyle\int_{0}^{1}\int_{0}^{\sqrt{1-x^2}} e^{-(x^2+y^2)}\ dy\ dx$

24. $\displaystyle\int_{0}^{\infty}\int_{0}^{\infty} e^{-(x^2+y^2)}\ dy\ dx$

For Problems 25 to 29, find the average value of the function on the given region R.

25. $f(x,y) = 7 + 3x + 2y$ with $R = \{(x,y):\ x^2 + y^2 \le 9\}$.

26. $f(x,y) = 2 + xy$ for $R = \{\text{top half of the disk } x^2 + y^2 \le 4\}$.

27. $f(x,y) = 5 + 2x + 3y$ with $R = \{(x,y):\ x^2 + y^2 \le 16\}$.

28. $f(x,y) = x^2 + y^2$ for R={part of the disk $x^2 + y^2 \le 9$ in the first quadrant}

29. A sprinkler (located at the origin) sprays water so after one hour the depth at location (x,y) feet is $f(x,y) = K \cdot e^{-(x^2+y^2)}$ feet.

(a) How much water reaches the annulus $2 \le r \le 4$ and the annulus $8 \le r \le 10$ in one hour?

(b) What is the average amount of water (depth per square foot) of water at (x,y) in each annulus after one hour?

(c) Why is this a poor design for a sprinkler?

30. $f(x) = K \cdot e^{-\left(\frac{x^2}{2}\right)}$ is the normal probability distribution for a population with mean 0 and standard deviation 1, and is extremely important in probability theory and applications. Unfortunately, it does not have a "nice" antiderivative in terms of elementary functions, but we can use double integrals in polar coordinates to evaluate $\int_{-\infty}^{\infty} e^{-x^2}\, dx$.

(a) The rectangular coordinate double integral of $f(x,y) = e^{-(x^2+y^2)} = e^{-x^2} \cdot e^{-x^2}$ with

R={square -C≤x≤C, -C≤y<C} is

$$\int_{-C}^{C}\int_{-C}^{C} e^{-(x^2+y^2)}\, dx\cdot dy = \int_{-C}^{C}\int_{-C}^{C} e^{-x^2}\cdot e^{-y^2}\, dx\cdot dy$$

$$= \int_{-C}^{C} e^{-y^2}\left(\int_{-C}^{C} e^{-x^2} dx\right) dy = \left(\int_{-C}^{C} e^{-x^2} dx\right)\left(\int_{-C}^{C} e^{-y^2} dy\right) = \left(\int_{-C}^{C} e^{-x^2} dx\right)^2.$$

So the

$$\iint_{xy-plane} e^{-(x^2+y^2)}\, dA = \lim_{C\to\infty} \int_{-C}^{C}\int_{-C}^{C} e^{-(x^2+y^2)}\, dx\cdot dy = \lim_{C\to\infty} \left(\int_{-C}^{C} e^{-x^2} dx\right)^2 = \left(\int_{-\infty}^{\infty} e^{-x^2} dx\right)^2$$

But from Practice 2 we know that $\iint_{xy-plane} e^{-(x^2+y^2)}\, dA = \pi$ so $\int_{-\infty}^{\infty} e^{-x^2} dx = \sqrt{\pi}$.

Finally, changing the variable to $u = \frac{x}{\sqrt{2}}$ we get $\int_{-\infty}^{\infty} e^{-\left(\frac{x^2}{2}\right)} dx = \int_{-\infty}^{\infty} e^{-u^2}\sqrt{2}\cdot du = \sqrt{2\pi}$.

That was a lot of work, but this is a very important integral.

Practice Answers

Practice 1: The only difference from Example 1 is that the domain angle θ goes from 0 to π instead of from 0 to 2π so the calculation is the same until we get to the final evaluation:

$$\text{volume} = \iint_R f(x,y)\cdot dA = \int_{\theta=0}^{\theta=2\pi}\int_{r=0}^{r=2} \{5+r\cdot\cos(\theta)+2\cdot r\cdot\sin(\theta)\}\cdot r\cdot dr\cdot d\theta$$

$$= 10\theta+\frac{8}{3}\cdot\sin(\theta)-\frac{16}{3}\cdot\cos(\theta)\Big|_{\theta=0}^{\theta=\pi} = \left\{10\pi+0+\frac{16}{3}\right\}-\left\{0+0-\frac{16}{3}\right\} = 10\pi+\frac{32}{3}$$

The new domain may be half the area of the original domain, but the function is larger over the new domain than over the original domain. Symmetry is very powerful, but we need to be careful and only use it when it is justified.

Practice 2: (a) The only change from Example 2 is that now the radius r goes from 0 to A instead of from 0 to 2 so most of the calculations are the same:

$$\int_{\theta=0}^{\theta=2\pi}\int_{r=0}^{r=A} e^{-r^2}\cdot r\cdot dr\cdot d\theta = \frac{1}{2}\left(1-e^{-A^2}\right)\theta\Big|_{\theta=0}^{\theta=2\pi} = \left(1-e^{-A^2}\right)\pi\ .$$

(b) $$\lim_{A\to\infty} \text{Volume} = \lim_{A\to\infty}\left(1-e^{-A^2}\right)\pi = \pi$$

Practice 3: From Example 1 we know the volume is 20π, and the area of the circular base is 4π so the average value is 5. (Fig. 7)

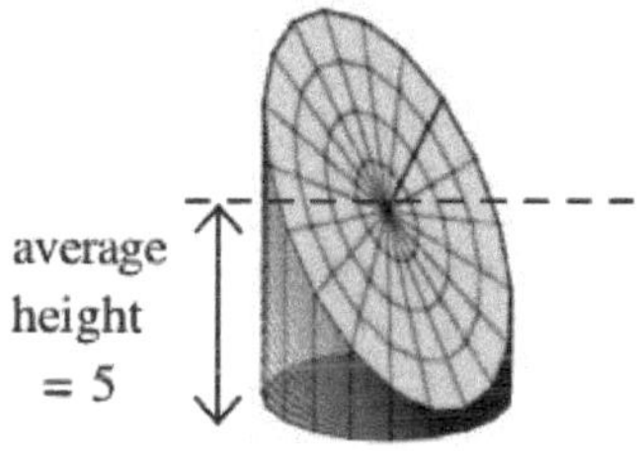

Fig. 7: f(x,y)=5+x+2y

14.4 APPLICATIONS OF DOUBLE INTEGRALS

In Section 5.4 we used single integrals to determine the mass, moments about each axis, and the center of mass of a thin plate (lamina) with uniform density δ. In this section we will use double integrals to extend those ideas and calculations to thin plates with varying densities.

Uniform density for a 2D region between f and the x-axis for $a \le x \le b$

Total Mass: $M = \delta \cdot \int_a^b f(x)\, dx$

Moments: about y-axis $M_y = \delta \cdot \int_a^b x \cdot f(x)\, dx$ about x-axis $M_x = \delta \cdot \int_a^b \frac{1}{2} f^2(x)\, dx$

Center of mass: $\bar{x} = \frac{M_y}{M}$, $\bar{y} = \frac{M_x}{M}$

If the density of the plate depends on the location (x,y) on the plate, then δ is a function of x and y: $\delta(x,y)$ with units of the form (mass)/area such as kg/m^2 .

Nonuniform density for a 2D region R in the xy-plane

Area: $Area = \iint_R 1\, dA$

Total Mass: $M = \iint_R \delta(x,y)\, dA$

Moments: about y-axis $M_y = \iint_R x \cdot \delta(x,y)\, dA$ about x-axis $M_x = \iint_R y \cdot \delta(x,y)\, dA$

Center of mass: $\bar{x} = \frac{M_y}{M}$, $\bar{y} = \frac{M_x}{M}$

Note: Be careful to use the x with M_y since the little dA piece is x units from the y-axis. Similarly, use the y with the M_x formula since the little dA piece is y units from the x-axis.

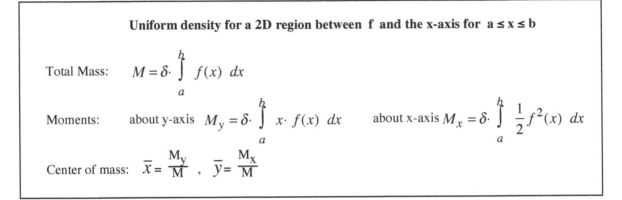

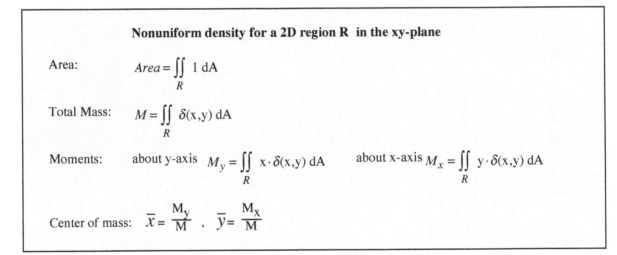

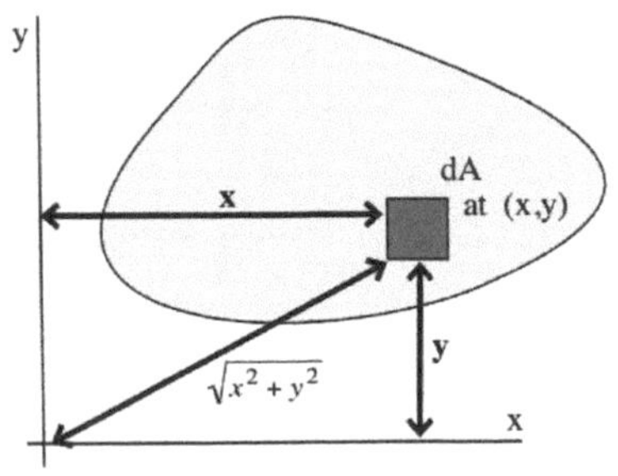

Fig. 1

Example 1: $R = \{(x,y) : 0 \le x \le 2 \text{ and } 0 \le y \le 4 - x^2\}$ (Fig. 2).

(a) Determine the center of mass of R if the region has uniform density δ.

(b) Determine the center of mass of R if $\delta(x,y) = x \cdot y^2$.

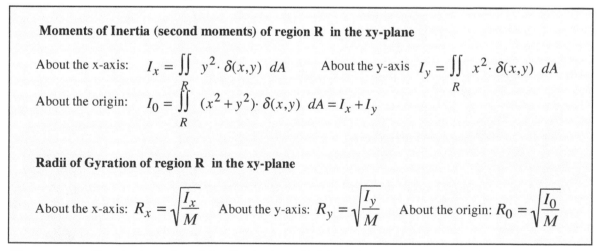

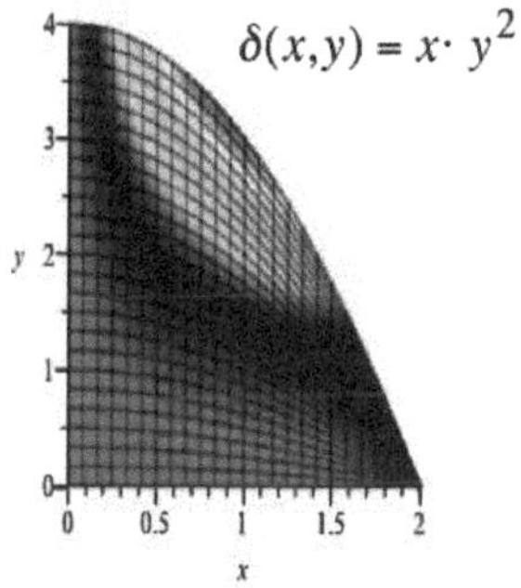

Fig. 2

Solution: (a) Using double integrals,

$$M = \iint_R \delta(x,y)\ dAM = \int_{x=0}^{2}\int_{y=0}^{4-x^2} 1\ dy\ dx = \frac{16}{3}.$$

$$M_y = \int_{x=0}^{2}\int_{y=0}^{4-x^2} x\ dy\ dx = 4 \text{ and } M_x = \int_{x=0}^{2}\int_{y=0}^{4-x^2} y\ dy\ dx = \frac{128}{15}$$

so $\bar{x} = \frac{3}{4}$ and $\bar{y} = \frac{8}{5}$.

(b) $\delta(x,y) = x \cdot y^2$, $M = \int_{x=0}^{2}\int_{y=0}^{4-x^2} x \cdot y^2\ dy\ dx = \frac{32}{3}$, $M_y = \int_{x=0}^{2}\int_{y=0}^{4-x^2} x \cdot x \cdot y^2\ dy\ dx = \frac{8192}{945}$

and $M_x = \int_{x=0}^{2}\int_{y=0}^{4-x^2} y \cdot x \cdot y^2\ dy\ dx = \frac{128}{5}$ so $\bar{x} = \frac{256}{315}$ and $\bar{y} = \frac{12}{5}$.

Practice 1: $R = \{(x,y) : 0 \le x \le 2 \text{ and } 0 \le y \le 4 - x^2\}$. Determine the center of mass of R if $\delta(x,y) = x^2 \cdot y$.

Moments of Inertia (second moments) and Radii of Gyration

Moments of inertia are needed to calculate the kinetic energy of rotating objects and also for formulas for stiffness of beams. Note that the second moments formulas are very similar to the first moment formulas but that they use the squares x^2 and y^2 of the lever arms distances of the dA piece from the axes instead of just x and y.

> **Moments of Inertia (second moments) of region R in the xy-plane**
>
> About the x-axis: $I_x = \iint_R y^2 \cdot \delta(x,y)\ dA$ About the y-axis $I_y = \iint_R x^2 \cdot \delta(x,y)\ dA$
>
> About the origin: $I_0 = \iint_R (x^2 + y^2) \cdot \delta(x,y)\ dA = I_x + I_y$
>
> **Radii of Gyration of region R in the xy-plane**
>
> About the x-axis: $R_x = \sqrt{\frac{I_x}{M}}$ About the y-axis: $R_y = \sqrt{\frac{I_y}{M}}$ About the origin: $R_0 = \sqrt{\frac{I_0}{M}}$

The moment of inertia about the origin, I_0, is also called the **polar moment**, and that integral uses the square of the distance of the dA piece from the origin.

The radius of gyration R_x is the distance from the x-axis that a point with mass M must be in order to give the same moment of inertia I_x: $I_x = M \cdot R_x^2$.

Example 2: If the units of x and y are meters and the units of $\delta(x,y)$ are kg/m^2 , determine the units for I_x and R_x.

Solution: dA = $dx \cdot dy$ has units m^2, so $I_x = \iint_R y^2 \cdot \delta(x,y)\ dA$ has units $m^2 \cdot \frac{kg}{m^2} \cdot m^2 = kg \cdot m^2$.

$$R_x = \sqrt{\frac{I_x}{M}} = \sqrt{\frac{kg \cdot m^2}{kg}} = m.$$

Practice 2: If the units of x and y are feet and the units of $\delta(x,y)$ are $\frac{slug}{ft^2}$, determine the units for I_x and R_x. (One slug of mass in the British system is approximately 14.594 kg.)

Example 3: Suppose x and y are given in meters and the triangular region R={(x,y): 0≤x≤2 and 0≤y≤4-2x} (Fig. 3) has uniform density $\delta(x,y) = \delta\ \frac{kg}{m^2}$.

Calculate M, $\bar{x}$, I_x , and R_x.

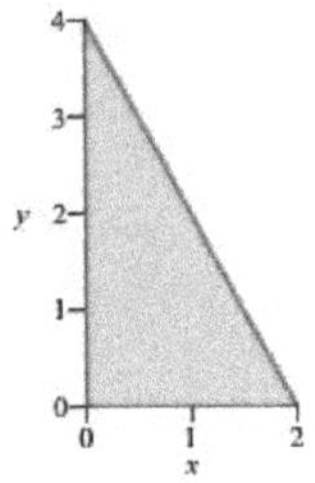

Fig. 3

Solution: $M = \iint_R \delta\ dA = \int_0^2 \int_0^{4-2x} \delta\ dy\ dx = \int_0^2 \delta \cdot (4-2x)\ dx = 4\delta$ kg.

$$M_y = \iint_R x \cdot \delta\ dA = \int_0^2 \int_0^{4-2x} x \cdot \delta\ dy\ dx = \int_0^2 \delta(4x - 2x^2)\ dx = \frac{8}{3}\delta\ kg \cdot m$$

$$I_x = \iint_R y^2 \cdot \delta\ dA = \int_0^2 \int_0^{4-2x} y^2 \cdot \delta\ dy\ dx = \int_0^2 \frac{\delta}{3}(4-2x)^3\ dx = \frac{32}{3}\ kg \cdot m^2$$

so $\bar{x} = \frac{M_y}{M} = \frac{2}{3}\ m$ and $R_x = \sqrt{\frac{I_x}{M}} = \sqrt{\frac{8}{3}\ m^2} \approx 1.63\ m$.

Practice 3: Calculate $\bar{y}$, I_y , and R_y for the region R in Example 3.

Example 4: R is the quarter circle of radius 2 in the first quadrant with density $\delta(r,\theta) = 1 + r$. (Fig. 4) Calculate M, $\bar{x}$, I_x , and R_x.

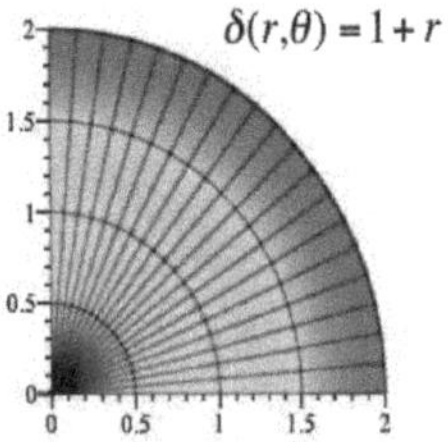

Fig. 4

Solution: $M = \iint_R \delta\ dA = \int_{\theta=0}^{\pi/2} \int_{r=0}^{2} (1+r) \cdot r\ dr\ d\theta = \int_0^{\pi/2} \frac{14}{3}\ d\theta = \frac{7}{3}\pi$

$$M_y = \iint_R x \cdot \delta \; dA = \int_{\theta=0}^{\pi/2} \int_{r=0}^{2} r \cdot \cos(\theta) \cdot (1+r) \cdot r \; dr \; d\theta = \int_0^{\pi/2} \frac{20}{3} \cos(\theta) \; d\theta = \frac{20}{3}$$

$$I_x = \iint_R y^2 \cdot \delta \; dA = \int_{\theta=0}^{\pi/2} \int_{r=0}^{2} r^2 \cdot \sin^2(\theta) \cdot (1+r) \cdot r \; dr \; d\theta = \int_0^{\pi/2} \frac{52}{5} \sin^2(\theta) \; d\theta = \frac{13}{5}\pi$$

so $\bar{x} = \frac{M_y}{M} = \frac{20}{7\pi} \approx 0.909$ and $R_x = \sqrt{\frac{I_x}{M}} = \sqrt{\frac{39}{35}} \approx 1.056$.

Note: Because of the symmetry of the R and δ, then $\bar{y} = \bar{x}$, $I_y = I_x$, and $R_y = R_x$.

Problems

In problems 1 to 8 use double integrals to calculate the area, total mass, the moments about each axis, and the center of mass of the region. Plot the location of the center of mass on the region.

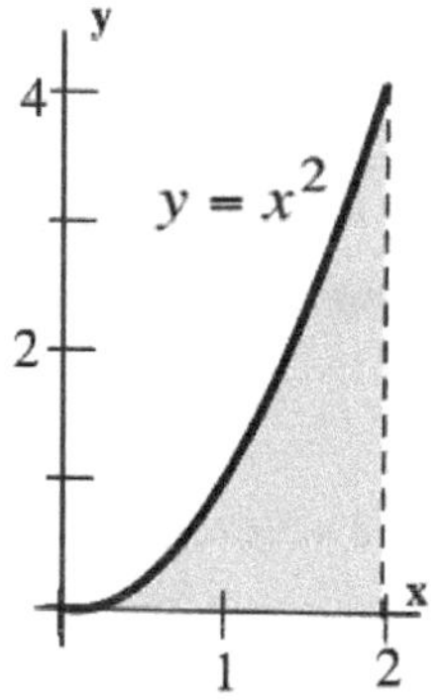

Fig. 5

1. R is the rectangular region bounded by the x and y axes and the lines x=2 and y=4 with $\delta = 1$.

2. R is the rectangular region bounded by the x and y axes and the lines x=2 and y=4 with $\delta = xy$.

3. R is the shaded region in Fig. 5 and $\delta(x,y) = 1 + x$.

4. R is the shaded region in Fig. 5 and $\delta(x,y) = 1 + y$.

5. R={(x,y): 0≤x≤3, 0≤y≤1+x} and $\delta(x,y) = x + y$.

6. R={(x,y): 0≤x≤3, 0≤y≤1+x} and $\delta(x,y) = 1 + y$.

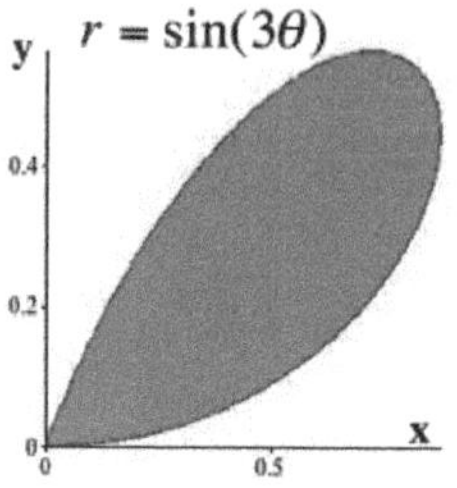

Fig. 6

7. R is the shaded region in Fig. 6 and $\delta(r,\theta) = 1$.

8. R is the shaded region in Fig. 6 and $\delta(r,\theta) = r$.

9. R={(r, θ): $0 \le \theta \le \pi$, $0 \le r \le 1 + \cos(\theta)$} and $\delta(r,\theta) = r$. (Fig. 7)

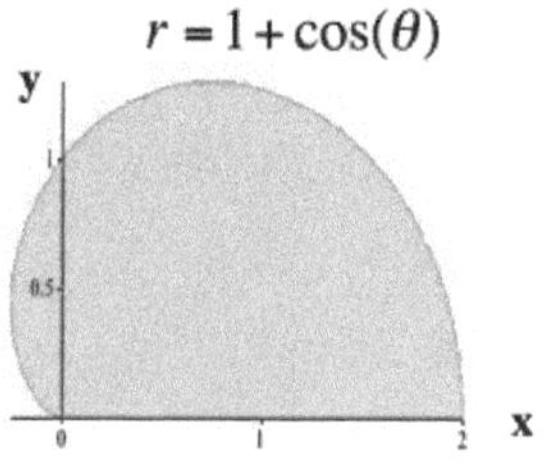

Fig. 7

10. R={(r, θ): $0 \le \theta \le \pi$, $0 \le r \le 1 + \cos(\theta)$} and $\delta(r,\theta) = 1 + r$. (Fig. 7)

11. Calculate the values of I_x and R_x for the region in Problem 1.

12. Calculate the values of I_y and R_y for the region in Problem 1.

13. Calculate the values of I_x and R_x for the region in Problem 2.

14. Calculate the values of I_y and R_y for the region in Problem 2.

In Problems 15 to 20, use the fact that the attractive gravitational force between two points with masses M and m at a distance of r is $\text{force} = \frac{\text{GMm}}{r^2}$. (It helps to sketch the regions.)

15. Represent the total force between a point mass of 10 kg at the origin and a bar from x=2 to x=4 m on the x-axis with a mass of 8 kg. (This is a single integral.)

16. 15. Represent the total force between a point mass of M kg at the origin and a bar from a to b m (0<a<b) m on the x-axis with a mass of m kg. (This is a single integral.)

17. Represent the total force between a bar along the x-axis from 0 to 2 with mass of 10 kg and another bar on the x-axis from 4 to 7 with a mass of 9 kg. (This is a double integral.)

18. Represent the total force between a bar along the x-axis from a to b with total mass M and another bar on the x-axis from c to d with a mass of m (a<b<c<d). (This is a double integral.)

Practice Solutions

Practice 1: $R = \{(x,y): 0 \le x \le 2 \text{ and } 0 \le y \le 4 - x^2\}$. Determine the center of mass of R if $\delta(x,y) = x^2 \cdot y$.

$$(\delta(x,y) = x^2 \cdot y \,,\ M = \int_{x=0}^{2} \int_{y=0}^{4-x^2} x^2 \cdot y \; dy \; dx = \frac{512}{105},\ M_y = \int_{x=0}^{2} \int_{y=0}^{4-x^2} x \cdot x^2 \cdot y \; dy \; dx = \frac{16}{3}$$

$$\text{and } M_x = \int_{x=0}^{2} \int_{y=0}^{4-x^2} y \cdot x^2 \cdot y \; dy \; dx = \frac{8192}{945} \quad \text{so } \bar{x} = \frac{35}{32} \text{ and } \bar{y} = \frac{16}{9} .$$

Practice 2: $dA = dx \cdot dy$ has units ft^2, so $I_x = \iint_R y^2 \cdot \delta(x,y) \; dA$ has units $ft^2 \cdot \frac{slug}{ft^2} \cdot ft^2 = slug \cdot ft^2$.

$$M = \iint_R \delta(x,y) \; dA = \frac{slug}{ft^2} \cdot ft^2 = slug \text{ so } R_x = \sqrt{\frac{I_x}{M}} = \sqrt{\frac{slug \cdot ft^2}{slug}} = feet .$$

Practice 3: $M_x = \iint_R y \cdot \delta \; dA = \int_0^2 \int_0^{4-2x} y \cdot \delta \; dy \; dx = \int_0^2 \delta \cdot \frac{1}{2}(4-2x)^2 \; dx = \frac{16}{3}\delta$

$$\text{so } \bar{y} = \frac{16/3 \ \delta}{4 \ \delta} = \frac{4}{3} \ m .$$

$$I_y = \iint_R x^2 \cdot \delta \; dA = \int_0^2 \int_0^{4-2x} x^2 \cdot \delta \; dy \; dx = \int_0^2 \delta \cdot x \cdot (4-2x) \; dx = \frac{8}{3}\delta \ kg \cdot m^2$$

$$\text{so } R_y = \sqrt{\frac{I_y}{M}} = \sqrt{\frac{8/3 \ \delta}{4 \ \delta} \ m^2} = \sqrt{\frac{2}{3}} \approx 0.816 \ m$$

surface of revolution

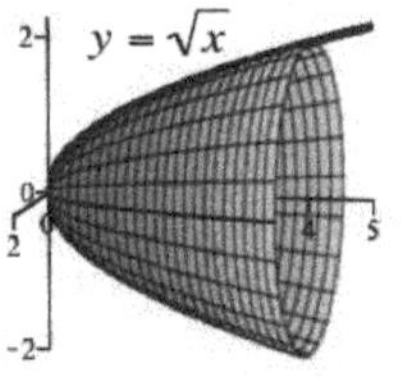

Fig. 1

14.5 SURFACE AREAS USING DOUBLE INTEGRALS

In Section 5.2 we determined a method for calculating the area of a surface of revolution (Fig. 1). Here we will build a way to calculate the area of a surface of the form $z = f(x,y)$ over a region R (Fig. 2). Sometimes both methods can be used and in that case they both give the same result. There is another type of surface called a parametric surface (Fig. 3) that is even more general, and we will build a way to determine those surface areas in Section 15.7.

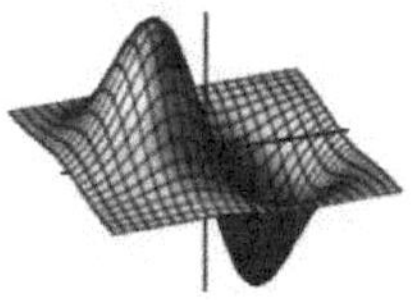

Fig. 2

To make the derivation and figures simpler, we assume that R is a rectangle in the xy-plane and that $z=f(x,y)\geq 0$ in R. As we have done before, we start by partitioning the domain into small Δx by Δy rectangles and note that the area ΔS of the tangent plane (Fig. 4) above one of these little ΔA rectangles has approximately the same area as the surface area above the ΔA rectangle:. If we can represent the sides of the tilted tangent plane above the ΔA rectangle as vectors, then we can use the cross product of those vectors to determine the area of the rectangle.

parametric surface

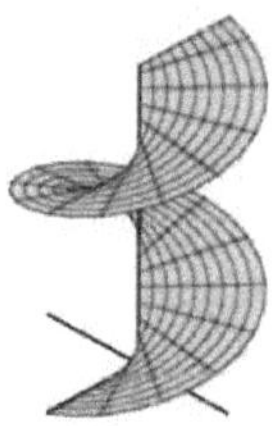

Fig. 3

Starting at one corner (x,y) of a little rectangle in the domain, and moving Δx in the x direction, the vector A along the tangent plane is $A = \langle \Delta x,\ 0,\ f_x(x,y)\cdot\Delta x \rangle$. Similarly, starting at (x,y) and moving in the y direction, the vector B along the tangent plane is $B = \langle 0,\ \Delta y,\ f_y(x,y)\cdot\Delta y \rangle$. Then the area of the tangent plane above the ΔA rectangle is the magnitude of the cross product of A and B:

$$\boxed{\Delta S = \{\text{tangent plane area}\} = |\, A\text{x}B \,|\,.}$$

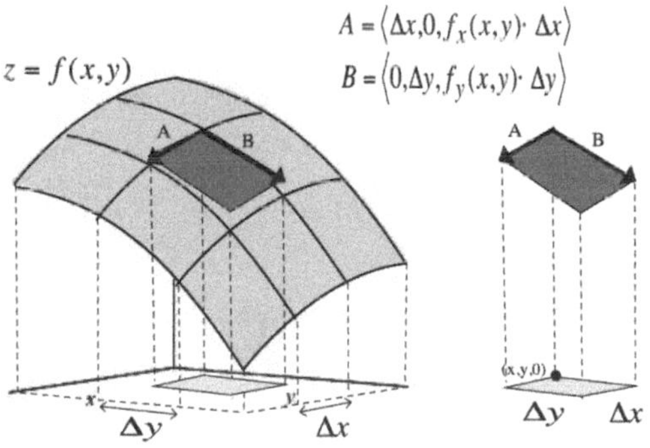

Fig. 4

Example 1: The function in Fig, 4 is $f(x,y) = 7 - \frac{1}{2}x^2 - y^2$. Determine the area of the little rectangle at the point (2,1,4) with $\Delta x = 0.3$ and $\Delta y = 0.1$.

Solution: $f_x(2,1) = -2$ and $f_y(2,1) = -2$ so $A = \langle 0.3,\ 0,\ -0.6\rangle$ and $B = \langle 0,\ 2,\ -0.2\rangle$.

$$A\text{x}B = \begin{vmatrix} i & j & k \\ 0.3 & 0 & -0.6 \\ 0 & 0.1 & -0.2 \end{vmatrix} = \langle 0.06,\ 0.06,\ 0.03\rangle \text{ so area} = |\, A\text{x}B \,| = 0.09\ ,$$

Practice 1: Determine the area of the little rectangle for the Example 1 function at the point (1, 1, 5.5) with $\Delta x = 0.2$ and $\Delta y = 0.3$.

In the general case $A = \langle \Delta x,\ 0,\ f_x(x,y) \rangle$ and $B = \langle 0,\ \Delta y,\ f_y(x,y) \rangle$ so

$$AxB = \begin{vmatrix} i & j & k \\ \Delta x & 0 & f_x(x,y)\cdot \Delta x \\ 0 & \Delta y & f_y(x,y)\cdot \Delta y \end{vmatrix} = \langle -\Delta y\cdot f_x \cdot \Delta x,\ -\Delta x\cdot f_y \cdot \Delta y,\ \Delta x\cdot \Delta y \rangle = \langle -f_x,\ -f_y,\ 1 \rangle \cdot \Delta A$$

and $|AxB| = \left(\sqrt{(f_x)^2 + (f_y)^2 + 1} \right) \cdot \Delta A = \Delta S$.

The total surface area is the accumulation of all of these little areas:

Surface area $A \approx \sum \sum \Delta S = \sum \sum \left(\sqrt{(f_x)^2 + (f_y)^2 + 1} \right) \cdot \Delta A \to \iint_R \left(\sqrt{(f_x)^2 + (f_y)^2 + 1} \right) \cdot dA$

For z=f(x,y), Surface Area = $\displaystyle \iint_R \sqrt{1 + \left(\frac{\partial z}{\partial x}\right)^2 + \left(\frac{\partial z}{\partial y}\right)^2} \cdot dA$

You should recognize the similarity of this formula to the formula for arc length from Section 5.2:

$$L = \int \sqrt{1 + \left(\frac{dy}{dx}\right)^2}\ dx ,$$

Example 2: Determine the surface area of the plane $f(x,y) = 1 + 2x + y$ over the rectangular region $R = \{(x,y):\ 0 \le x \le 2,\ 0 \le y \le 3\}$. (Fig. 5)

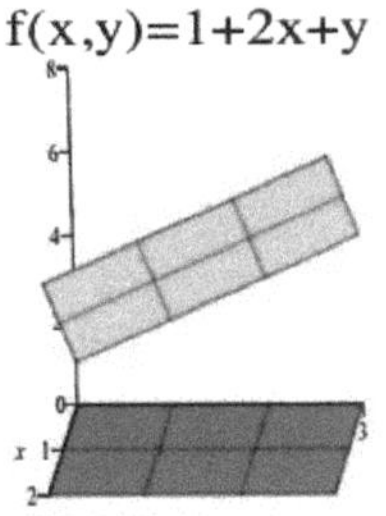

Fig. 5

Solution: $\dfrac{\partial z}{\partial x} = 2$ and $\dfrac{\partial z}{\partial y} = 1$ so

Surface area $= \displaystyle \int_{x=0}^{2} \int_{y=0}^{3} \sqrt{1 + (2)^2 + (1)^2}\ dy\ dx = \int_{x=0}^{2} 3\sqrt{6}\ dx = 6\sqrt{6}$

Practice 2: Determine the surface area of the plane $f(x,y) = 3 + 8x + 4y$ over the rectangular region $R = \{(x,y):\ 0 \le x \le 4,\ 0 \le y \le 3\}$.

The surface area formula also works for domains that are not rectangular, and sometimes polar coordinates make the evaluation easier.

Example 3: Determine the surface area of the plane $f(x,y) = 10 - 2x - 3y$ over the circular disk $R = \{(x,y):\ 0 \le x^2 + y^2 \le 4\}$. (Fig. 6)

f(x,y)=10-2x-3y

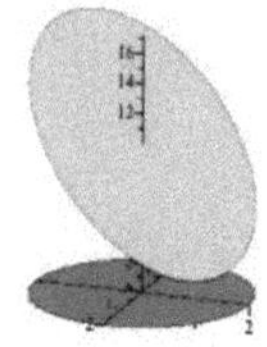

Fig. 6

Solution: $\frac{\partial z}{\partial x} = -2$ and $\frac{\partial z}{\partial y} = -3$ so Surface area = $\iint\limits_R \sqrt{1 + (-2)^2 + (-3)^2}\cdot dA$

Because of the symmetry of R, this is easier to evaluate using polar coordinates:

$$\text{Surface area} = \int_{\theta=0}^{2\pi}\int_{r=0}^{2} \sqrt{14}\ r\cdot dr\cdot d\theta = \int_{\theta=0}^{2\pi} 2\sqrt{14}\cdot d\theta = 4\pi\sqrt{14}\,.$$

$f(x,y) = 5 + x^2 - y^2$

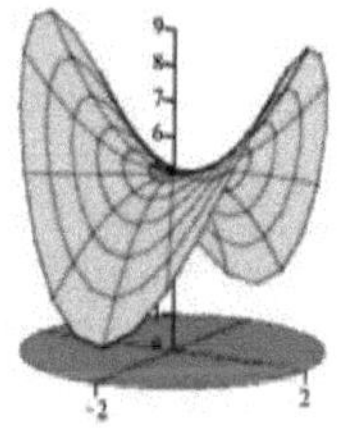

Fig. 7

Example 4: Determine the surface area of the saddle $f(x,y) = 5 + x^2 - y^2$ over the circular disk $R = \{(x,y):\ 0 \le x^2 + y^2 \le 4\}$. (Fig. 7)

Solution: $\frac{\partial z}{\partial x} = 2x$ and $\frac{\partial z}{\partial y} = -2y$ so

$$\text{Surface area} = \iint\limits_R \sqrt{1 + (2x)^2 + (-2y)^2}\cdot dA = \iint\limits_R \sqrt{1 + 4(x^2 + y^2)}\cdot dA$$

Again, polar coordinates make this easier:

$$\text{Surface area} = \int_{\theta=0}^{2\pi}\int_{r=0}^{2} \sqrt{1 + 4r^2}\cdot r\cdot dr\cdot d\theta$$

$$= \int_{\theta=0}^{2\pi} \frac{1}{12}(17)^{3/2} - \frac{1}{12}\ d\theta = \left(\frac{1}{12}(17)^{3/2} - \frac{1}{12}\right)\cdot 2\pi$$

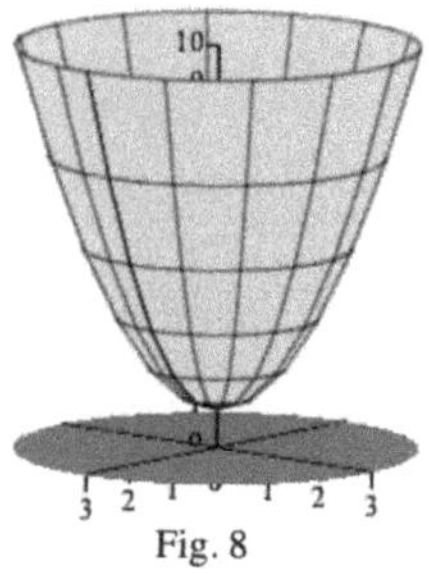

Fig. 8

Practice 3: Determine the surface area of the paraboloid $f(x,y) = 1 + x^2 + y^2$ over the circular disk $R = \{(x,y):\ 0 \le x^2 + y^2 \le 9\}$. (Fig. 8)

These examples and practice problems were chosen so that it was possible to evaluate the integrals "by hand." Unfortunately that is rarely the situation. Once we take partial derivatives, square them, add those squares and a 1, and then take a square root the result is usually not an integral that we can evaluate by hand, at least not easily If we just change the Example 4 function slightly to be $f(x,y) = 5 + 2x - y^2$ then $f_x(x,y) = 2$ and $f_y(x,y) = -2y$ so the surface area integral is $\iint\limits_R \sqrt{5 + 4y^2}\ dA$. This is a rather difficult antidrivative which involves the inverse hyperbolic sine function. In many cases the surface area involves integrals that do not have antiderivatives involving only elementary functions and we need to resort to software such as Maple or Mathematica.

Example 5: The formula for Fig. 2 is $f(x,y) = x \cdot e^{-x^2-y^2}$ and the graph of f is over the rectangle $-2 \le x \le 2$ and $-2 \le y \le 2$. Represent the surface area using double integrals.

Solution: $f_x(x,y) = x \cdot \left(e^{-x^2-y^2}\right)(-2x) + \left(e^{-x^2-y^2}\right) = (-2x^2+1) \cdot e^{-x^2-y^2}$ and

$$f_y(x,y) = x \cdot \left(e^{-x^2-y^2}\right)(-2y) = -2xy \cdot e^{-x^2-y^2} \text{ so}$$

$$\text{Surface area} = \int_{x=-2}^{2} \int_{y=-2}^{2} \sqrt{1 + \left((-2x^2+1)^2 + 4x^2y^2\right)\left(e^{-x^2-y^2}\right)^2} \, dy \, dx \quad \text{Yuck.}$$

But the Maple command

int(sqrt(1+ ((1-2*x^2)^2+4*x^2*y^2)*(exp(-x^2-y^2))^2), x=-2 .. 2, y=-2 ..2);

quickly gives the result 16.72816232 .

Problems

1. Find the area of the surface $f(x,y) = x^2 + y$ over the triangular domain bounded by the x-axis, the line x=2 and the line y=2x.
2. Find the area of the surface $f(x,y) = x^2 + 3y$ over the triangular domain bounded by the x-axis, the line x=2 and the line y=x.
3. Find the area of the surface $f(x,y) = 4x + y^2$ over the triangular domain with vertices (0,0), (0,4) and (2,4).
4. Find the area of the surface $f(x,y) = x + 3y^2$ over the triangular domain with vertices (0,0), (0,4) and (2,4).
5. Find the area of the surface $f(x,y) = 1 + 3x + 4y$ over the domain bounded by the x-axis and the parabola $y = 4 - x^2$.
6. Find the area of the surface $f(x,y) = 2 + 12x + y$ over the domain bounded by the x-axis and the parabola $y = 1 - x^2$.
7. Find the area of the surface $f(x,y) = 5 + 4x + 3y$ over the circular domain $R = \{(x,y):\ x^2 + y^2 \le 9\}$.
8. Find the area of the surface $f(x,y) = 1 + x + y$ over the circular domain $R = \{(x,y):\ x^2 + y^2 \le 4\}$.
9. Find the area of the surface $f(x,y) = 1 + x + y$ over the domain $R = \{(x,y):\ 1 \le x^2 + y^2 \le 9\}$.
10. Find the area of the cone $f(x,y) = \sqrt{x^2 + y^2}$ over the circular domain $R = \{(x,y):\ x^2 + y^2 \le 4\}$.
11. Find the area of the cone $f(x,y) = \sqrt{x^2 + y^2}$ over the domain $R = \{(x,y):\ 1 \le x^2 + y^2 \le 9\}$.
12. Find the area of the paraboloid $f(x,y) = x^2 + y^2$ over the circular domain $R = \{(x,y):\ x^2 + y^2 \le 4\}$.

In problems 13 to 16 set up the integrals representing the surface area of the given function on the given domain. (These may be too difficult to evaluate by hand.)

13. $f(x,y) = x \cdot y^2$ on the domain $R = \{(x,y): -2 \le x \le 2, x^2 \le y \le 4\}$.

14. $f(x,y) = x^2 + y^3$ on the domain $R = \{(x,y): -2 \le x \le 2, x^2 \le y \le 4\}$.

15. $f(x,y) = 2 + \sin(x) + \cos(y)$ on the domain $R = \{(x,y): 0 \le x \le 2, 0 \le y \le 3\}$.

16. $f(x,y) = 2 + x^3 - y^2$ on the domain $R = \{(x,y): 0 \le x \le 2, 0 \le y \le 1\}$.

Practice Answers

Practice 1: $f_x(1,1) = -1$ and $f_y(1,1) = -2$ so $A = \langle 0.2, 0, -0.2 \rangle$ and $B = \langle 0, 0.3, -0.6 \rangle$

.

$$AxB = \begin{vmatrix} i & j & k \\ 0.2 & 0 & -0.2 \\ 0 & 0.3 & -0.6 \end{vmatrix} = \langle 0.06, 0.12, 0.06 \rangle \text{ so area} = |AxB| \approx 0.146 .$$

Practice 2: $f(x,y) = 3 + 8x + 4y$ so $\frac{\partial z}{\partial x} = 8$ and $\frac{\partial z}{\partial y} = 4$ then

$$\text{Surface area} = \int_{x=0}^{4} \int_{y=0}^{3} \sqrt{1 + (8)^2 + (4)^2} \; dy \; dx = \int_{x=0}^{4} 27 \; dx = 108$$

Practice 3: $\frac{\partial z}{\partial x} = 2x$ and $\frac{\partial z}{\partial y} = 2y$ so

$$\text{Surface area} = \iint_R \sqrt{1 + (2x)^2 + (2y)^2} \cdot dA = \iint_R \sqrt{1 + 4(x^2 + y^2)} \cdot dA$$

Again, polar coordinates make this easier (with a $u = 1 + 4r^2$ substitution).

$$\text{Surface area} = \int_{\theta=0}^{2\pi} \int_{r=0}^{3} \sqrt{1+4r^2} \cdot r \cdot dr \cdot d\theta = \int_{\theta=0}^{2\pi} \frac{1}{12}(37)^{3/2} - \frac{1}{12} \; d\theta = \left(\frac{1}{12}(37)^{3/2} - \frac{1}{12} \right) \cdot 2\pi$$

14.6 TRIPLE INTEGRALS AND APPLICATIONS

Sometimes the value of a continuous function f(x,y,z) depends on the location in 3 dimensions: perhaps we know the density (kg/m^3) at each location (x,y,z) of a 3D object and want to determine the total mass (kg) of the object. Everything in this section is in rectangular coordinates. The next section considers triple integrals in cylindrical and spherical coordinates.

Our strategy is similar to that used to create double integrals, except now our region R is a 3D solid and we partition R into small rectangular cells (boxes) by cuts parallel to the coordinate planes. (Fig. 1) Then the volume of each little box is $\Delta V = \Delta x \cdot \Delta y \cdot \Delta z$. If we want the mass of the little box, it is approximately the density δ at some point (x*, y*, z*) inside each box times the volume of the box:

$\Delta M = \delta(x^*, y^*, z^*) \cdot \Delta V = \delta(x^*, y^*, z^*) \cdot \Delta x \cdot \Delta y \cdot \Delta z$. By adding the approximate masses of all of the boxes together, a triple sum, we can approximate the total mass of the solid:

$$M \approx \sum_{\Delta z} \sum_{\Delta y} \sum_{\Delta x} \delta(x^*,y^*,z^*) \cdot \Delta x \cdot \Delta y \cdot \Delta z$$

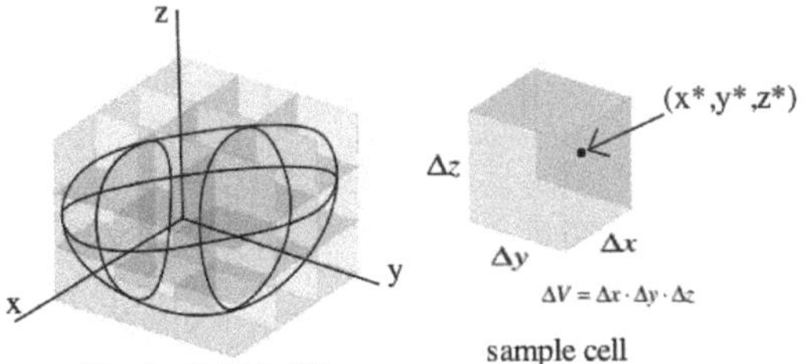

Fig. 1: Solid in 3D

sample cell

As before, letting all of the side lengths of the boxes approach 0, we get a triple integral:

$$\lim_{\Delta x, \Delta y, \Delta z \to 0} \sum_{\Delta z} \sum_{\Delta y} \sum_{\Delta x} \delta(x^*,y^*,z^*) \cdot \Delta x \cdot \Delta y \cdot \Delta z \to \iiint_R \delta(x,y,z)\, dV \cdot$$

Triple integrals have all of the properties that you might expect, and these properties follow from the properties of finite sums.

1) $\iiint_R k \cdot f\, dV = k \cdot \iiint_R f\, dV$
2) $\iiint_R \{f \pm g\}\, dV = \iiint_R f\, dV \pm \iiint_R g\, dV$
3) If f≥g for all points in R, then $\iiint_R f\, dV \geq \iiint_R g\, dV$
4) If the R is partitioned into two subregions R1 and R2 by a smooth surface, then $\iiint_R f\, dV = \iiint_{R1} f\, dV + \iiint_{R2} f\, dV$
5) Volume of R = $\iiint_R 1\, dV$
6) Units of $\iiint_R f\, dV$ are (units of f) . (units of x). (units of y). (units of z)

Evaluating Triple Integrals

Triple integrals are rarely evaluated as limits of triple sums. Instead, we evaluate single integrals, working from the inside out just as we did with double integrals.

Example 1: Evaluate $\iiint_R \text{f dV}$ for $f(x,y,z) = 2x + y + z^2$ on the solid

$$R = \{(x,y,z):\ 0 \le x \le 2,\ 1 \le y \le 4,\ 0 \le z \le 1\}$$

Solution: $\iiint_R \text{f dV} = \int_{z=0}^{1} \int_{y=1}^{4} \int_{x=0}^{2} 2x + y + z^2 \text{ dx dy dz}$

Starting on the inside, $\int_{x=0}^{2} 2x + y + z^2 \ dx = x^2 + xy + xz^2 \Big|_{x=0}^{2} = 4 + 2y + 2z^2$

Next, $\int_{y=1}^{4} \int_{x=0}^{2} 2x + y + z^2 \ dx \ dy = 4y + y^2 + 2yz^2 \Big|_{y=1}^{4} = 12 + 15 + 6z^2$

Finally, $\int_{z=0}^{1} 27 + 6z^2 \ dz = 27z + 2z^3 \Big|_{z=0}^{1} = 29$.

If the x, y and z units are meters, and the f units are kg/m^3 , then $\iiint_R \text{f dV} = 35$ kg.

> 3D Fubini's Theorem
>
> If f is continuous on the domain R,
> then the triple integral can be evaluated in any order that describes R.

In the case of a box [a,b]x[c,d]x[e,f], that means any order of dx, dy and dz gives the same result as long as the end points match the variable: a≤x≤b, c<y≤d and e≤z≤f.

Practice 1: Evaluate $\iiint_R x^3 + y^2 + z \ dV$ for $R = \{(x,y,z):\ 0 \le x \le 2,\ 0 \le y \le,\ 1 \le z \le 3\}$

using two different orders of integration.

Often the most difficult part of working with triple integrals is setting up the order and endpoints of integration . **Keep in mind that the outer integral must have constant endpoints, the middle integral can have at most one variable in the endpoints, and the inside integral can have at most two variables in the end points.**

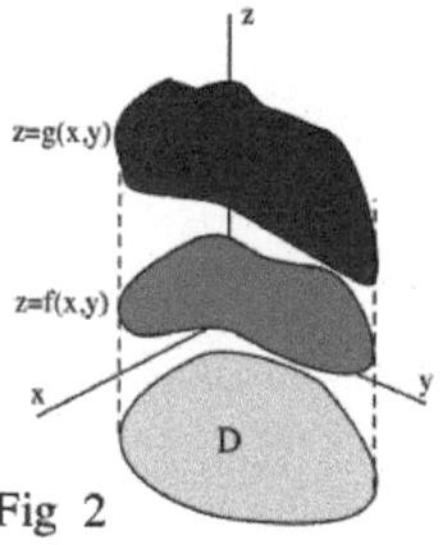

Fig 2

If D is a region in the xy-plane, and $R = \{(x,y,z):\ (x,y) \text{ is in D and } f(x,y) \le z \le g(x,y)\}$

as in Fig. 2, then $\iiint_R F(x,y,z)\, dV = \iint_D \left\{ \int_{z=f}^{z=g} F(x,y,z)\ dz \right\} dA$

Example 2: Determine the volume of the region bounded by $0 \le x \le 1$, $0 \le y \le 3$ and z between plane $z = f(x,y) = 5 - 2x - y$ and the surface $z = g(x,y) = 13 - 3x^2 - y^2$ as shown in Fig. 3.

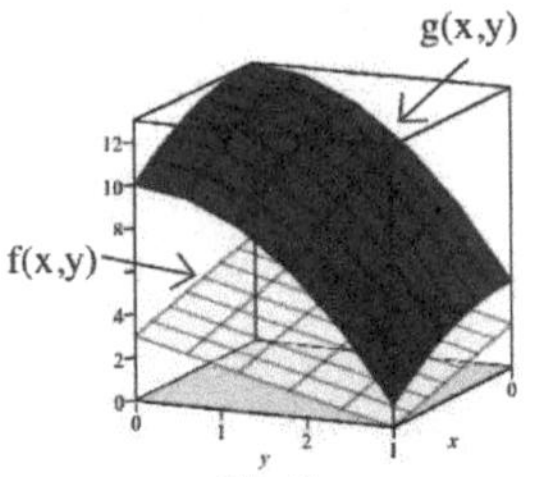

Fig. 3

Solution: The inside integral must be **dz** with z from $z = f(x,y) = 5 - 2x - y$ to $z = g(x,y) = 13 - 3x^2 - y^2$. The outer integral can be either **dx** or **dy** since each of them is bounded by constants.

$$\text{volume} = \iiint_R 1 \, dV = \int_{x=0}^{1} \int_{y=0}^{3} \int_{z=5-2x-y}^{z=13-3x^2-y^2} 1 \, dz \, dy \, dx = \int_{x=0}^{1} \int_{y=0}^{3} (8 - 3x^2 - y^2 + 2x + y) \, dy \, dx$$

$$= \int_{x=0}^{1} (-9x^2 + \frac{39}{2} + 6x) \, dx = \frac{39}{2}$$

Practice 2: Write the triple integral that represents the volume of the region bounded by $0 \le x \le 1$, $0 \le y \le 3-3x$ and z between plane $z = f(x,y) = 5 - 2x - y$ and the surface $z = g(x,y) = 13 - 3x^2 - y^2$. How does this region differ from the one in Example 2?

Example 3: Write an iterated triple integral for f(x,y,z) in the solid bounded by the paraboloid $z = x^2 + y^2$ and the plane z=4 (Fig. 4).

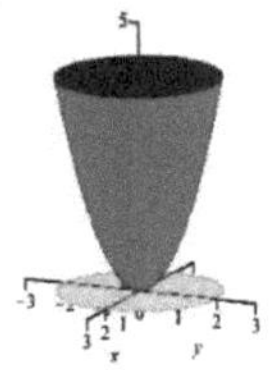
Fig. 4: Solid paraboloid

Solution: The domain in the xy-plane is the circle $x^2 + y^2 \le 4$ which can be described as $-2 \le x \le 2$ and $-\sqrt{4-x^2} \le y \le \sqrt{4-x^2}$. And z goes from $x^2 + y^2$ to 4.

Putting this information together, the triple integral is

$$\iiint_R f(x,y,z) \, dV = \int_{x=-2}^{2} \int_{y=-\sqrt{4-x^2}}^{\sqrt{4-x^2}} \int_{z=x^2+y^2}^{4} f(x,y,z) \, dz \, dy \, dx$$

Note that the endpoints of the outside integral had no variables, the middle endpoints have one variable, and the inside endpoints have two variables.

Practice 3: (a) Write an iterated triple integral for f(x,y,z) in the solid bounded below by the cone $z = \sqrt{x^2 + y^2}$ and above by the sphere $x^2 + y^2 + z^2 = 8$ (Fig. 5).

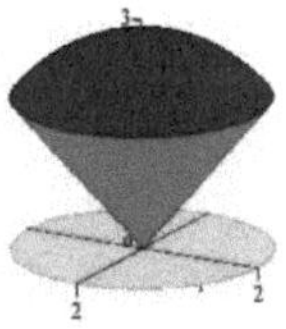
Fig. 5: Cone topped by sphere

(b) Write an iterated triple integral for this function and solid that is only in the first octant.

Applications of Triple Integrals

These are similar to the applications of single and double integrals and are very useful in some applications.

Volume of a solid region R = $\iiint_R 1\, dV$

Average value of f on a solid region R $= \frac{1}{\text{volume of R}} \cdot \iiint_R f\, dV$

If $\delta = \delta(x,y,z)$ is the density of a solid region R at the location (x,y,z) then Mass = M = $\iiint_R \delta\, dV$

First moments about the coordinate planes:

$M_{yz} = \iiint_R x \cdot \delta\, dV$ $\quad M_{xz} = \iiint_R y \cdot \delta\, dV$ $\quad M_{xy} = \iiint_R z \cdot \delta\, dV$

Center of Mass: $\bar{x} = \frac{M_{yz}}{M}$ $\quad \bar{y} = \frac{M_{xz}}{M}$ $\quad \bar{z} = \frac{M_{xy}}{M}$

Second moments (moments of inertia):

$I_x = \iiint_R (y^2+z^2) \cdot \delta\, dV$ $\quad I_y = \iiint_R (x^2+z^2) \cdot \delta\, dV$ $\quad I_z = \iiint_R (x^2+y^2) \cdot \delta\, dV$

about line L: $I_L = \iiint_R r^2 \cdot \delta\, dV$ where r(x,y,z) = distance of (x,y,z) from line L

Radius of gyration about a line L: $R_L = \sqrt{I_L / M}$

Example 4: A 1 cm by 1 cm ($0 \le y \le 1$, $0 \le z \le 1$) bar along the x-axis has a length of 6 cm ($0 \le x \le 6$). The density of the bar is $\delta(x,y,z) = 1 + x$ g/ cm^3. Determine (a) the mass M, (b) M_{yz}, (c) $\bar{x}$, (d) I_z and (e) R_z.

Solution: (a) mass = $\int_{z=0}^{1} \int_{y=0}^{1} \int_{x=0}^{6} (1+x)\, dx\, dy\, dz = 24$ g

Fig. 6: Bar along the x-axis

(b) $M_{yz} = \iiint_R x \cdot \delta\, dV = \int_{z=0}^{1} \int_{y=0}^{1} \int_{x=0}^{6} x \cdot (1+x)\, dx\, dy\, dz = 90\ g \cdot cm$

(c) $\bar{x} = \frac{M_{yz}}{M} = \frac{90\ g \cdot cm}{24} = \frac{15}{4} = 3.75\ cm$ The bar balances on a fulcrum at location x=3.75 cm.

(d) $I_z = \iiint_R (x^2+y^2) \cdot \delta\, dV = \int_{z=0}^{1} \int_{y=0}^{1} \int_{x=0}^{6} (x^2+y^2) \cdot (1+x)\, dx\, dy\, dz = 404\ g \cdot cm^2$

(e) $R_z = \sqrt{I_z / M} = \sqrt{404/24} \approx 4.10\ cm$ The kinetic energy of this bar rotating around the z-axis is the same as a point mass of 24 g located at x=4.1 cm rotating around the z-axis with the same angular speed.

Based on each integral in the example, you should be able to justify the units attached to the numerical result.

Practice 4: A cube ($0 \le x \le 2$, $0 \le y \le 2$, $0 \le z \le 2$ cm) has density $\delta(x,y,z) = 1 + x + y + z^2$ g/ cm^3.

Determine (a) the mass M, (b) the moment about each coordinate plane, and (c) the center of mass of the cube.

Example 5: $\iiint\limits_R \sqrt{x^2+y^2}\ dV$ where R is the solid bounded by the paraboloid $z = x^2 + y^2$ and the plane z=4 (Fig. 4).

Solution: This is the solid from Example 2 so

$$\iiint\limits_R f(x,y,z)\ dV = \int_{x=-2}^{2} \int_{y=-\sqrt{4-x^2}}^{\sqrt{4-x^2}} \int_{z=x^2+y^2}^{4} \sqrt{x^2+y^2}\ dz\ dy\ dx$$

$$= \int_{x=-2}^{2} \int_{y=-\sqrt{4-x^2}}^{\sqrt{4-x^2}} (4-(x^2+y^2)) \cdot \sqrt{x^2+y^2}\ dy\ dx$$

But the domain of this remaining double integral is the disk $0 \le x^2 + y^2 \le 2$ so it is useful to switch to polar coordinates. Then we have

$$= \int_{\theta=0}^{2\pi} \int_{r=0}^{2} \left((4-r^2) \cdot \sqrt{r^2} \right) r\ dr\ d\theta = \int_{\theta=0}^{2\pi} \int_{r=0}^{2} \left(4r^2 - r^4 \right) dr\ d\theta = \int_{\theta=0}^{2\pi} \frac{64}{15}\ d\theta = \frac{128}{15}\pi\ .$$

Problems

In Problems 1 to 6, set up the appropriate iterated integrals for $\iiint\limits_R f\ dV$ on the indicated domains.

1. R is the solid prism in Fig. 7.

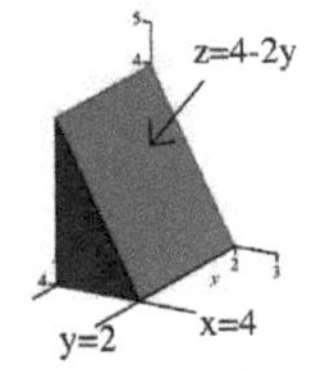

Fig. 7: Solid prism

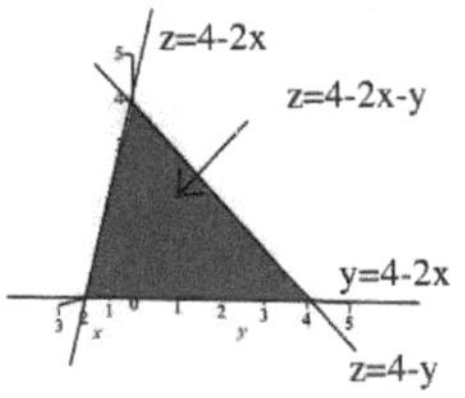

Fig. 8: Solid tetrahedron

2. R is the solid tetrahedron in Fig. 8.

3. R is the solid cone in Fig. 9.

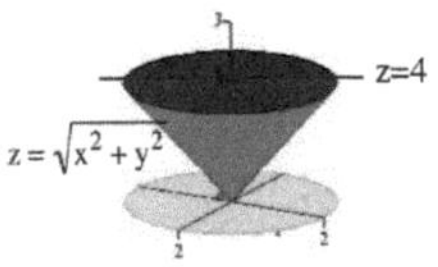

Fig. 9: Solid cone

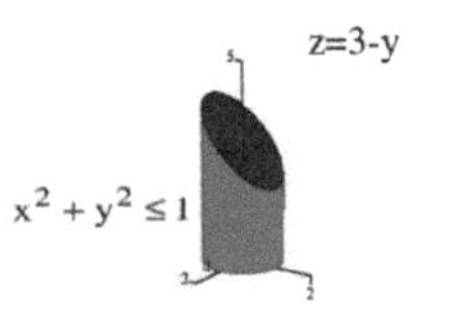

Fig. 10: Truncated cylinder

4. R is the solid sliced cylinder in Fig. 10.

5. R is the solid in Fig. 11.

6. R is the solid in Fig. 12.

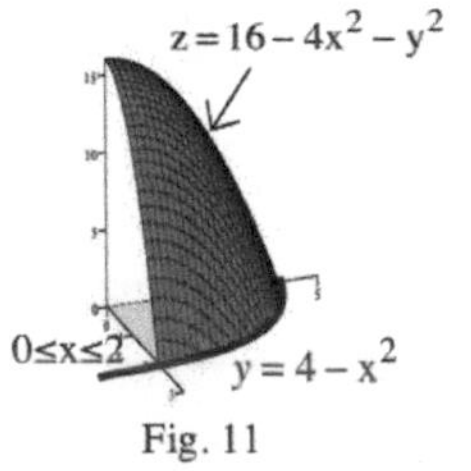

Fig. 11

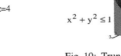

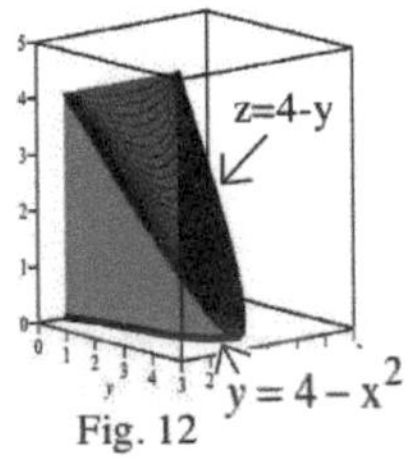

Fig. 12

In problems 7 to 18, evaluate the integrals.

7. $\int_{x=0}^{1} \int_{y=0}^{1} \int_{z=0}^{1} \left(x^2 + y^2 + z\right) \, dz \, dy \, dx$

8. $\int_{z=1}^{e} \int_{y=1}^{e} \int_{x=1}^{e} \frac{1}{xyz} \, dx \, dy \, dz$

9. $\int_{0}^{1} \int_{0}^{x-2} \int_{0}^{3-x-y} dz \, dy \, dx$

10. $\int_{0}^{1} \int_{0}^{\pi} \int_{0}^{\pi} y \cdot \cos(z) \, dx \, dy \, dz$

11. $\int_{0}^{1} \int_{0}^{2} \int_{0}^{x+z} (12xz) \, dy \, dx \, dz$

12. $\int_{0}^{2} \int_{x}^{2x} \int_{0}^{y} (6xyz) \, dz \, dy \, dx$

13. $\int_{0}^{3} \int_{0}^{1} \int_{0}^{1-z} z \cdot e^{y} \, dx \, dz \, dy$

14. $\int_{0}^{3} \int_{0}^{1} \int_{0}^{1-z} x \cdot e^{y} \, dx \, dz \, dy$

15. $\int_{0}^{2} \int_{0}^{\sqrt{4-x^2}} \int_{0}^{\sqrt{4-x^2}} 1 \, dy \, dz \, dx$

16. $\int_{0}^{4} \int_{0}^{\ln(y)} \int_{\ln(y)}^{\ln(3y)} e^{x+y-z} \, dx \, dz \, dy$

17. $\int_{0}^{2} \int_{0}^{3} \int_{0}^{y} (2x + 4y + 6z) \, dz \, dy \, dx$

18. $\int_{0}^{3} \int_{0}^{1} \int_{0}^{y} (4xy) \, dx \, dz \, dy$

For problems 19 to 22 write the Maple command to evaluate the triple integral.

19. f(x,y)=2x+3 on the domain of Problem 19.

20. f(x,y)=sin(xy)+z on the domain of Problem 20.

21. f(x,y)=xyz on the domain of Problem 21.

22. f(x,y)=1 on the domain of Problem 22.

Practice Answers

Practice 1: Evaluate $\iiint\limits_R x^3+y^2+z\ dV$ for $R=\{(x,y,z):\ 0\le x\le 2,\ 0\le y\le,\ 1\le z\le 3\}$.

$$\iiint\limits_R x^3+y^2+z\ dV = \int_{z=1}^{3}\int_{y=0}^{1}\int_{x=0}^{2} x^3+y^2+z\ dx\ dy\ dz$$

$$\int_{x=0}^{2} x^3+y^2+z\ dx = 4+2y^2+2z, \quad \int_{y=0}^{1} 4+2y^2+2z\ dy = 4+\frac{2}{3}+2z,$$

and finally $\displaystyle\int_{z=1}^{3} \frac{14}{3}+2z\ dz = \frac{28}{3}+8.$

Check that integrating in some other order gives the same result.

Try $\displaystyle\int_{x=0}^{2}\int_{z=1}^{3}\int_{y=0}^{1} (x^3+y^2+z)\ \text{dy dz dx}$ and $\displaystyle\int_{y=0}^{1}\int_{x=0}^{2}\int_{z=1}^{3} (2x+y+z^2)\ \text{dz dx dy}$

Practice 2: $\text{volume} = \displaystyle\int_{x=0}^{1}\int_{y=0}^{3-3x}\int_{z=5-2x-y}^{z=13-3x^2-y^2} 1\ \text{dz dy dx}$ (= 23/2)

The domain is now a triangle in the xy-plane (Fig. P2).

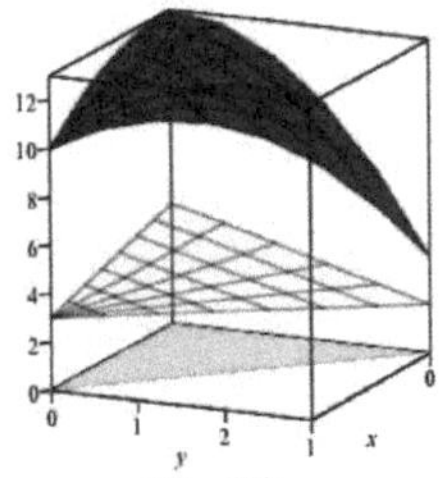

Fig. P2

Practice 3: (a) $\displaystyle\iiint\limits_R \text{f dV} = \int_{x=-2}^{2}\int_{y=-\sqrt{4-x^2}}^{\sqrt{4-x^2}}\int_{z=\sqrt{x^2+y^2}}^{\sqrt{8-x^2-y^2}} \text{f(x,y,z) dz dy dx}$

(b) $\displaystyle\int_{x=0}^{2}\int_{y=0}^{\sqrt{4-x^2}}\int_{z=\sqrt{x^2+y^2}}^{\sqrt{8-x^2-y^2}} \text{f(x,y,z) dz dy dx}$

Practice 4: Since x, y and z have constant endpoints, the integration can be done in any order.

(a) $\text{Mass} = \displaystyle\int_{x=0}^{2}\int_{y=0}^{2}\int_{z=0}^{2} \left(1+x+y+z^2\right) \text{dz dy dx} = \frac{104}{3}\ g$

(b) $M_{yz} = \displaystyle\int_{x=0}^{2}\int_{y=0}^{2}\int_{z=0}^{2} x\cdot\left(1+x+y+z^2\right) \text{dz dy dx} = \frac{112}{3}\ g\cdot cm,\ M_{xz} = M_{yz}$

$M_{xy} = \displaystyle\int_{x=0}^{2}\int_{y=0}^{2}\int_{z=0}^{2} z\cdot\left(1+x+y+z^2\right) \text{dz dy dx} = 40\ g\cdot cm$

(c) $\bar{x} = \dfrac{M_{yz}}{M} = \dfrac{14}{13}\ cm,\ \bar{y} = \bar{x} = \dfrac{14}{13}\ cm,\ \bar{z} = \dfrac{M_{xy}}{M} = \dfrac{15}{13}\ cm$

14.7 TRIPLE INTEGRALS IN CYLINDRICAL AND SPHERICAL COORDINATES

"In physics everything is straight, flat or round." Statement by a physics teacher

Maybe not, but lots of applications have pieces of round or spherical domains, and using cylindrical or spherical coordinates can make many triple integrals much easier to evaluate.

$\iiint f\,dV$ in Cylindrical Coordinates

If our domain of integration is round or is easily described using polar coordinates (r, θ), then a triple integral in cylindrical coordinates (r, θ, z) is often the best method, and it begins with the form of the ΔV. Fig. 1 illustrates that if we partition each of r, θ and z, then the volume of each little cell is $\Delta V = (r\cdot\Delta\theta)\cdot\Delta r\cdot\Delta z$. Then the triple Riemann sum is

$\sum\sum\sum f(r^*,\theta^*,z^*)\cdot r\cdot\Delta r\cdot\Delta\theta\cdot\Delta z$. If f is continuous, then the limit of this Riemann sum as $\Delta r,\ \Delta\theta,\ \Delta z\to 0$ is the triple integral in cylindrical coordinates

$$\iiint_R f\,dV = \int_z\int_\theta\int_r f(r,\theta,z)\ r\ dr\ d\theta\ dz\ .$$

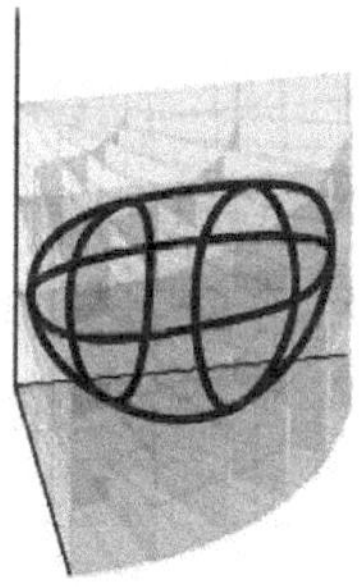
Partition of r, θ and z

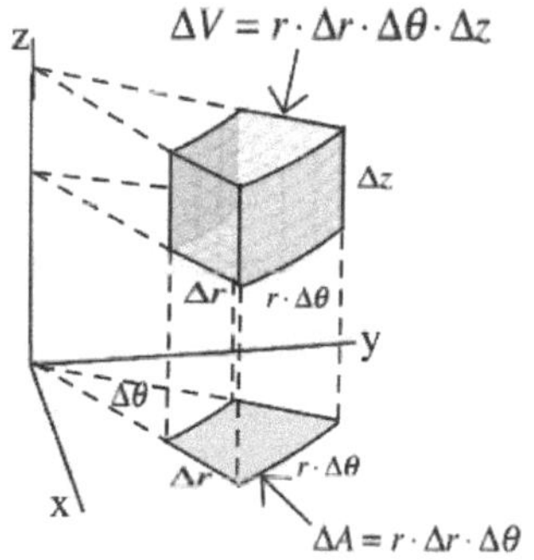

Fig. 1

As before, we can alter the order of the integrals as long as we accurately describe the domain of integration. Also, the outer integral endpoints must be constants, the middle integral endpoints can have only one variable, and the inner integral endpoints can have two variables.

(Note: Recall that $x = r\cdot\cos(\theta)$, $y = r\cdot\sin(\theta)$ so $x^2+y^2=r^2$ and z=z .)

Example 1: Evaluate (a) $\iiint_R f\,dV$ for $f(r,\theta,z) = r\cdot z$ with $R = \{1\le r\le z,\ 0\le\theta\le\pi,\ 0\le z\le 2\}$

and (b) $\displaystyle\int_{r=0}^{3}\int_{\theta=0}^{2\pi}\int_{z=0}^{e^{-r^2}} 1\ r\ dz\ d\theta\ dr$

Solution: (a) $\displaystyle\iiint_R f\,dV = \int_{z=0}^{2}\int_{\theta=0}^{\pi}\int_{r=1}^{z}(r\cdot z)\ r\ dr\ d\theta\ dz = \int_{z=0}^{2}\int_{\theta=0}^{\pi}\left(\frac{1}{3}r^3\right)\Big|_{r=1}^{z}\ d\theta\ dz = \int_{z=0}^{2}\frac{1}{3}\left(z^4-z\right)\pi\ dz = \frac{22}{15}\pi$

(b) $\displaystyle\int_{r=0}^{3}\int_{\theta=0}^{2\pi}\int_{z=0}^{e^{-r^2}} 1\ r\ dz\ d\theta\ dr = \int_{r=0}^{3}\int_{\theta=0}^{2\pi}\left(e^{-r^2}\right)r\ d\theta\ dr = \int_{r=0}^{3}2\pi\left(e^{-r^2}\right)r\ dr = -\pi e^{-r^2}\Big|_{r=0}^{3} = \pi\left(1-e^{-9}\right)$

Practice 1: Evaluate (a) $\iiint\limits_R dV$ for $R=\{0 \le r \le 2,\ 0 \le \theta \le \pi/2,\ 0 \le z \le 8 - r^3\}$ and

(b) the volume of the solid cylinder above the disk $x^2+y^2 \le 4$ and below the plane z=4-y.

Example 2: R is the solid bounded by the paraboloid $z = x^2 + y^2$ and the plane z=4 (Fig. 2).

(a) Write and evaluate an iterated triple integral for the volume of R.

(b) Write and evaluate an iterated triple integral for the mass of R if the density is $\delta(x,y,x) = 1 + z$.

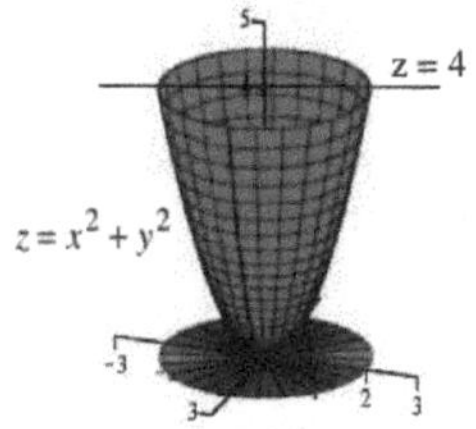

Fig. 2: Solid paraboloid

Solution: (a) The domain of this integral is the circle $x^2+y^2 \le 4$ so $r^2 \le 4$ and $0 \le \theta \le 2\pi$:

$$\iiint\limits_R f\, dV = \int_{\theta=0}^{2\pi}\int_{r=0}^{2}\int_{z=r^2}^{4} (1)\ r\ dz\ dr\ d\theta = \int_{\theta=0}^{2\pi}\int_{r=0}^{2} (z\cdot r)\Big|_{z=r^2}^{4}\ dr\ d\theta = \int_{\theta=0}^{2\pi}\int_{r=0}^{2} \left(4r - r^3\right)\ dr\ d\theta = 8\pi$$

(b) $$\text{mass} = \int_{\theta=0}^{2\pi}\int_{r=0}^{2}\int_{z=r^2}^{4} (1+z)\cdot r\ dz\ dr\ d\theta = \frac{88}{3}\pi$$

Practice 2: R is the solid hemisphere $x^2+y^2+z^2 \le 4$ with $z \ge 0$ (Fig. 3).

(a) Write and evaluate an iterated triple integral for the volume of R.

(b) Write and evaluate an iterated triple integral for the mass of R if the density is $\delta(x,y,x) = 1 + z$.

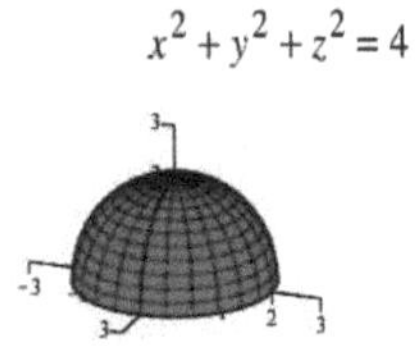

Fig. 3: Solid hemisphere

Example 3: Find the centroid of the solid that is bounded below by the disk $x^2+y^2 \le 9$ and above by the paraboloid $z = x^2 + y^2$.

Solution: $x^2+y^2 \le 9$ means $0 \le r \le 3$, and $z = x^2 + y^2$ means $z = r^2$, the domain is

$$R = \{(r,\theta,z):\ 0 \le r \le 3,\ 0 \le \theta \le 2\pi,\ 0 \le z \le r^2\}$$

$$\text{mass} = \int_{\theta=0}^{2\pi}\int_{r=0}^{3}\int_{z=0}^{r^2} r\cdot dz\cdot dr\cdot d\theta = \int_{\theta=0}^{2\pi}\int_{r=0}^{3} r\cdot z\Big|_{z=0}^{r^2}\ dr\cdot d\theta = \int_{\theta=0}^{2\pi}\int_{r=0}^{3} r^3\ dr\cdot d\theta = \frac{81}{2}\pi$$

$$M_{xy} = \int_{\theta=0}^{2\pi}\int_{r=0}^{3}\int_{z=0}^{r^2} z\cdot r\ dz\cdot dr\cdot d\theta = \int_{\theta=0}^{2\pi}\int_{r=0}^{3} r\cdot\frac{z^2}{2}\Big|_{z=0}^{r^2}\ dr\cdot d\theta = \int_{\theta=0}^{2\pi}\int_{r=0}^{3} \frac{r^5}{2}\ dr\cdot d\theta$$

$$= \int_{\theta=0}^{2\pi}\int_{r=0}^{3} \frac{r^6}{12}\Big|_{t=0}^{3}\ d\theta = 2\pi\left(\frac{729}{12}\right) = \frac{243}{2}\pi\ .$$

Then $\bar{z} = \dfrac{(243/2)\pi}{(81/2)\pi} = 3$. Because of the symmetry about the z-axis, $\bar{x}$ and $\bar{y}$ are both 0 so the centroid is (0, 0, 3).

$\iiint$ f dV **in Spherical Coordinates**

This development is very similar to what was done for cylindrical coordinates. First we partition our domain R into $(\Delta\rho, \Delta\theta, \Delta\varphi)$ cells, pick a representative point $(\rho^*, \theta^*, \varphi^*)$ in each cell, form the triple Riemann sum $\sum \sum \sum f(\rho^*, \theta^*, \varphi^*)\ \Delta V$, and, finally, take the limit as all of the cell dimensions approach 0 in order to form a triple integral: $\lim_{\Delta \to 0} \sum \sum \sum f(\rho^*, \theta^*, \varphi^*)\ \Delta V \to \iiint_R f(\rho, \theta, \varphi) \cdot dV$

But before we can actually use this idea, we first need to determine dV in terms of the variables ρ, θ and φ. That is a bit complicated and is derived in the Appendix of this section as well as in the next section when Jacobeans are introduced . In either case, $dV = \rho^2 \cdot \sin(\varphi) \cdot d\rho \cdot d\theta \cdot d\varphi$, and then

$$\iiint_R f(\rho, \theta, \varphi) \cdot dV = \iiint_R f(\rho, \theta, \varphi) \cdot \rho^2 \cdot \sin(\varphi) \cdot d\rho \cdot d\theta \cdot d\varphi$$

As before, the we can use any order of integration that describes the domain as long as the outside integral has constant endpoints, and the middle integral has at most one variable endpoint The inside integral can have two variable endpoints.

Example 4: Represent each domain R using iterated triple integrals.

(a) R is shown in Fig. 4a, (b) R is shown in Fig. 4b.

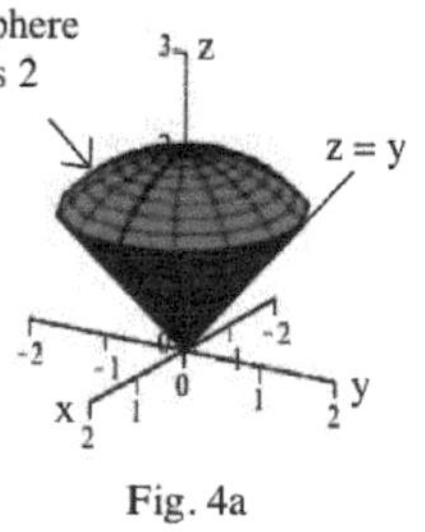

Fig. 4a

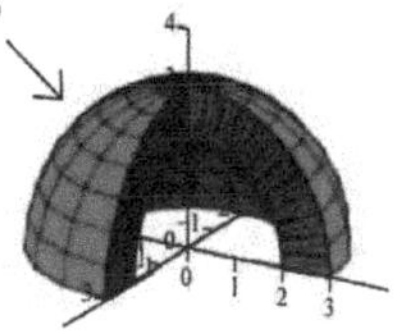

Fig. 4b

Solution: (a) $R = \{(\rho, \theta, \varphi):\ 0 \le \rho \le 2,\ 0 \le \theta \le 2\pi,\ 0 \le \varphi \le \pi/4\}$

$$\iiint_R f\ dV = \int_{\varphi=0}^{\pi/4} \int_{\theta=0}^{2\pi} \int_{\rho=0}^{2} f(\rho, \theta, \varphi)\ \rho^2 \cdot \sin(\varphi) \cdot d\rho \cdot d\theta \cdot d\varphi$$

(b) $R = \{(\rho, \theta, \varphi):\ 2 \le \rho \le 3,\ \pi/2 \le \theta \le 2\pi,\ 0 \le \varphi \le \pi/2\}$

$$\iiint_R f\ dV = \int_{\varphi=0}^{\pi/2} \int_{\theta=\pi/2}^{2\pi} \int_{\rho=2}^{3} f(\rho, \theta, \varphi)\ \rho^2 \cdot \sin(\varphi) \cdot d\rho \cdot d\theta \cdot d\varphi$$

All of these integral endpoints are constants so the integrals can done be in any order.

Practice 3: Represent each domain R using iterated triple integrals.

(a) R is shown in Fig. 5a, (b) R is shown in Fig. 5b.

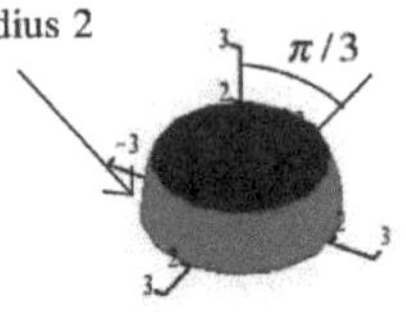

Fig. 5a

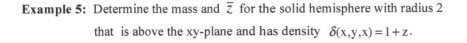

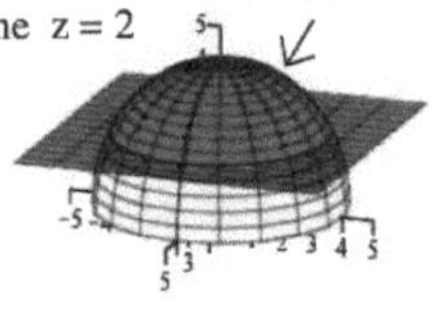

Fig. 5b

Example 5: Determine the mass and $\bar{z}$ for the solid hemisphere with radius 2 that is above the xy-plane and has density $\delta(x,y,x) = 1+z$.

Solution: $R = \{(\rho,\theta,\varphi): 0 \le \rho \le 2,\ 0 \le \theta \le 2\pi,\ 0 \le \varphi \le \pi/2\}$ so

$$\text{mass} = \int_{\varphi=0}^{\pi/2}\int_{\theta=0}^{2\pi}\int_{\rho=0}^{2} (1+z)\ \rho^2\cdot\sin(\varphi)\cdot d\rho\cdot d\theta\cdot d\varphi$$

$$= \int_{\varphi=0}^{\pi/2}\int_{\theta=0}^{2\pi}\int_{\rho=0}^{2} (1+\rho\cdot\cos(\varphi))\ \rho^2\cdot\sin(\varphi)\cdot d\rho\cdot d\theta\cdot d\varphi$$

$$= \int_{\varphi=0}^{\pi/2}\int_{\theta=0}^{2\pi}\int_{\rho=0}^{2} \left(\rho^2\cdot\sin(\varphi)+\rho^3\cdot\cos(\varphi)\cdot\sin(\varphi)\right)\ d\rho\cdot d\theta\cdot d\varphi$$

$$= \int_{\varphi=0}^{\pi/2}\int_{\theta=0}^{2\pi} \left(\frac{8}{3}\cdot\sin(\varphi)+4\cdot\cos(\varphi)\cdot\sin(\varphi)\right)\cdot d\theta\cdot d\varphi$$

$$= \int_{\varphi=0}^{\pi/2} 2\pi\left(\frac{8}{3}\cdot\sin(\varphi)+4\cdot\cos(\varphi)\cdot\sin(\varphi)\right)\cdot d\varphi = 2\pi\left(2\cdot\sin^2(\varphi)-\frac{8}{3}\cdot\cos(\varphi)\right)\Bigg|_{\varphi=0}^{\pi/2}$$

$$= 2\pi(2) + 2\pi\left(\frac{8}{3}\right) = \frac{28}{3}\pi$$

$$M_{xy} = \int_{\varphi=0}^{\pi/2}\int_{\theta=0}^{2\pi}\int_{\rho=0}^{2} z\cdot(1+z)\ \rho^2\cdot\sin(\varphi)\cdot d\rho\cdot d\theta\cdot d\varphi = \frac{124}{15}\pi \quad \text{(using Maple)}$$

$$\text{so } \bar{z} = \frac{(124/15)\pi}{(28/3)\pi} = \frac{32}{35} \approx 0.914$$

Conclusion: Even a simple looking problem can take a long time.

These conversion formulas for cylindrical and spherical coordinates are useful.

Coordinate conversion formulas

Cylindrical to Rectangular	Spherical to Cylindrical	Spherical to Rectangular
$x = r\cdot\cos(\theta)$	$r = \rho\cdot\sin(\varphi)$	$x = \rho\cdot\sin(\varphi)\cdot\cos(\theta)$
$y = r\cdot\sin(\theta)$	$z = \rho\cdot\cos(\varphi)$	$y = \rho\cdot\sin(\varphi)\cdot\sin(\theta)$
$z = z$	$\theta = \theta$	$z = \rho\cdot\cos(\varphi)$

$$dV = dx\cdot dy\cdot dz = r\cdot dr\cdot d\theta\cdot dz = \rho^2\cdot\sin(\varphi)\cdot d\rho\cdot d\theta\cdot d\varphi$$

Problems

In problems 1-8, evaluate the triple integrals in cylindrical coordinates.

1. $\int_0^{\pi}\int_0^{1}\int_0^{\sqrt{2-r^2}} r\ dz\cdot dr\cdot d\theta$

2. $\int_0^{4}\int_0^{\pi}\int_r^{6} r\ dz\cdot d\theta\cdot dr$

3. $\int_0^{\pi}\int_0^{\theta/\pi}\int_0^{\sqrt{4-r^2}} z\ dz\cdot dr\cdot d\theta$

4. $\int_0^{\pi}\int_0^{1}\int_{-1/2}^{1/2} \left(r^2\cdot\sin(\theta)+z^2\right)\ dz\cdot dr\cdot d\theta$

5. $\int_0^{2\pi}\int_0^{3}\int_0^{\pi/3} r^3\ dr\cdot dz\cdot d\theta$

6. $\int_{-1}^{1}\int_0^{2\pi}\int_0^{1+\sin(\theta)} 2r\ dr\cdot d\theta\cdot dz$

7. $\int_0^{2}\int_{r-2}^{\sqrt{4-r^2}}\int_0^{2\pi} (r\cdot\sin(\theta)+1)\cdot r\ d\theta\cdot dz\cdot dr$

8. $\int_0^{\pi}\int_r^{2r}\int_0^{\pi} r\cdot\cos(\theta)\ d\theta\cdot dz\cdot dr$

In problems 9 to 12, evaluate the integrals in cylindrical coordinates.

9. $\int_0^{4}\int_0^{\sqrt{2}/2}\int_x^{\sqrt{1-x^2}} e^{-(x^2+y^2)}\ dy\cdot dx\cdot dz$

10. $\int_0^{4}\int_{-1}^{1}\int_{-\sqrt{1-x^2}}^{\sqrt{1-x^2}} 1\ dy\cdot dx\cdot dz$

11. $\int_0^{1}\int_0^{\sqrt{1-x^2}}\int_0^{4-y} 1\ dz\cdot dy\cdot dx$

12. $\int_{-2}^{2}\int_0^{\sqrt{4-x^2}}\int_0^{1} \cos(x^2+y^2)\ dz\cdot dy\cdot dx$

In problems 13 to 18, set up and evaluate the triple integrals in cylindrical coordinates.

13. $f(x,y)=\sqrt{x^2+y^2}$. R is the region inside the cylinder $x^2+y^2=9$ and between the planes z=3 and z=5.

14. $f(x,y)=(x^3+xy^2)$. R is the region in the first octant and under the paraboloid $z=4-x^2-y^2$.

15. $f=e^z$. R is the region enclosed by paraboloid $z=1+x^2+y^2$, the cylinder $x^2+y^2=7$, and the xy-plane.

16. $f=x^2$. R is the region inside the cylinder $x^2+y^2=4$, below the cone $z^2=9x^2+9y^2$ and above the xy-plane.

17. Find the volume of the region R in first octant below the $z=x^2+y^2$ and above $z=36-3x^2-3y^2$.

18. $f(x,y)=6+4x^2+4y^2$. R is the region in the 2nd, 3rd and 4th octants, inside the cylinder $x^2+y^2=1$, and between the planes z=2 and z=3.

Now spherical

In problems 19 to 26 evaluate the integrals in spherical coordinates.

19. $\int_0^{\pi}\int_0^{\pi}\int_0^{2\cdot\cos(\varphi)} \rho^2\cdot\sin(\varphi)\ d\rho\cdot d\varphi\cdot d\theta$

20. $\int_0^{2\pi}\int_0^{\pi/4}\int_0^{2} \rho^3\cdot\sin^2(\varphi)\ d\rho\cdot d\varphi\cdot d\theta$

21. $\int_0^{\pi}\int_0^{\pi}\int_0^{1} 5\rho^3\cdot\sin^3(\varphi)\ d\rho\cdot d\varphi\cdot d\theta$

22. $\int_0^{\pi}\int_0^{\pi/3}\int_{\sec(\varphi)}^{1} 3\rho^2\cdot\sin(\varphi)\ d\rho\cdot d\varphi\cdot d\theta$

23. $\int_0^{\pi/2}\int_0^{\pi}\int_1^{2} \rho^2\cdot\sin(\varphi)\ d\rho\cdot d\theta\cdot d\varphi$

24. $\int_0^{\pi}\int_0^{\pi}\int_0^{1} e^{(\rho^2)}\cdot\rho\cdot\sin(\varphi)\ d\rho\cdot d\theta\cdot d\varphi$

25. $\int_0^{\pi}\int_0^{\pi/3}\int_0^{\cos(\varphi)} 4\rho^3\cdot\sin(\varphi)\ d\rho\cdot d\varphi\cdot d\theta$

26. $\int_0^{2\pi}\int_0^{\pi/2}\int_0^{\csc(\varphi)} \sin(\varphi)\ d\rho\cdot d\varphi\cdot d\theta$

Practice Answers

Practice 1: (a) $\int_{r=0}^{2}\int_{\theta=0}^{\pi/2}\int_{z=0}^{8-r^3} r\ dz\ d\theta\ dr = \int_{z=0}^{2}\int_{\theta=0}^{\pi/2} (z\cdot r)\Big|_{z=0}^{8-r^3}\ d\theta\ dr = \int_{z=0}^{2}\int_{\theta=0}^{\pi/2} 8r-r^4\ d\theta\ dr$

$= \int_{z=0}^{2} \frac{\pi}{2}(8r-r^4)\ dr = \frac{\pi}{2}\left(\frac{48}{5}\right)$

(b) $y = r\cdot\sin(\theta)$ so

volume $= \int_{r=0}^{2}\int_{\theta=0}^{2\pi}\int_{z=0}^{4-r\cdot\sin(\theta)} 1\ r\ dz\ d\theta\ dr = \int_{r=0}^{2}\int_{\theta=0}^{2\pi} (4-r\cdot\sin(\theta)\cdot r\ d\theta\ dr = \int_{r=0}^{2} 2\pi(4r)\ dr = 16\pi$

Practice 2: $z^2 \le 4-(x^2+y^2) = 4-r^2$, and, as in Example 1, $0\le r\le 2$ and $0\le\theta\le 2\pi$.

(a) volume $= \int_{\theta=0}^{2\pi}\int_{r=0}^{2}\int_{z=0}^{\sqrt{4-r^2}} (1)\cdot r\ dz\ dr\ d\theta = \int_{\theta=0}^{2\pi}\int_{r=0}^{2} z\cdot r\Big|_{z=0}^{\sqrt{4-r^2}}\ dr\ d\theta = \int_{\theta=0}^{2\pi}\int_{r=0}^{2} \sqrt{4-r^2}\cdot r\ dr\ d\theta$

$= \int_{\theta=0}^{2\pi}\int_{r=0}^{2} \sqrt{4-r^2}\cdot r\ dr\ d\theta = \int_{\theta=0}^{2\pi}\left(-\frac{1}{3}\left(4-r^2\right)^{3/2}\right)\Big|_{r=0}^{2}\ d\theta = \int_{\theta=0}^{2\pi}\frac{8}{3}\ d\theta = \frac{16}{3}\pi$

(b) mass $= \int_{\theta=0}^{2\pi}\int_{r=0}^{2}\int_{z=0}^{\sqrt{4-r^2}} (1+z)\cdot r\ dz\ dr\ d\theta = \frac{28}{3}\pi$

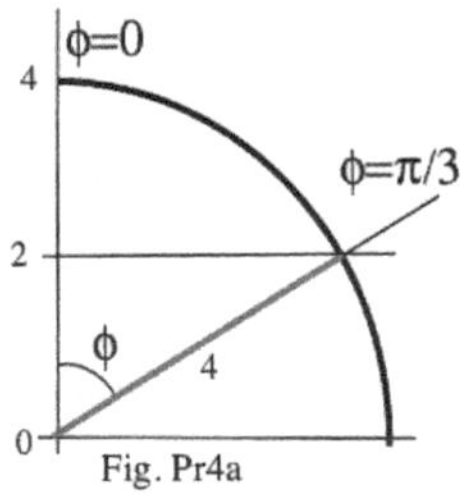

Fig. Pr4a

Practice 3: (a) $R = \{(\rho,\theta,\varphi):\ 0\le\rho\le 2,\ 0\le\theta\le 2\pi,\ \pi/3\le\varphi\le\pi/2\}$

$\int_{\varphi=\pi/3}^{\pi/2}\int_{\theta=0}^{2\pi}\int_{\rho=0}^{2} f(\rho,\theta,\varphi)\ \rho^2\cdot\sin(\varphi)\cdot d\rho\cdot d\theta\cdot d\varphi$

(b) Clearly $0\le\theta\le 2\pi$, but φ and ρ require a bit of work (Fig. Pr4):

$R = \{(\rho,\theta,\varphi):\ 2\sec(\varphi)\le\rho\le 4,\ 0\le\theta\le 2\pi,\ 0\le\varphi\le\pi/3\}$

$\int_{\varphi=0}^{\pi/3}\int_{\theta=0}^{2\pi}\int_{\rho=2\sec(\varphi)}^{4} f(\rho,\theta,\varphi)\ \rho^2\cdot\sin(\varphi)\cdot d\rho\cdot d\theta\cdot d\varphi$

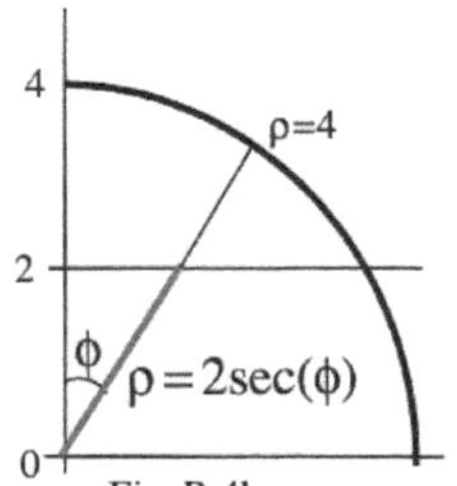

Fig. Pr4b

Appendix: Why $dV = \rho^2 \cdot \sin(\varphi) \cdot d\rho \cdot d\theta \cdot d\varphi$

The figures A1-A3 are an attempt to explain where this strange equation comes from. We need to partition the space by partitioning each of the three variables ρ, θ and φ. This results in cells as shown greatly magnified in Fig. A1. Using different views of a typical cell in Fig. A2, it is possible to determine the lengths of the sides of this cell. Putting all of this together in Fig. A3, the volume of the cell, ΔV ia the product of the lengths of the sides:

$$dV = \rho^2 \cdot \sin(\varphi) \cdot d\rho \cdot d\theta \cdot d\varphi$$

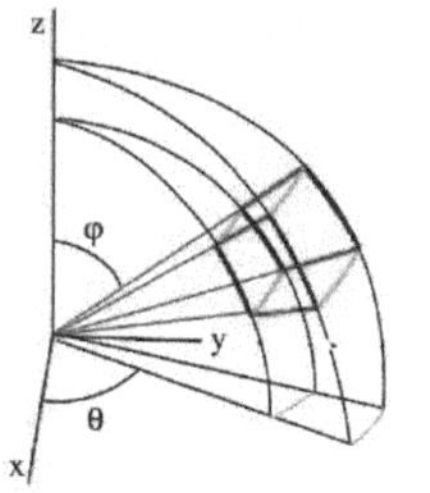

Fig. A1

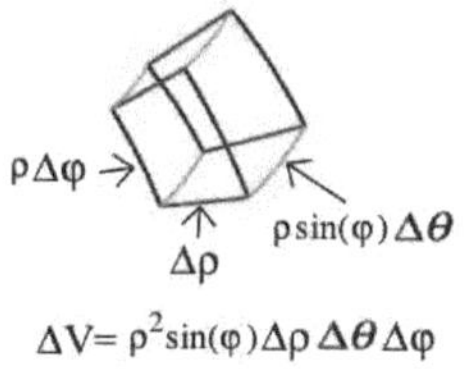

Fig. A3

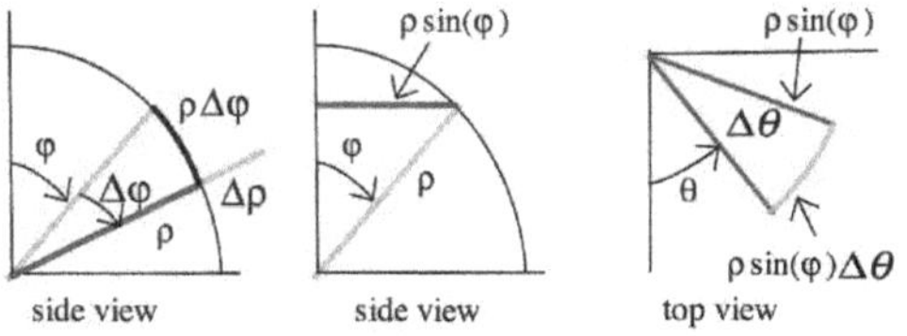

Fig. A2

14.8 Changing Variables in Double and Triple Integrals

In section 4.5 we saw how a change of variables could make some integrals easier to evaluate: the integral $\int x\sqrt{3-x^2}\ dx$ is made easier by using the substitution $u = 3 - x^2$. Similarly, in section 14.3 we saw that converting some integrals from rectangular to polar coordinates can make them easier: $\iint \sqrt{x^2+y^2}\ dx\ dy$ is made easier by replacing x, y and dx dy with $r\cdot\cos(\theta)$, $r\cdot\sin(\theta)$ and $r\cdot dr\cdot d\theta$ respectively. Then $\iint_R \sqrt{x^2+y^2}\ dA = \iint_G r\cdot r\cdot dr\ d\theta$. There are other substitutions that can make double and triple integrals easier, and this section shows how to make some of those transformations.

Changing Variables in Double Integrals

Since a double integral depends on x and y, we will typically need two substitution variables, u and v, and formulas that replace x=x(u,v) and y=y(u,v). Then the domain S in the uv-plane will be mapped into a region R in the xy-plane. Reversing the transformation, we can map R from the xy-plane to S in the uv- plane. The goal is to map a complicated xy domain to an easier uv domain.

Example 1: Suppose $S = \{u,v): 0 \le u \le 2 \text{ and } 1 \le v \le 2\}$ in the uv-plane, and that the transformation T is given by $x = x(u,v) = 2u + v$ and $y = y(u,v) = u - v$. (a) What is the R= T(S) region in the xy-plane? (b) What is the inverse transformation u=u(x,y), v=v(x,y) that maps R back onto S?

Solution: (a) S is shown in Fig. 1a. The corners of S, moving counterclockwise, are (0,1), (2,1), (2,2) and (0,2) and these are mapped by T to the (x,y) points (1,–1), (5,1), (6,0) and (2,–2) respectively, and these become the corners of R = T(S) in the xy-plane as shown in Fig. 1b. Since the transformation T is linear, the straight line boundaries of S are mapped to straight lines of R.

(b) Solving x=2u+v and y=u-v for u and v, we get u=(x+y)/3 and v=(x–2y)/3 . It is difficult to describe the domain of integration for R in terms of x and y, but quite easy for S in terms of u and v.

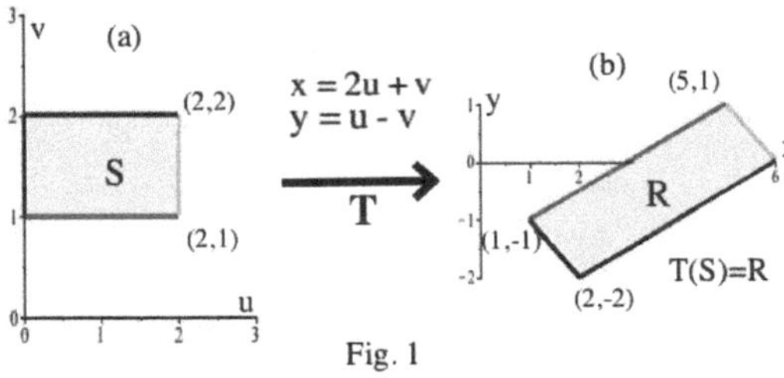

Fig. 1

Practice 1: Suppose $S = \{u,v): 1 \le u \le 3 \text{ and } 0 \le v \le 2\}$ in the uv-plane, and T is given by $x = x(u,v) = u + 2v$ and $y = y(u,v) = 2u - v$. (a) What is the R= T(S) region in the xy-plane? (b) What is the inverse transformation u=u(x,y), v=v(x,y) that maps R back onto S?

Example 2: Suppose $S = \{u,v): 0 \le u \le 1$ and $1 \le v \le 2\}$ in the uv-plane, and T is given by x=u+v and y=u/v . (a) What is the R= T(S) region in the xy-plane?

(b) What is the inverse transformation u=u(x,y), v=v(x,y) that maps R back onto S?

Solution: (a) S is shown in Fig. 2. The corners of S, moving counterclockwise, are (0,1), (1,1), (1,2) and (0,2) and these are mapped by T to the (x,y) points (1,0), (2,1), (3, 1/2) and (2,0) respectively, and these become the corners of R = T(S) in the xy-plane as shown in Fig. 2.

The boundary: If v=1 and 0≤u≤1, then y=x-1 for 1≤x≤2. If v=2 and 0≤u≤1, then y=x/2-1 for 2≤x≤3. If u=0 and 1≤v≤2, then y=0 and 1≤x≤2. Finally, the interesting boundary: if u=1 and 1≤v≤2, then y=1/(x-1) for 2≤x≤3.

(b) The inverse transformation (solving x=u+v and y=u/v for u and v) is u=xy/(1+y) and v=x/(1+y).

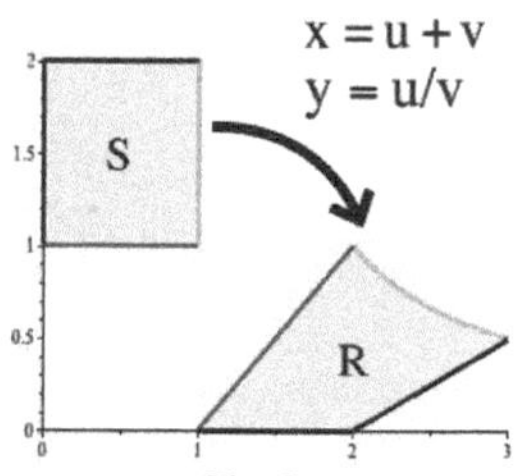

Fig. 2

Practice 2: Suppose $S = \{u,v): 0 \le u \le \pi$ and $1 \le v \le 2\}$ in the uv-plane, and T is given by and $x = 1 + v \cdot \sin(u)$ and y=u . (a) What is the R= T(S) region in the xy-plane? (b) What is the inverse transformation u=u(x,y), v=v(x,y) that maps R back onto S?

Our goal in this section is not to simply map regions into other regions, but it is to do substitutions that make double integrals easier, either by making the integral domain easier or by making the integrand function easier or both. However, we need one more piece, the Jacobian.

Definition

The **Jacobian** of the transformation T:(u,v)->(x,y) by the substitution x=x(u,v) and y=y(u,v) is

$$J(u,v) = \frac{\partial(x,y)}{\partial(u,v)} = \begin{vmatrix} \frac{\partial x}{\partial u} & \frac{\partial x}{\partial v} \\ \frac{\partial y}{\partial u} & \frac{\partial y}{\partial v} \end{vmatrix} = \frac{\partial x}{\partial u}\cdot\frac{\partial y}{\partial v} - \frac{\partial y}{\partial u}\cdot\frac{\partial x}{\partial v}$$

Note: partials are with respect to u and v

The Jacobian is the determinant of the 2x2 matrix of partial derivatives. It is required that x(u,v) and y(u,v) have continuous first partial derivatives on the region S in the uv-plane.

The Jacobian of the inverse transformation with u=u(x,y) and v=v(x,y) is

$$J(x,y) = \frac{\partial(u,v)}{\partial(x,y)} = \begin{vmatrix} \frac{\partial u}{\partial x} & \frac{\partial u}{\partial y} \\ \frac{\partial v}{\partial x} & \frac{\partial v}{\partial y} \end{vmatrix} = \frac{\partial u}{\partial x}\cdot\frac{\partial v}{\partial y} - \frac{\partial v}{\partial x}\cdot\frac{\partial u}{\partial y}$$

The derivation of this Jacobian formula is intricate and is given in the Appendix.

Example 3: (a) Calculate the Jacobian J(u,v) of the transformation $x = x(u,v) = 2u + v$ and $y = y(u,v) = u - v$ from Example 1.

(b) Also calculate the Jacobian J(x,y) of the inverse transformation.

Solution: (a) $\frac{\partial x}{\partial u} = 2,\ \frac{\partial x}{\partial v} = 1,\ \frac{\partial y}{\partial u} = 1$ and $\frac{\partial y}{\partial v} = -1$ so $J(u,v) = \frac{\partial(x,y)}{\partial(u,v)} = \begin{vmatrix} 2 & 1 \\ 1 & -1 \end{vmatrix} = -3$.

(b) The inverse transformation (using algebra) is u=(x+y)/3 and v=(x-2y)/3 so

$$\frac{\partial u}{\partial x} = \frac{1}{3},\ \frac{\partial u}{\partial y} = \frac{1}{3},\ \frac{\partial v}{\partial x} = \frac{1}{3} \text{ and } \frac{\partial v}{\partial y} = -\frac{2}{3} \text{ and } J(x,y) = \frac{\partial(u,v)}{\partial(x,y)} = \begin{vmatrix} 1/3 & 1/3 \\ 1/3 & -2/3 \end{vmatrix} = -\frac{1}{3} .$$

Practice 3: (a) Calculate the Jacobian J(u,v) of the transformation x=u+2v and y=2u-v from Practice 1.

(b) Also calculate the Jacobian J(x,y) of the inverse transformation.

Fact: You might have noticed that $J(x,y) = \frac{1}{J(u,v)}$ in the previous Examples and Practices. That is true in general and can make some computations much easier. A proof of this for the 2D Jacobian is given in the Appendix.

Now we finally get to change variables in double integrals.

Change of Variables Theorem

If the region G in the (u,v) plane is transformed into the region R in the xy-plane by x=x(u,v) and y=y(u,v), and if x(u,v) and y(u,v) have continuous first partial derivatives,

then $$\iint_R F(x,y)\,dx\,dy = \iint_S F(x(u,v),y(u,v))\ |J(u,v)|\ du\,dv$$

for any continuous function F.

Example 4: (a) Calculate the area of the region R in Example 1.

(b) Calculate the mass of the region in Example 1 when the density is $\delta(x,y) = 5 + y$.

Solution: x=2u+v and y=u-v so $J(u,v) = \frac{\partial(x,y)}{\partial(u,v)} = \begin{vmatrix} \frac{\partial x}{\partial u} & \frac{\partial x}{\partial v} \\ \frac{\partial y}{\partial u} & \frac{\partial y}{\partial v} \end{vmatrix} = \begin{vmatrix} 2 & 1 \\ 1 & -1 \end{vmatrix} = -3$.

(a) $$\{\text{area of R}\} = \iint_R 1\,dA = \iint_S 1\ |J(u,v)|\ du\,dv = \int_{v=1}^{2}\int_{u=0}^{2} 1\cdot(3)\ du\ dv = \int_{v=1}^{2} 6\ dv = 6$$

(b) $$\text{Mass} = \iint_R \delta\,dA = \iint_S (5+y)\ |J(u,v)|\ du\,dv = \int_{v=1}^{2}\int_{u=0}^{2} (5+u-v)\cdot(3)\ du\ dy = \int_{v=1}^{2} 36-6v\ dv = 27$$

Practice 4: (a) Calculate the area of the region R in Practice 1.

(b) Calculate the mass of the region in Practice 1 when the density is $\delta(x,y) = x + y$.

Example 5: Let $\iint_R \frac{2x-y}{2}\, dA = \int_{y=0}^{4} \int_{x=y/2}^{(y/2)+1} \frac{2x-y}{2}\, dx\ dy$

(a) Sketch the region R.

(b) Make the substitutions $u = \frac{2x-y}{2}$, $v = \frac{y}{2}$ and solve for x=x(u,v) and y=y(u,v).

(c) Calculate the Jacobian J(u,v) and (d) Rewrite the xy-integral in terms of u and v and evaluate this integral.

Solution: (a) Since this is a linear transformation, corners get mapped to corners and straight line boundaries get mapped to straight line boundaries. The corners in the xy-plane of R are (0,0), (2,4), (1,0) and (3,4) as shown in Fig. 3a. Under the change of variables, (0,0)-->(0,0), (2,4)-->(0,2), (1,0)-->(1,0), and (3,4)-->(1,2) so the integration region in the uv-plane is the rectangle S in Fig. 3b.

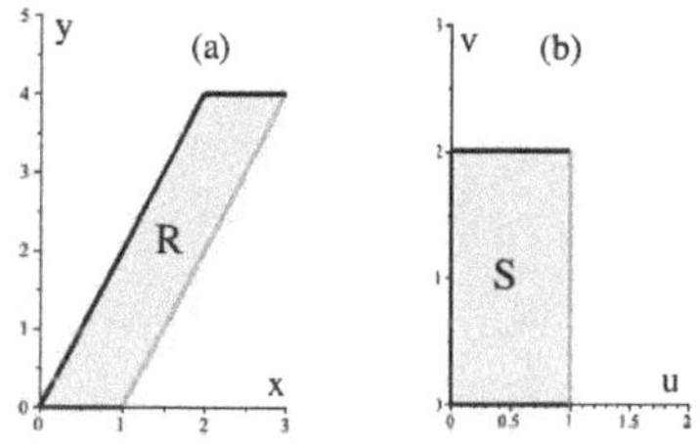

Fig. 3

(b) Simple algebra gives x=u+v and y=2v. (c) $J(u,v) = \frac{\partial(x,y)}{\partial(u,v)} = \begin{vmatrix} \frac{\partial x}{\partial u} & \frac{\partial x}{\partial v} \\ \frac{\partial y}{\partial u} & \frac{\partial y}{\partial v} \end{vmatrix} = \begin{vmatrix} 1 & 1 \\ 0 & 2 \end{vmatrix} = 2$

(d) $\int_{y=0}^{4} \int_{x=y/2}^{(y/2)+1} \frac{2x-y}{2}\, dx\ dy = \int_{v=0}^{2} \int_{u=0}^{1} u\ |J(u,v)|\ du\ dv = \int_{v=0}^{2} \int_{u=0}^{1} u \cdot 2\ du\ dv = \int_{v=0}^{2} 1\ dv = 2$.

Practice 5: Suppose the area of S is $\iint_S 1\ dV = 20$, and $J(u,v) = 2$ for the transformation T(S)=R.

Determine the area of region R.

In Section 14.3 we discussed double integrals in polar coordinates and used geometry to find the conversion formula between the two types of double integrals. The conversion from polar to rectangular coordinates is simply the transformation $x = r \cdot \cos(\theta)$ and $y = r \cdot \sin(\theta)$. Then

$$J(r,\theta) = \frac{\partial(x,y)}{\partial(r,\theta)} = \begin{vmatrix} \frac{\partial x}{\partial r} & \frac{\partial x}{\partial \theta} \\ \frac{\partial y}{\partial r} & \frac{\partial y}{\partial \theta} \end{vmatrix} = \begin{vmatrix} \cos(\theta) & -r \cdot \sin(\theta) \\ \sin(\theta) & r \cdot \cos(\theta) \end{vmatrix} = r \cdot \cos^2(\theta) + r \cdot \sin^2(\theta) = r \text{ and}$$

$\iint_R F(x,y)\ dx\ dy = \iint_S F(r \cdot \cos(\theta), r \cdot \sin(\theta))\ r\ dr\ d\theta$, the same result we got in this special case in 14.3.

Creating the Desired Transformation for Double Integrals

So far in this section all of the transformations have been given. But often we have a "strange" xy-domain and need to pick a transformation that maps it to something nice such as a rectangle in the uv-plane. Sometimes that is very difficult, but there are a few situations and ideas that are much easier. Technology and software can help us evaluate double integrals once they are set up, but it is usually up to us to set up those integrals first.

(1) Region R is bounded by parallel lines

In Example 1 the region R (Fig. 4) is bounded by the pair of parallel lines y=6–x and y=–x which can be rewritten as x+y=**6** and x+y=**0**. This suggests that we might set x+y=**u** and let u vary from 0 to 6. Similarly with the parallel pair 2y=x–3 and 2y=x–6, rewritten as x-2y=**3** and x–2y=**6**, suggesting that we put x–2y=**v** with v going from 3 to 6. It is straight forward to verify that the transformation u=x+y, v=x-2y transforms R into the uv-plane rectangle shown in Fig. 5. (The transformation in Example 1 was u=(x+y)/3 and v=(x-2y)/3 which leads to a smaller rectangle in the uv plane.)

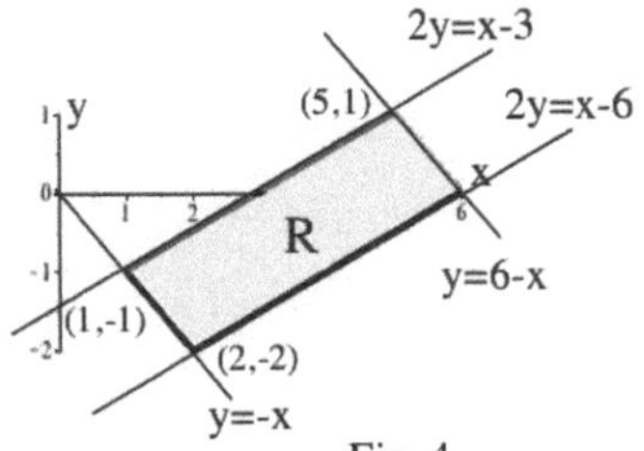

Fig. 4

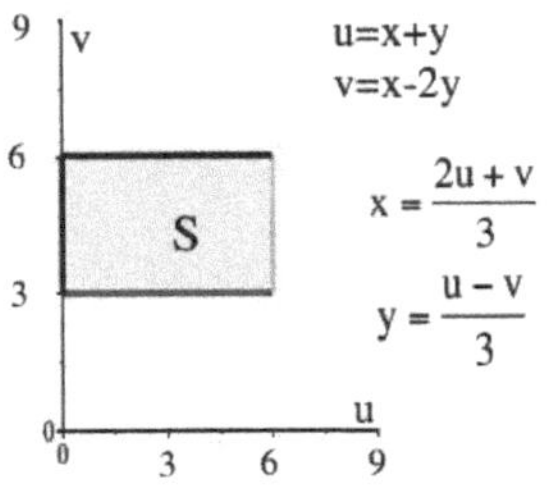

Fig. 5

Practice 6: Find a transformation of the region R bounded by the lines y=x, y=x+2, y=6–2x and y=9–2x (Fig. 6) into a rectangle S in the uv-plane.

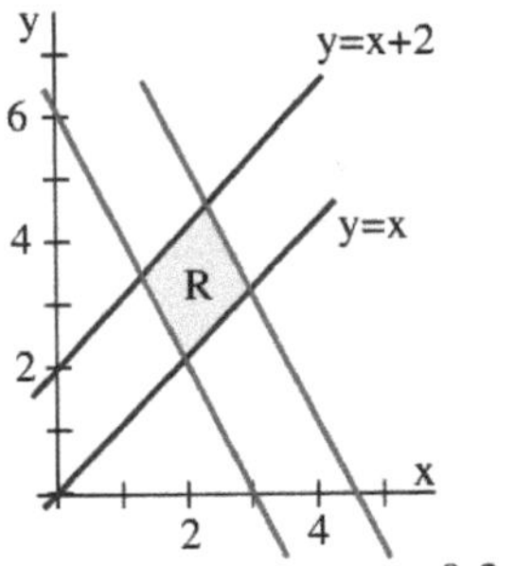

Fig. 6 y=6-2x y=9-2x

(2) Region R is bounded by shifted curves

Example 6: R is the region in the xy plane bounded by the lines y=x and y=x+3 and the parabolas $y = 9 - x^2$ and $y = 16 - x^2$ (Fig. 7). Find a transformation of R into a rectangle S in the uv-plane.

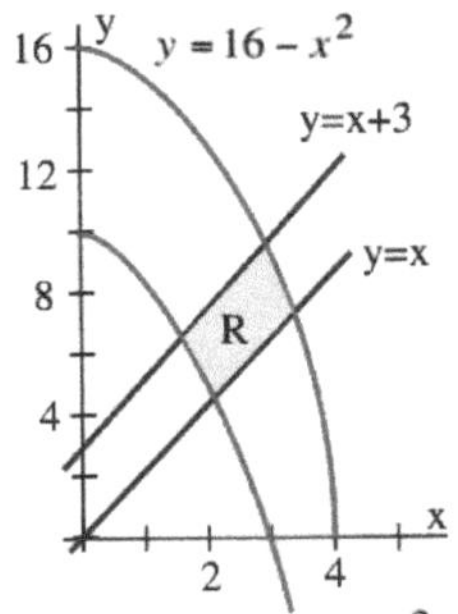

Fig. 7 $y = 9 - x^2$

Solution: y-x=0 and y-x=3 suggests putting u=y-x (so u goes from 0 to 3). The parabolas $x^2 + y = 9$ and $x^2 + y = 16$ suggests putting $v = x^2 + y$ (so v goes from 9 to 16). You can verify that this works.

Practice 7: R is the region in the xy-plane bounded by the four parabolas $y = 9 - x^2$, $y = 16 - x^2$, $y = x^2$ and $y = x^2 + 4$. Sketch the region R and find a transformation of R into a rectangle in the uv-plane.

(3) Two parameter family of curves

If the bounding curves of the region R can be written using the two parameters u and v, then we can usually transform R into a rectangle in the uv-plane.

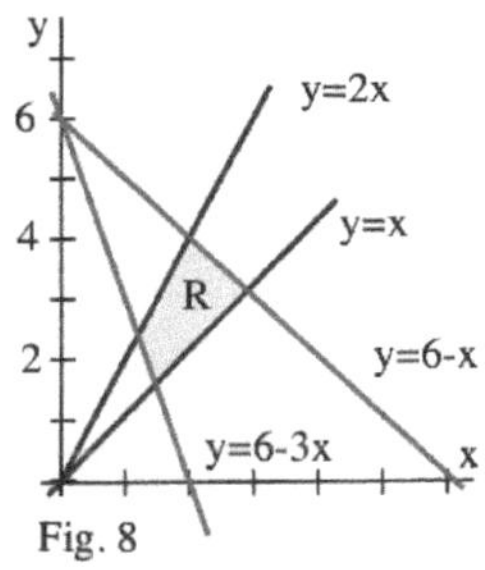

Fig. 8

Example 7: R is the region in the xy-plane bounded by the four lines y=x, y=2x, y=6–x and y=6–3x in Fig. 8. Find a transformation of R into a rectangle S in the uv-plane.

Solution: y=1x and y=2x can be rewritten with a single parameter u as y=ux for u from 1 to 2. y=6–1x and y=6–3x can be rewritten with the single parameter y=6–vx with v going from 1 to 3. Solving for u and v, we get the transformation u=y/x and v=(6–y)/x which takes R to a rectangle in the uv-plane. The inverse transformation is x=6/(u+v) and y=6u/(u+v) .

Changing Variables in Triple Integrals

Changing variables for triple integrals is very similar to the situation for double integrals. If T is a transformation from an uvw-space region S to an xyz-space region R (so x, y and z are each differentiable functions of u, v and w: x=g(u,v,w), y=h(u,v,w) and z=k(u,v,w)) then the

3D Jacobian is $$J(u,v,w)=\begin{vmatrix} \frac{\partial x}{\partial u} & \frac{\partial x}{\partial v} & \frac{\partial x}{\partial w} \\ \frac{\partial y}{\partial u} & \frac{\partial y}{\partial v} & \frac{\partial y}{\partial w} \\ \frac{\partial z}{\partial u} & \frac{\partial z}{\partial v} & \frac{\partial z}{\partial w} \end{vmatrix}$$

The 3D change of variables formula is

$$\iiint_R f(x,y,z)\, dx \cdot dy \cdot dz = \iiint_S f(g(u,v,w),\ h(u,v,w),\ k(u,v,w)) \cdot |J(u,v,w)|\ du \cdot dv \cdot dw$$

Example 8: Suppose f(x,y,z)=2x+4z on the box-like region (Fig. 9)

$R = \{(x,y,z):\ x \le y \le x+3,\ x \le z \le x+2 \text{ and } 1-x \le z \le 3-x\}$. Evaluate

$\iiint_R f(x,y,z)\, dx \cdot dy \cdot dz$ by using the transformation T: u=y-x (0≤u≤3),

v=z–x (0≤v≤2) w=z+x (1≤w≤3) and then evaluating the new integral.

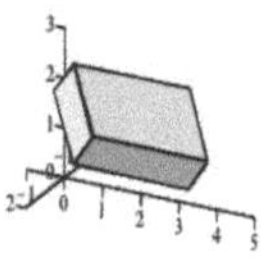
Fig. 9

Solution: We need several pieces. The inverse transformation (after a bit of algebra) is x=–v/2+w/2, y= u-v/2+w/2 , z= v/2+w/2 , f(x,y,z)=2x+4z=v+3w , S={(u,v,w): 0≤u≤3, 0≤v≤2, 1≤w≤3}, and the Jacobean is $J(u,v,w)=\begin{vmatrix} 0 & -\frac{1}{2} & \frac{1}{2} \\ 1 & -\frac{1}{2} & \frac{1}{2} \\ 0 & \frac{1}{2} & \frac{1}{2} \end{vmatrix} = \frac{1}{2}$

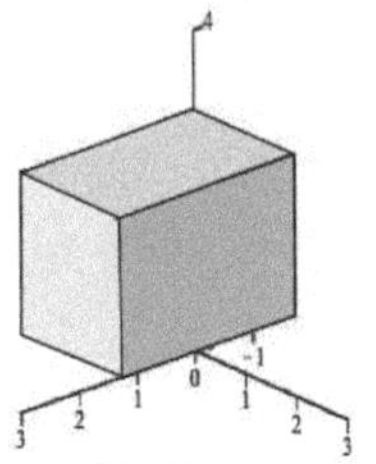
Fig. 10

The new domain in uvw-space is shown in Fug. 10. Finally,

$$\iiint_R f(x,y,z)\, dx \cdot dy \cdot dz \quad \int_1^3 \int_0^2 \int_0^3 (v+3w)\left(\frac{1}{2}\right) du \cdot dv \cdot dw$$

$$\int_1^3 \int_0^2 3(v+3w)\left(\frac{1}{2}\right) dv \cdot dw = \int_1^3 3+9w\, dw = 42\ .$$

Practice 8: Evaluate $\iiint_R (x+z)\, dx \cdot dy \cdot dz$ on the region

$R = \{(x,y,z):\ x+1 \le y \le x+3,\ x \le z \le 2x \text{ and } 0 \le z \le 3\}$ by using the transformation

T: $u=y-x\ (1 \le u \le 3)$, $v=x/z\ (1 \le v \le 2)$ $w=z\ (0 \le w \le 3)$.

Rectangular to Spherical: In section 14.7 we transformed some integrals in rectangular coordinates into ones in spherical coordinates, and we derived the $dV = \rho^2 \cdot \sin(\varphi)\ d\rho \cdot d\theta \cdot d\varphi$ formula for the transformation geometrically. Instead, we can use the Jacobean. For spherical coordinates $x = \rho \cdot \sin(\varphi) \cdot \cos(\theta)$, $y = \rho \cdot \sin(\varphi) \cdot \sin(\theta)$, and $z = \rho \cdot \cos(\varphi)$. Then the Jacobean J(u,v,w) is

$$J(\rho,\theta,\varphi) = \begin{vmatrix} \frac{\partial x}{\partial \rho} & \frac{\partial x}{\partial \theta} & \frac{\partial x}{\partial \varphi} \\ \frac{\partial y}{\partial \rho} & \frac{\partial y}{\partial \theta} & \frac{\partial y}{\partial \varphi} \\ \frac{\partial z}{\partial \rho} & \frac{\partial z}{\partial \theta} & \frac{\partial z}{\partial \varphi} \end{vmatrix} = \begin{vmatrix} \sin(\varphi)\cdot\cos(\theta) & -\rho\cdot\sin(\varphi)\cdot\sin(\theta) & \rho\cdot\cos(\varphi)\cdot\cos(\theta) \\ \sin(\varphi)\cdot\sin(\theta) & \rho\cdot\sin(\varphi)\cdot\cos(\theta) & \sin(\varphi)\cdot\sin(\theta) \\ \cos(\varphi) & 0 & -\rho\cdot\sin(\varphi) \end{vmatrix} = \ldots = \rho^2 \cdot \sin(\varphi)\ d\rho \cdot d\theta \cdot d\varphi .$$

The " ..." is simply a matter of carefully calculating the determinant and then using some fundamental trigonometric identities to simplify the result.

Practice 9: Calculate $\sin(\varphi)\cdot\cos(\theta)\cdot \begin{vmatrix} \rho\cdot\sin(\varphi)\cdot\cos(\theta) & \sin(\varphi)\cdot\sin(\theta) \\ 0 & -\rho\cdot\sin(\varphi) \end{vmatrix}$

Other triple integral transformations are possible, but rectangular to cylindrical or spherical are the most common.

Problems

In problems 1 to 4, (a) sketch the given set S and the image of S under the given transformation, (b) calculate the Jacobians, J(x,y) and J(u,v), and rewrite $\iint_R f(x,y)\,dx\,dy$ as $\iint_S$ ____ du dv.

1. $S=\{(u,v):\ 0\le u\le 2,\ 1\le v\le 4\}$ under $x=u+v$ and $y=2u-v$.
2. $S=\{(u,v):\ 1\le u\le 2,\ 0\le v\le 2\}$ under $x=2u-3v$ and $y=u+2v$.
3. $S=\{(u,v):\ 0\le u\le 1,\ 0\le v\le 1\}$ under $x=au+bv$ and $y=cu+dv$.
4. $S=\left\{(u,v):\ 2\le u\le 4,\ \frac{\pi}{6}\le v\le\frac{\pi}{2}\right\}$ under $x=u\cdot\cos(v)$ and $y=u\cdot\sin(v)$.

In problems 5 to 10, (a) sketch the given set S and the image of S under the given transformation, and (b) calculate the Jacobians, J(x,y) and J(u,v).

5. $S=\{(x,y):\ 0\le x\le 2,\ 1\le y\le 3\}$ under $u=\frac{3x-3y}{4}$ and $v=\frac{y}{3}$.
6. $S=\{(x,y):\ 0\le x\le 2,\ 1\le y\le 3\}$ under $u=x+y$ and $v=\frac{x}{y}$.
7. $S=\{(x,y):\ 0\le x\le 1,\ 0\le y\le 1\}$ under $u=x^2-y^2$ and $v=2xy$.
8. $S=\{$region bounded by the hyperbolas $y=1/x$ and $y=4/x$ and the lines $y=x$ and $y=4x\}$ with $x=u/v$ and $y=uv$.
9. $S=\{$trapezoid with vertices $(x,y)=(1,0)$, $(2,0)$, $(0,-2)$ and $(0,-1)\}$ under $u=x-y$ and $v=x+y$.
10. $S=\{$triangle with vertices $(x,y)=(1,0)$, $(3,1)$ and $(0,4)\}$ under $u=x-y$ and $v=x+y$.

11. Suppose $\iint_S 1\,du\,dv=14$, and $J(u,v)=7$ for the transformation T(S)=R. Evaluate $\iint_R 1\,dV$.
12. Suppose $\iint_S 1\,du\,dv=30$, and $J(u,v)=5$ for the transformation T(S)=R. Evaluate $\iint_R 1\,dA$.
13. Suppose $\iint_S 1\,du\,dv=15$, and $J(x,y)=3$ for the transformation T(S)=R. Evaluate $\iint_R 1\,dA$.
14. Suppose $\iint_S 1\,du\,dv=24$, and $J(x,y)=4$ for the transformation T(S)=R. Evaluate $\iint_R 1\,dA$.

In problems 15-24, use the given transformation to evaluate the integral.

15. $\iint_R (x+3y)\,dA$. R is the triangular region with vertices (0,0), (2,1) and (1,2). Use x=2u+v and y=u+2v (so u=(2x–y)/3 and v=(2y–x)/3).
16. $\iint_R x\,dA$. R is the ellipse $\frac{x^2}{9}+\frac{y^2}{4}\le 1$. Use x=3u and y=2u.
17. $\iint_R \frac{x+y}{3}\,dA$ where R is the region shown in Fig. 4.

18. $\iint_R (x-2y)\,dA$ where R is the region shown in Fig. 4.

19. $\iint_R (3x+6y)\,dA$ where R is the region shown in Fig. 6.

20. $\iint_R (6x-3y)\,dA$ where R is the region bounded by the lines y=x, y=2x, y=6-2x and y=9-2x (Fig. 6).

21. $\iint_R \sqrt{\frac{y}{x}}+\sqrt{xy}\,dA$ where R={region bounded by the hyperbolas y=1/x and y=4/x and the lines y=x and y=4x} Use the substitution x=u/v and y=uv.

22. $\iint_R (6x-3y)\,dA$ where R is the region bounded by the lines y=x, y=2x, y=6-x and y=6-3x (Fig. 8).

Use u=y/x and v=(6-y)/x.

23. The area of the ellipse $\frac{x^2}{a^2}+\frac{y^2}{b^2}\le 1$ can be found using earlier methods, but the integral requires a trigonometric substitution. Instead, transform the ellipse into a circle using the substitution x=au and y=bv, write the new uv integral and evaluate it to show that the area of the ellipse is πab .

24. Evaluate $\iint_R xy\,dA$ where R is the square with vertices at (0,0), (1,1), (2,0) and (1,-1).

Practice Solutions

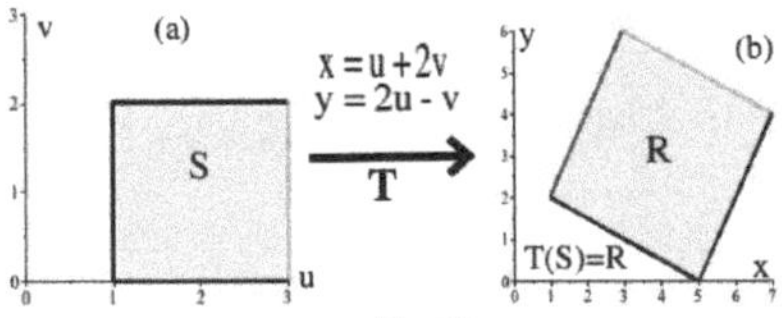

Fig. P1

Practice 1: See Fig. P1 for the S and R regions. u=(x+2y)/5 and v=(2x–y)/5 .

Practice 2: See Fig. P2 for the S and R regions. u=y and v=x/(1+sin(y)).

Practice 3: $\frac{\partial x}{\partial u}=1,\ \frac{\partial x}{\partial v}=2,\ \frac{\partial y}{\partial u}=2$ and $\frac{\partial y}{\partial v}=-1$ so $J(u,v)=\frac{\partial(x,y)}{\partial(u,v)}=\begin{vmatrix}1 & 2\\ 2 & -1\end{vmatrix}=-5$

The inverse transformation is u=(x+2y)/5 and v=(2x-y)/5 so

$\frac{\partial u}{\partial x}=\frac{1}{5},\ \frac{\partial u}{\partial y}=\frac{2}{5},\ \frac{\partial v}{\partial x}=\frac{2}{5}$ and $\frac{\partial v}{\partial y}=-\frac{1}{5}$ and $J(x,y)=\frac{\partial(u,v)}{\partial(x,y)}=\begin{vmatrix}1/5 & 2/5\\ 2/5 & -1/5\end{vmatrix}=-\frac{1}{5}$

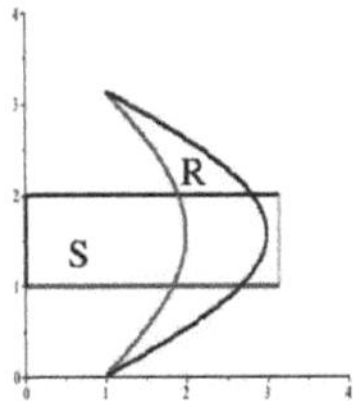

Fig. P2

Practice 4: $\delta(x,y)=x+y$. x = x(u,v) = u+2v and y = y(u,v) = u - v so J(u,v)=5 (see Practice 3)

(a) $\{\text{area of R}\} = \iint_R 1\,dA = \iint_S 1\cdot |J(u,v)|\,du\,dv = \int_{v=0}^{2}\int_{u=1}^{3} (5)\,du\,dv = \int_{v=1}^{2} 10\,dv = 10$

(b) $\text{Mass} = \iint_R \delta\,dA = \iint_S (x+y)\cdot |J(u,v)|\,du\,dv = \int_{v=0}^{2}\int_{u=1}^{3} (2u+v)(5)\,du\,dv = \int_{v=1}^{2} 40+10v\,dv = 65$

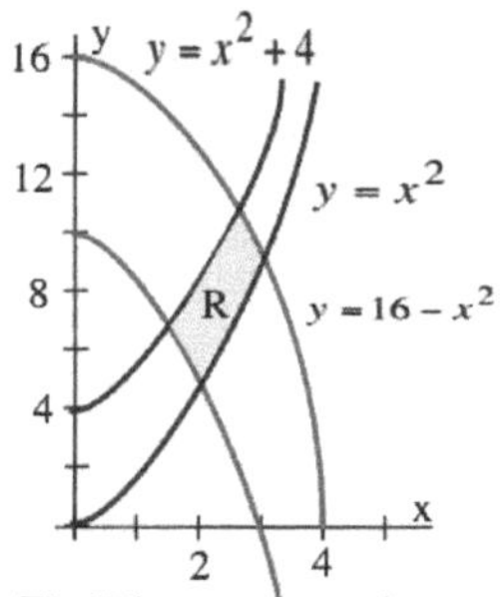

Fig. P6

Practice 5: The area of R is $\iint\limits_R 1 \, dV = \iint\limits_S 1 \cdot |J(u,v)| \, dV \iint\limits_S 2 \, dV = 2(20) = 40 \cdot$

Practice 6: The first two line equations can be written as y-x=0 and y-x=2 so put u=y-x (then u goes from 0 to 2). The other two parallel lines can be written as 2x+y=6 and 2x+y=9 so put v=2x+y (then v goes from 6 to 9). This transformation takes R in the xy-plane to the 0≤u≤2 and 6≤v≤9 rectangle in the uv-plane.

Practice 7: R is shown in Fig. P6. $x^2 + y = 9$ and $x^2 + y = 16$ so put $u = x^2 + y$ (then u goes from 9 to 16). Similarly, $x^2 - y = 0$ and $x^2 - y = 4$ so put $u = x^2 - y$ (then v goes from 0 to 4).

Practice 8: The inverse transformation is x=uv, y=vw+u and z=w. x+z=uv+w and

$J(u,v,w) = \begin{vmatrix} v & u & 0 \\ 1 & w & 0 \\ 0 & 0 & 1 \end{vmatrix} = vw$ so the new integral is

$$\iiint\limits_R (x+z) \, dx \cdot dy \cdot dz = \int_0^3 \int_1^2 \int_1^3 (uv+w)(vw) \, du \cdot dv \cdot dw = \int_0^3 \int_1^2 \int_1^3 (uv^2w + vw^2) \, du \cdot dv \cdot dw$$

$$= \int_0^3 \int_1^2 \left(4v^2w + 2vw^2\right) dv \cdot dw = \int_0^3 \left(\frac{28}{3}w + 3w^2\right) dw = 69$$

Practice 9: $\sin(\varphi) \cdot \cos(\theta) \cdot \begin{vmatrix} \rho \cdot \sin(\varphi) \cdot \cos(\theta) & \sin(\varphi) \cdot \sin(\theta) \\ 0 & -\rho \cdot \sin(\varphi) \end{vmatrix} = \sin(\varphi) \cdot \cos(\theta) \left[-\rho^2 \cdot \sin^2(\varphi) \cdot \cos(\theta)\right] = -\rho^2 \cdot \sin^3(\varphi) \cdot \cos^2(\theta)$

Two Transformations, Same Result

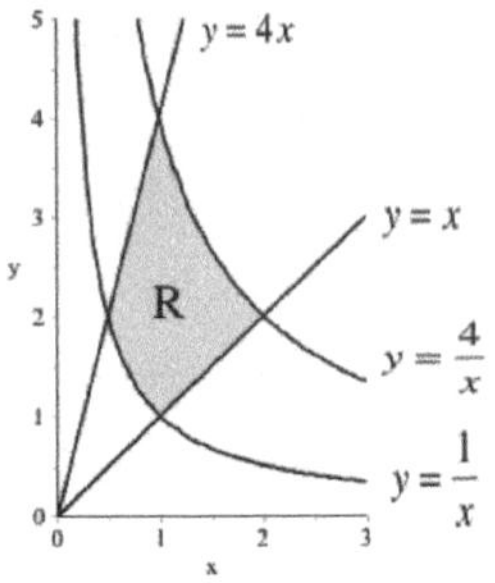

Integral of $f(x,y) = \sqrt{\dfrac{y}{x}} + \sqrt{xy}$ on

$R = \{(x,y)$ bounded by $y = 1/x$, $y = 4/x$, $y = x$ and $y = 4x\}$ (see figure)

(A) The "natural" transformation (pages 5 and 6):

$y=vx$ for $v=1..4$ so $v=y/x$. $y=u/x$ for $u=1..4$ so $u=xy$.

Then $x = \sqrt{\dfrac{u}{v}}$ and $y = \sqrt{uv}$.

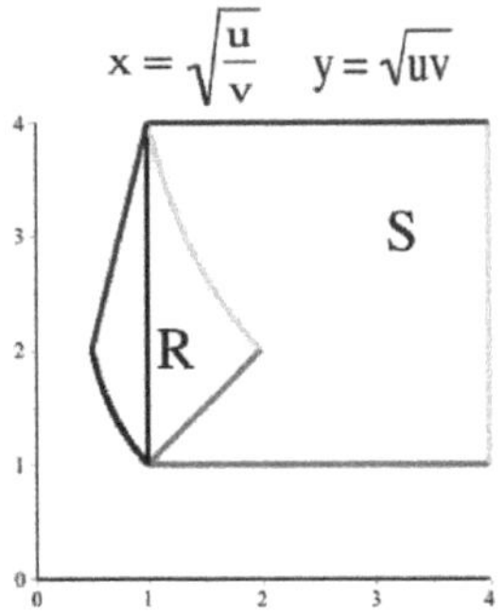

The Jacobian is $J(x,y) = \begin{vmatrix} \dfrac{\partial u}{\partial x} & \dfrac{\partial u}{\partial y} \\ \dfrac{\partial u}{\partial x} & \dfrac{\partial u}{\partial x} \end{vmatrix} = \begin{vmatrix} y & x \\ \dfrac{-y}{x^2} & \dfrac{1}{x} \end{vmatrix} = \dfrac{2y}{x} = 2v$.

$$\iint_R \sqrt{\frac{y}{x}} + \sqrt{xy}\, dx\, dy = \int_{u=1}^{4} \int_{v=1}^{4} [\sqrt{v} + \sqrt{u}]\left(\frac{1}{2v}\right) dv\, du$$

$$= \int_{u=1}^{4} \int_{v=1}^{4} \left[\frac{1}{2\sqrt{v}} + \frac{\sqrt{u}}{2v}\right] dv\, du = \int_{u=1}^{4} \left[\sqrt{v} + \frac{\sqrt{u}}{2}\ln(v)\right] \Big|_{v=1}^{v=4} du$$

$$= \int_{u=1}^{4} \left[1 + \frac{\sqrt{u}}{2}\ln(4)\right] \Big|_{v=1}^{v=4} du = u + \frac{1}{2}\cdot\frac{2}{3}(u)^{3/2}\ln(4) \Big|_{u=1}^{u=4} = \mathbf{3 + \frac{7}{3}\ln(4)}$$

(B) The "suggested" transformation (Problem 8):

$x=u/v$ and $y=uv$. Then $u=xv$ so

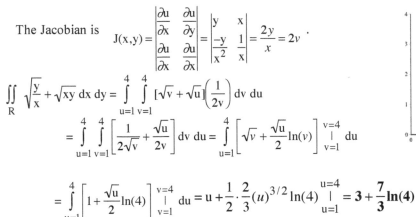

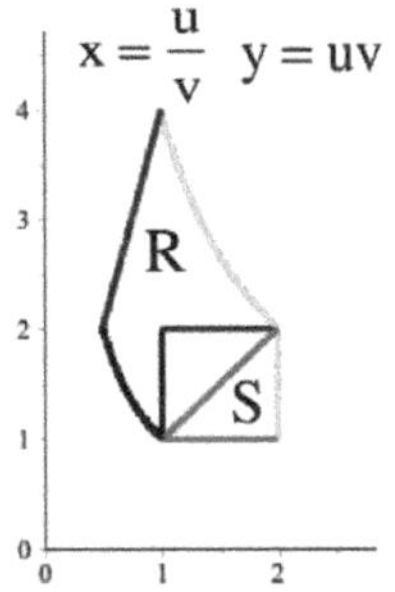

$y = (xv)v = x \cdot v^2$ with $\mathbf{v^2 = 1..4}$ **and** $\mathbf{v = 1..2}$ and $v = \sqrt{\dfrac{y}{x}}$.

$xy = \left(\dfrac{u}{v}\right)(uv) = u^2$ with $\mathbf{u^2 = 1..4}$ **and** $\mathbf{u = 1..2}$ and $u = \sqrt{xy}$.

The Jacobian is $J(u,v) = \begin{vmatrix} \dfrac{\partial x}{\partial u} & \dfrac{\partial x}{\partial v} \\ \dfrac{\partial y}{\partial u} & \dfrac{\partial y}{\partial v} \end{vmatrix} = \begin{vmatrix} \dfrac{1}{v} & -\dfrac{u}{v^2} \\ v & u \end{vmatrix} = \dfrac{2u}{v}$.

$$\iint_R \sqrt{\frac{y}{x}} + \sqrt{xy}\, dx\, dy = \int_{u=1}^{2} \int_{v=1}^{2} [v + u]\left(\frac{2u}{v}\right) dv\, du = \int_{u=1}^{2} \int_{v=1}^{2} \left[2u + \frac{2u^2}{v}\right] dv\, du$$

$$= \int_{u=1}^{2} \left[2uv + 2u^2\ln(v)\right] \Big|_{v=1}^{v=2} du = \int_{u=1}^{2} \left[2u + 2u^2\ln(2)\right] du = u^2 + \frac{2}{3}u^3\ln(2) \Big|_{u=1}^{u=2}$$

$$= u^2 + \frac{2}{3}u^3\ln(2) \Big|_{u=1}^{u=2} = 3 + \frac{14}{3}\ln(2) = \mathbf{3 + \frac{7}{3}\ln(4)} \ .$$

Conclusion: These two transformations have different Jacobians (magnifications) and lead to different S regions ([1,4]x[1,4] verses [1,2]x[1,2]) but the resulting integral values are the same.

Appendix: Derivation of the 2D Jacobian Formula

Suppose T is a transformation from a rectangular region S in the uv-plane to a region R in the xy-plane given by x=x(u,v) and y=y(u,v) as in Fig A1.

The corners of the rectangular region S are (u,v) , $(u+\Delta u,v)$, $(u,v+\Delta v)$ and $(u+\Delta u,v+\Delta v)$.

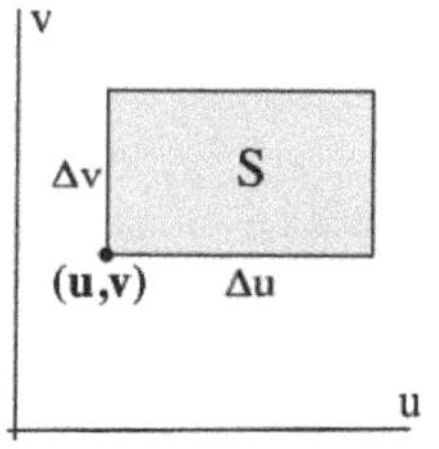

Fig. A1

Then T:(u,v)-> (x(u,v), y(u,v)) = A, a point in the xy-plane (Fig. A2)
T: $(u+\Delta u,v)$->($x(u+\Delta u,v)$, $y(u+\Delta u,v)$) = B
T: $(u,v+\Delta v)$->($x(u,v+\Delta v)$, $y(u,v+\Delta v)$) = C

Next we want to approximate the area of the region R using the cross product.
Put P = vector AB $=\langle x(u+\Delta u,v) - x(u,v),\ y(u+\Delta u,v) - y(u,v)\rangle$

$$=\left\langle \frac{x(u+\Delta u,v)-x(u,v)}{\Delta u},\ \frac{y(u+\Delta u,v)-y(u,v)}{\Delta u}\right\rangle \Delta u=\left\langle \frac{\Delta x}{\Delta u},\ \frac{\Delta y}{\Delta u}\right\rangle \Delta u$$

and Q = vector AC = $\langle x(u,v+\Delta v) - x(u,v),\ y(u,v+\Delta v) - y(u,v)\rangle$

$$=\left\langle \frac{x(u,v+\Delta v)-x(u,v)}{\Delta v},\ \frac{y(u,v+\Delta v)-y(u,v)}{\Delta v}\right\rangle \Delta v=\left\langle \frac{\Delta x}{\Delta v},\ \frac{\Delta y}{\Delta v}\right\rangle \Delta v$$

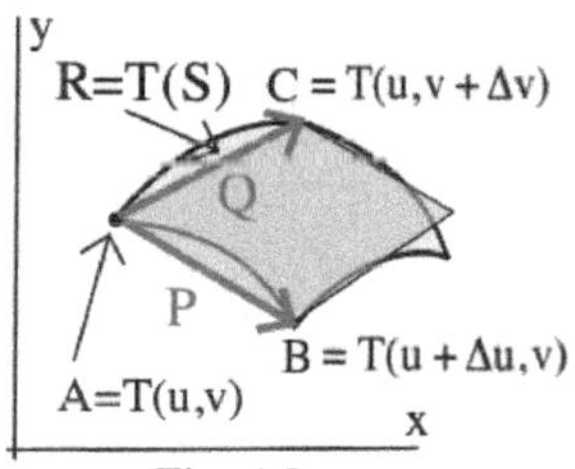

Fig. A2

Then the area of region R is approximately $|PxQ|=\begin{vmatrix} \frac{\Delta x}{\Delta u} & \frac{\Delta y}{\Delta u} \\ \frac{\Delta x}{\Delta v} & \frac{\Delta y}{\Delta v} \end{vmatrix}\Delta u\Delta v$

As $\Delta u,\ \Delta v \to 0$, the fractions in the determinant approach the partial derivatives.

So $dA = dx\cdot dy = \begin{vmatrix} \frac{\partial x}{\partial u} & \frac{\partial x}{\partial v} \\ \frac{\partial y}{\partial u} & \frac{\partial y}{\partial v} \end{vmatrix} du\cdot dv = |J(u,v)|du\cdot dv$

Appendix: Simple proof that $J(x,y)=\dfrac{1}{J(u,v)}$ for a linear transformation in 2D (could be a student assignment)

Theorem; If T:(u,v)-->(x,y) is a linear transformation (x=au+bv and y=cu+dv) then J(u,v)=1/J(x,y).

Proof: $J(x,y)=\begin{vmatrix} \frac{\partial x}{\partial u} & \frac{\partial x}{\partial v} \\ \frac{\partial y}{\partial u} & \frac{\partial y}{\partial v} \end{vmatrix}=\begin{vmatrix} a & b \\ c & d \end{vmatrix}=ad-bc$.

Using elementary algebra and solving for u and v we get $u=\dfrac{dx-by}{ad-bc}$ and $v=\dfrac{-cx+ay}{ad-bc}$ so

$$J(u,v)=\begin{vmatrix} \frac{\partial u}{\partial x} & \frac{\partial u}{\partial y} \\ \frac{\partial v}{\partial x} & \frac{\partial v}{\partial y} \end{vmatrix}=\begin{vmatrix} \frac{d}{ad-bc} & \frac{-b}{ad-bc} \\ \frac{-c}{ad-bc} & \frac{a}{ad-bc} \end{vmatrix}=\frac{ad-bc}{(ad-bc)^2}=\frac{1}{ad-bc}=\frac{1}{J(x,y)}.$$

Appendix: Proof that $J(x,y)=\dfrac{1}{J(u,v)}$ for a general linear transformation in 2D.

The following results are true for larger matrices and the proofs are similar.

Lemma: For 2x2 matrices S and T, $|S\cdot T| = |S|\cdot|T|$

Proof: If $S=\begin{pmatrix} a & b \\ c & d \end{pmatrix}$ and $T=\begin{pmatrix} A & B \\ C & D \end{pmatrix}$ then $|S| = ad - bc$ and $|T| = AD - BC$.

By matrix multiplication, $S\cdot T = \begin{pmatrix} aA+bC & aB+bD \\ cA+dC & cB+dD \end{pmatrix}$ so

$$|S\cdot T| = (aA+bC)(cB+dD) - (cA+dC)(aB+bD)$$

$$= (aAcB + aAdD + bCcB + bCdD) - (cAaB + cAbD + dCaB + dCbD)$$

$$= acAB + adAD + bcBC + bdCD - acAB - bcAD - adBC - bdCD$$

$$= adAD + bcBC - bcAD - adBC \;\; = (ad - bc)(AD - BC) = |S|\cdot|T|$$

Lemma: If $S:(x,y)\to(u,v)$ and $T:(u,v)\to(w,z)$ then $J(T\circ S) = J(S)\cdot J(T)$.

Proof. $J(S)=\begin{vmatrix} \frac{\partial x}{\partial u} & \frac{\partial x}{\partial v} \\ \frac{\partial y}{\partial u} & \frac{\partial y}{\partial v} \end{vmatrix}$ and $J(S)=\begin{vmatrix} \frac{\partial x}{\partial u} & \frac{\partial x}{\partial v} \\ \frac{\partial y}{\partial u} & \frac{\partial y}{\partial v} \end{vmatrix}$.

Then $T\circ S:(x,y)\to(u,v)\to(w,z)$ so $J(T\circ S)=\begin{vmatrix} \frac{\partial x}{\partial w} & \frac{\partial x}{\partial z} \\ \frac{\partial y}{\partial w} & \frac{\partial y}{\partial z} \end{vmatrix}$.

But by the Chain Rule for functions of several variables,

$$\frac{\partial x}{\partial w} = \frac{\partial x}{\partial u}\cdot\frac{\partial u}{\partial w} + \frac{\partial x}{\partial v}\cdot\frac{\partial v}{\partial w} \qquad \frac{\partial y}{\partial w} = \frac{\partial y}{\partial u}\cdot\frac{\partial u}{\partial w} + \frac{\partial y}{\partial v}\cdot\frac{\partial v}{\partial w}$$

$$\frac{\partial x}{\partial z} = \frac{\partial x}{\partial u}\cdot\frac{\partial u}{\partial z} + \frac{\partial x}{\partial v}\cdot\frac{\partial v}{\partial z} \qquad \frac{\partial y}{\partial z} = \frac{\partial y}{\partial u}\cdot\frac{\partial u}{\partial z} + \frac{\partial y}{\partial v}\cdot\frac{\partial v}{\partial z}$$

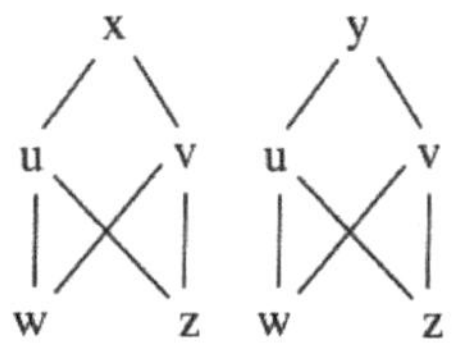

so $$J(T\circ S)=\begin{vmatrix} \frac{\partial x}{\partial w} & \frac{\partial x}{\partial z} \\ \frac{\partial y}{\partial w} & \frac{\partial y}{\partial z} \end{vmatrix} = \begin{vmatrix} \frac{\partial x}{\partial u}\cdot\frac{\partial u}{\partial w} + \frac{\partial x}{\partial v}\cdot\frac{\partial v}{\partial w} & \frac{\partial x}{\partial u}\cdot\frac{\partial u}{\partial z} + \frac{\partial x}{\partial v}\cdot\frac{\partial v}{\partial z} \\ \frac{\partial y}{\partial u}\cdot\frac{\partial u}{\partial w} + \frac{\partial y}{\partial v}\cdot\frac{\partial v}{\partial w} & \frac{\partial y}{\partial u}\cdot\frac{\partial u}{\partial z} + \frac{\partial y}{\partial v}\cdot\frac{\partial v}{\partial z} \end{vmatrix}$$

$$= \left|\begin{pmatrix} \frac{\partial x}{\partial u} & \frac{\partial x}{\partial v} \\ \frac{\partial y}{\partial u} & \frac{\partial y}{\partial v} \end{pmatrix}\begin{pmatrix} \frac{\partial u}{\partial w} & \frac{\partial u}{\partial z} \\ \frac{\partial v}{\partial w} & \frac{\partial v}{\partial z} \end{pmatrix}\right| = \begin{vmatrix} \frac{\partial x}{\partial u} & \frac{\partial x}{\partial v} \\ \frac{\partial y}{\partial u} & \frac{\partial y}{\partial v} \end{vmatrix}\cdot\begin{vmatrix} \frac{\partial u}{\partial w} & \frac{\partial u}{\partial z} \\ \frac{\partial v}{\partial w} & \frac{\partial v}{\partial z} \end{vmatrix} = J(S)\cdot J(T)$$

Theorem: If S and T are inverse transformations $\left(T\circ S = I:(x,y)\to(x,y) = \begin{pmatrix} 1 & 0 \\ 0 & 1 \end{pmatrix}\right)$, then $J(S)\cdot J(T)=1$.

Proof: Since $T\circ S = I$ then $J(T\circ S)=J(I)=\begin{vmatrix} 1 & 0 \\ 0 & 1 \end{vmatrix}=1$ and $J(S)\cdot J(T)=J(T\circ S)=1$.

14.0 Odd Answers

1. v=(24)(54/6)=216 (actual value is 368 – find out how in section 14.1)
3. (a) v=(32)(64/8)=256 (b) v=(32)(112/8)=448 (actual = 1184/3 = 393.667)
5. sum=39 inches, avg=39/16=2.4375 inches= 0.203125 ft, total = (avg)(area)=125.075 ft^3
7. v=(16/4)(4)(5)=80 ft^3
9. v=(19/5)(PI*2^2)=(76/5)*Pi=47.752 f^3
11. v=(26.7/15)*90=160.2 m^3
13. crude avg ht= 2, so vol=(2)(6)(10)(Pi)/4=30Pi = 94.25 m^3
15. avg =30/6=5, v=5*6*4= 120

14.1 Odd Answers

1. $\frac{8}{3}y^3$, $\frac{1}{4}x^2$ 3. $e^{2+y} + e^y = e^y(e^2 + 1)$, $xe^{x+1} - xe^x = xe^x(e - 1)$ 5. 60

7. $3+e^4+4\sin(1) \approx 60.96$ 9. 26 11. 32/3 13. $\frac{4}{15}(31 - 9\sqrt{3})$ 15. 6

17. Easier $\int y \cdot \sqrt{x^2 + y^2}\ dy = \frac{1}{3}(x^2 + y^2)^{3/2} + C(x)$

19. $\int \sin(x) \cdot e^{x+y}\ dy = \int \sin(x) \cdot e^x \cdot e^y\ dy = \sin(x) \cdot e^x \cdot e^y + C(x)$

21. $\int \sqrt{x + y^2}\ dx = \frac{2}{3}(x + y^2)^{3/2} + C(y)$ 23. $\int e^{(y^2)}\ dx = x \cdot e^{(y^2)} + C(y)$

25. $-\frac{585}{8}$ 27. $\frac{\pi}{12} - \frac{3}{2} + \sqrt{3} - \frac{\pi}{12}\sqrt{3} \approx 0.0404$ 29 $\frac{1}{6}(60) = 10$ 31. $\frac{1}{6}(26) = \frac{13}{3}$

33. One reasonable approximation is 33 m^3. Your approximation should be close to this number, and you should be able to justify why your method is reasonable.

35. Using midpoints, one reasonable estimate is 88 m^3.

14.2 Odd Answers

1. 9 3. 63 5. $\frac{1}{2} - \frac{1}{2}\cos(1)$

7. $\int_{x=-2}^{x=2} \int_{y=0}^{y=4-x^2} f\ dy\ dx$, $\int_{y=0}^{y=4} \int_{x=-\sqrt{4-y}}^{x=\sqrt{4-y}} f\ dx\ dy$ 9. $\int_1^4 \int_x^4 f\ dy\ dx$, $\int_1^4 \int_1^y f\ dx\ dy$

11. $\int_0^2 \int_{x^2}^{6-x} f\ dy\ dx$, $\int_0^4 \int_0^{\sqrt{y}} f\ dx\ dy + \int_4^6 \int_0^{6-y} f\ dx\ dy$

13. $\frac{1}{12}$ 15. $\frac{3}{4}$ 17. $\frac{32}{3}$ 19. $-\frac{72}{5}$

21. avg. value = volume/area = $\frac{44}{4} = 11$ 23. volume= $\frac{45}{4}$. area = $\frac{9}{2}$, avg. value = $\frac{5}{2}$

25. $\int_0^1 \int_y^1 f(x,y)\ dx\ dy$

27. $\int_0^3 \int_0^{6-2x} f(x,y)\ dy\ dx$

29. $\int_0^{\ln(2)} \int_{e^y}^2 f(x,y)\ dx\ dy$

31. $\int_0^4 \int_0^{2-y/2} f\ dx\ dy$

33. $\int_0^2 \int_0^{y^2} f\ dx\ dy$

35. $\int_1^e \int_{\ln(y)}^1 f\ dx\ dy$

37. One reasonable approximation of the volume is $24\ m^3$. Your approximation should be close to this.

14.3 Odd Answers

1. A: rectangular B: polar C: polar

3. A: rectangular B: rectangular (since the circle is not centered at the origin) C: polar

5. 63π 7. $A\cdot D^2\cdot\pi$

9. V= $\iint_R \sqrt{9-x^2-y^2}\ dA$ where $R=\{(x,y):\ x^2+y^2\le 9\}$.

$$V=\int_{\theta=0}^{2\pi}\int_{r=0}^{3}\sqrt{9-r^2}\ r\cdot dr\cdot d\theta=\int_{\theta=0}^{2\pi} 9\,d\theta=18\pi\ .$$

11. V= $\iint_R 2+xy\ dA$ where R is the region in inside the circle $x^2+y^2=4$ and above the x-axis..

$$V=\int_{\theta=0}^{\pi}\int_{r=0}^{2}\{2+r\cdot\cos(\theta)\cdot r\cdot\sin(\theta)\}\cdot r\cdot dr\cdot d\theta=\int_{\theta=0}^{\pi}\{4+4\cos(\theta)\sin(\theta)\}\ d\theta=4\pi\ .$$

13. Under the plane $z=5+2x+3y$ and above the disk $x^2+y^2\le 16$.

$$V=\int_{\theta=0}^{2\pi}\int_{r=0}^{4}\{5+2\cos(\theta)+3\sin(\theta)\}\ r\cdot dr\cdot d\theta=\int_{\theta=0}^{2\pi}\{40+16\cos(\theta)+24\sin(\theta)\}\ d\theta=80\pi$$

15. Between the surfaces $z=1+x+y$ and $z2=8+2x+3y$ for $x^2+y^2\le 1$ and $0\le y$.

$$V=\iint_R\{z2-z1\}\ dA=\int_{\theta=0}^{\pi}\int_{r=0}^{1}\{z2-z1\}\ dA=\int_{\theta=0}^{\pi}\int_{r=0}^{1}\{7+x+2y\}\ r\cdot dr\cdot d\theta$$

$$=\int_{\theta=0}^{\pi}\int_{r=0}^{1}\{7+r\cdot\cos(\theta)+2r\cdot\sin(\theta)\}\ r\cdot dr\cdot d\theta=\int_{\theta=0}^{\pi}\left\{\frac{7}{2}+\frac{1}{3}\cos(\theta)+\frac{2}{3}\sin(\theta)\right\}\ d\theta=\frac{4}{3}+\frac{7}{2}\pi$$

17. Between the surface $z=f(x,y)=\dfrac{1}{1+x^2+y^2}$ and the xy-plane for (x,y) in the first quadrant and $1\le x^2+y^2\le 4$. The domain is $R=\{(x,y):1\le x^2+y^2\le 4\}=\left\{(r,\theta):1\le r\le 2\ \ 0\le\theta\le\frac{\pi}{2}\right\}$.

$$\text{Volume}=\iint_R f\cdot dA=\int_{\theta=0}^{\pi/2}\int_{r=1}^{2}\frac{1}{1+r^2}\cdot r\cdot dr\cdot d\theta$$

$$\int_{r=1}^{2} \frac{1}{1+r^2}\cdot r\cdot dr = \frac{1}{2}\ln(1+r^2)\Big|_{r=1}^{r=2} = \frac{1}{2}\ln\left(\frac{5}{2}\right) \text{ and } \int_{\theta=0}^{\pi/2} \frac{1}{2}\ln\left(\frac{5}{2}\right)\, d\theta = \frac{\pi}{2}\cdot\frac{1}{2}\ln\left(\frac{5}{2}\right) \approx 0.72 .$$

19. $\int_0^1 \int_0^{\pi} r\ d\theta\ dr = \frac{\pi}{2}$

21. $\int_0^1 \int_0^{\pi/2} r\cdot \sin(\theta)\cdot r\ d\theta\ dr = \int_0^1 r^2\ dr = \frac{1}{3}$

23. $\int_0^{\pi/2} \int_0^1 e^{-r^2}\cdot r\ dr\ d\theta = \int_0^{\pi/2} \frac{1}{2}\left(1-\frac{1}{e}\right) d\theta = \frac{\pi}{4}\left(1-\frac{1}{e}\right)$

25. $f(x,y) = 7+3x+2y$ with $R = \{(x,y):\ x^2+y^2 \le 9\}$. (from Problem 5)

Area of R = 9π and $\iint_R$ f dA = 63π (from Problem 5) so the average value of f on R is 7.

27. $f(x,y) = 5+2x+3y$ with $R = \{(x,y):\ x^2+y^2 \le 16\}$. (from Problem 13)

Area of R = 16π and $\iint_R$ f dA = 80π (from Problem 5) so the average value of f on R is 5 .

29. A sprinkler (located at the origin) sprays water so after one hour the depth at location (x,y) feet is

$f(x,y) = K\cdot e^{-(x^2+y^2)}$ feet.

(a) How much water reaches the annulus $2 \le r \le 4$ and the annulus $8 \le r \le 10$ in one hour?

From Practice 2 we know that the total amount of the water in a circle of radius A is

$K\cdot\left(1-e^{-A^2}\right)\pi$. So the amount in the annulus $2 \le r \le 4$ is

$K\cdot\left(1-e^{-16}\right)\pi - K\cdot\left(1-e^{-4}\right)\pi = K\cdot\left(e^{-4}-e^{-16}\right)\pi$ and the amount in the annulus

$8 \le r \le 10$ is $K\cdot\left(e^{-64}-e^{-100}\right)\pi$.

(b) The area of the $2 \le r \le 4$ annulus is $(4^2-2^2)\pi = 12\pi$ square feet, and the area of the $8 \le r \le 10$

annulus is 36π square feet. The average depth for the first annulus is $\frac{K\cdot\left(e^{-4}-e^{-16}\right)\pi}{12\pi} \approx 0.00153K$

and it is $\frac{K\cdot\left(e^{-64}-e^{-100}\right)\pi}{36\pi} \approx 4.5\cdot 10^{-30}\cdot K$ (almost no water) for the second annulus.

14.4 Odd Answers

Author confession: Many of these integrals are very messy and take a long time. I used Maple to evaluate them.

1. Area $= \iint_R 1\ dA = \int_{x=0}^{2}\int_{y=0}^{4} 1\ dy\ dx = \int_{x=0}^{2} 4\ dx = 8 = M$

$$M_x = \iint_R y\cdot\delta\ dA = \int_{x=0}^{2}\int_{y=0}^{4} y\cdot(1)\ dy\ dx = \int_{x=0}^{2} 8\ dx = 16$$

$$M_y = \iint_R x\cdot\delta\ dA = \int_{x=0}^{2}\int_{y=0}^{4} x\cdot(1)\ dy\ dx = \int_{x=0}^{2} 4x\ dx = 4, \text{ and } \bar{x}=1,\quad \bar{y}=2$$

3. $$\text{Area} = \iint_R 1\ dA = \int_{x=0}^{2}\int_{y=0}^{x^2} 1\ dy\ dx = \int_{x=0}^{2} x^2\ dx = \frac{8}{3}$$

$$\text{M} = \iint_R \delta\ dA = \int_{x=0}^{2}\int_{y=0}^{x^2} (1+x)\ dy\ dx = \int_{x=0}^{2} x^3+x^2\ dx = \frac{20}{3}$$

$$M_x = \iint_R y\cdot\delta\ dA = \int_{x=0}^{2}\int_{y=0}^{x^2} y\cdot(1+x)\ dy\ dx = \int_{x=0}^{2} \frac{1}{2}(x^4+x^5)\ dx = \frac{128}{15}$$

$$M_y = \iint_R x\cdot\delta\ dA = \int_{x=0}^{2}\int_{y=0}^{x^2} x\cdot(1+x)\ dy\ dx = \int_{x=0}^{2} (x^3+x^4)\ dx = \frac{52}{5}$$

$$\bar{x} = \frac{M_y}{\text{M}} = \frac{39}{25} \text{ and } \bar{y} = \frac{M_x}{\text{M}} = \frac{32}{25}$$

5. Area $= \frac{15}{2}$, $M = 24$, $M_x = \frac{341}{8}$, $M_y = \frac{405}{8}$, $\bar{x} = \frac{135}{64}$, $\bar{y} = \frac{341}{192}$

7. $$\text{Area} = \iint_R 1\ dA = \int_{\theta=0}^{\pi/3}\int_{r=0}^{\sin(3\theta)} 1\ r\ dr\ d\theta = \frac{2}{3} = M ,$$

$$M_x = \iint_R y\cdot\delta\ dA = \int_{\theta=0}^{\pi/3}\int_{r=0}^{\sin(3\theta)} r\cdot\sin(\theta)\ r\ dr\ d\theta = \frac{9}{70}$$

$$M_y = \iint_R x\cdot\delta\ dA = \int_{\theta=0}^{\pi/3}\int_{r=0}^{\sin(3\theta)} r\cdot\cos(\theta)\ r\ dr\ d\theta = \frac{9}{70}\sqrt{3}$$

$$\bar{x} = \frac{27}{140}\sqrt{3},\quad \bar{y} = \frac{27}{140}$$

9. Area $= \pi$, $M = \frac{3}{4}\pi$, $M_x = \frac{3}{4}$, $M_y = \frac{5}{8}\pi$, $\bar{x} = \frac{5}{6}$, $\bar{y} = \frac{16}{9\pi}$

11. $\text{M} = 8,\quad \text{I}_x = \frac{128}{3},\quad \text{R}_x = \frac{4}{3}\sqrt{3}$

13. $\text{M} = 16,\quad \text{I}_x = 128,\quad \text{R}_x = 2\sqrt{2}$

15. The density of the bar is 4 kg/m. The force between the point mass and a small Δx piece of the bar at location x is $f_i = GM\frac{4\cdot\Delta x}{x^2}$. Forming the usual Riemann sum of the little forces and taking the limit,

total force $= G\cdot 10\cdot\int_2^4 \frac{4}{x^2}\ dx$,

17. The density is k = 5 kg/m for the first bar and K = 3 kg/m for the second bar. The force between a small Δx piece of the first bar at location x and a small Δy piece of the second bar at location y is $f_{ij} = G \cdot \frac{(5\Delta x)(3\Delta y)}{(y-x)^2}$.

Total force $= G \cdot \int_4^7 \int_0^2 \frac{15}{(y-x)^2} \, dx \, dy$

14.5 Odd Answers

1. $SA = \int_0^2 \int_0^{2x} \sqrt{1+(2x)^2+1} \, dy \, dx = \int_0^2 2x\sqrt{4x^2+2} \, dx = \frac{26}{3}\sqrt{2}$

3. $SA = \int_0^4 \int_0^{y/2} \sqrt{1+(4)^2+(2y)^2} \, dx \, dy = \int_0^4 \frac{y}{2}\sqrt{4y^2+17} \, dx = \frac{243}{8} - \frac{17}{24}\sqrt{17}$

5. $SA = \int_{-2}^{2} \int_0^{4-x^2} \sqrt{26} \, dy \, dx = \int_{-2}^{2} (4-x^2)\sqrt{26} \, dx = \frac{32}{3}\sqrt{26}$

7. $SA = \iint_R \sqrt{26} \, dA = 9\pi \cdot \sqrt{26}$ since R is a circle of radius 3

or $SA = \int_{\theta=0}^{2\pi} \int_{r=0}^{3} \sqrt{26} \cdot r \, dr \, d\theta = \int_{\theta=0}^{2\pi} \frac{9}{2}\sqrt{26} \, d\theta = 9\sqrt{26} \cdot \pi$

9. $8\pi\sqrt{3}$

11. R is an annulus ($1 \le r \le 3$). $SA = \int_{\theta=0}^{2\pi} \int_{r=1}^{3} \sqrt{2} \cdot r \, dr \, d\theta = 9\pi\sqrt{2} - 1\pi\sqrt{2} = 8\pi\sqrt{2}$

13. $SA = \int_{-2}^{2} \int_{x^2}^{4} \sqrt{1+(y^2)^2+(2xy)^2} \, dy \, dx$

15. $SA = \int_0^2 \int_0^3 \sqrt{1+\cos^2(x)+\sin^2(y)} \, dy \, dx$

14.6 Odd Answers

1. $\int_{x=0}^{4} \int_{y=0}^{2} \int_{z=0}^{4-2y} f \, dz \, dy \, dx$

3. $\int_{x=-2}^{2} \int_{y=-\sqrt{4-x^2}}^{\sqrt{4-x^2}} \int_{z=\sqrt{x^2+y^2}}^{4} f \, dz \, dy \, dx$

5. $\int_{x=0}^{2} \int_{y=0}^{4-x^2} \int_{z=0}^{16-4x^2-y^2} f \, dz \, dy \, dx$

7. 7/6

9. -5

11. 24

13. $(e^3-1)/6$ 15. 16/3 17. 144

19. int(2*x+3, y=0..2*x, x=0..2, z=0..3); 21. int(x*y*z, z=0..8-2*y, y=x..4, x=0..4);

14.7 Odd Answers

1. $\frac{\pi}{3}\left(2\sqrt{2}-1\right)$ 3. $\frac{23}{24}\pi$ 5. $\frac{1}{54}\pi^5$ 7. 8π

9. $\int_0^4\int_{\pi/4}^{\pi/2}\int_0^1 r\cdot e^{-r^2}\ dr\cdot d\theta\cdot dz=\frac{\pi}{2}\left(1-\frac{1}{e}\right)$

11. $\int_0^1\int_0^{\pi/2}\int_0^{4-r\cdot\sin(\theta)} r\ dz\cdot d\theta\cdot dr=\pi-\frac{1}{3}$

13. $\iiint_R \text{f dV}=\int_0^{2\pi}\int_0^3\int_3^5 r\cdot r\ dz\cdot dr\cdot d\theta=36\pi$

15. $\iiint_R \text{f dV}=\int_0^{2\pi}\int_0^{\sqrt{7}}\int_0^{1+r^2}\left(e^z\right)\cdot r\ dz\cdot dr\cdot d\theta=2\pi\left(\frac{1}{2}e^8-\frac{1}{2}e-\frac{7}{2}\right)$

17. $\iiint_R \text{f dV}=\int_0^{\pi/2}\int_0^3\int_{r^2}^{36-3r^2} r\ dz\cdot dr\cdot d\theta=\frac{81}{2}\pi$

19. 0 21. $\frac{5}{3}\pi$ 23. $\frac{7}{3}\pi$ 25. $\frac{31}{160}\pi$

14.8 Odd Answers

Problem 1

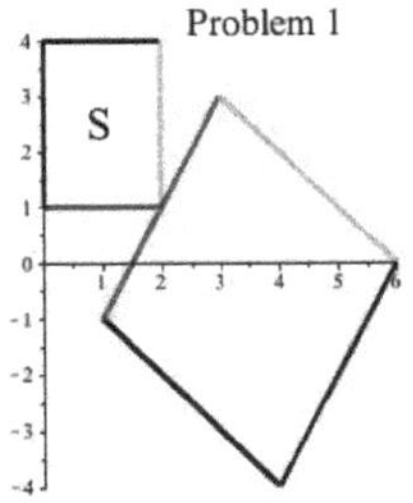

1. $S=\{(u,v):\ 0\le u\le 2,\ 1\le v\le 4\}$ under $x=u+v$ and $y=2u-v$.

$J(u,v)=\begin{vmatrix}1 & 1\\ 2 & -1\end{vmatrix}=-3,\ \ J(x,y)=-\frac{1}{3}$

$\iint_R f(x,y)\,dx\,dy=\int_1^4\int_0^2 f(u+v,\ 2u-v)(3)\ du\,dv$

Problem 3: a=1, b=3, c=3, d=2

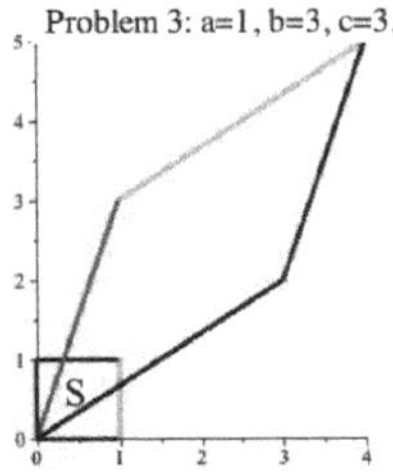

3. $S=\{(u,v):\ 0\le u\le 1,\ 0\le v\le 1\}$ under

$x=au+bv$ and $y=cu+dv$.

$J(u,v)=\begin{vmatrix}a & b\\ c & d\end{vmatrix}=ad-bc,\ \ J(x.y)=\frac{1}{ad-bc}$

$\iint_R f(x,y)\,dx\,dy=\int_0^1\int_0^1 f(au+bv,\ cu+dv)\,|\,ad-bc\,|\ du\,dv$

5. $S=\{(x,y):\ 0\le x\le 2,\ 1\le y\le 3\}$ under $u=\frac{3x-3y}{4}$ and $v=\frac{y}{3}$.

Then x=4u/3+3v and y=3v. $J(x,y)=\begin{vmatrix}\frac{3}{4} & -\frac{3}{4}\\ 0 & \frac{1}{3}\end{vmatrix}=\frac{1}{4},\ \ J(u,v)=4$

Problem 5

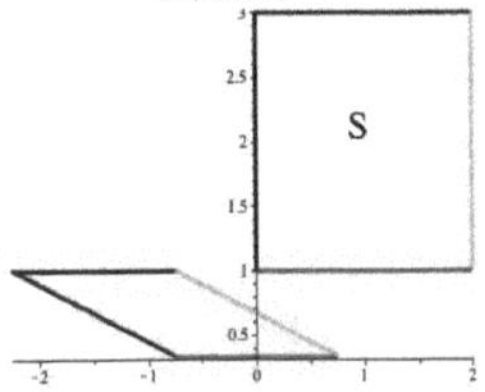

7. $S=\{(x,y):\ 0\le x\le 1,\ 0\le y\le 1\}$ under $u=x^2-y^2$ and $v=2xy$.

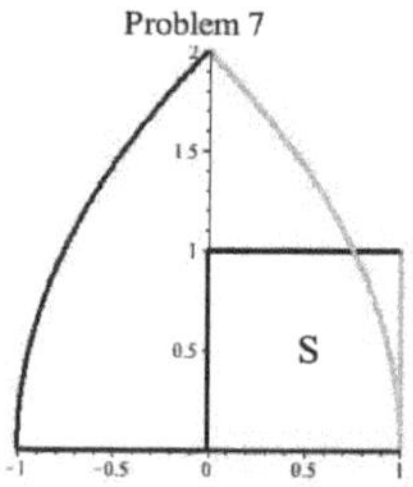

$$J(x,y)=\begin{vmatrix}2x & -2y\\ 2y & 2x\end{vmatrix}=4x^2+4y^2,\quad J(u,v)=\frac{1}{4x^2+4y^2}$$

(It is difficult/impossible to solve for x and y.)

9. $R=\{$trapezoid with vertices $(x,y)=(1,0),\ (2,0),\ (0,-2)$ and $(0,-1)\}$

under u=x-y and v=x+y . Then $x=\frac{u+v}{2}$ and $y=\frac{-u+v}{2}$.

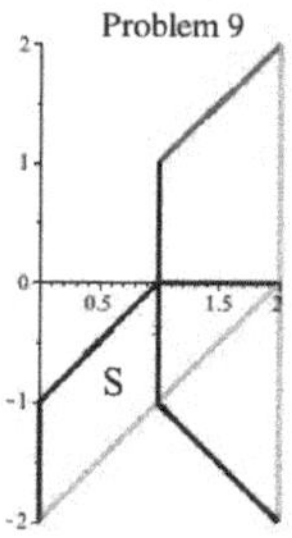

$$J(u,v)=\begin{vmatrix}1 & -1\\ 1 & 1\end{vmatrix}=2,\quad J(x,y)=\frac{1}{2}$$

11. $\iint_R 1\ dV=\iint_S 1\cdot|J(u,v)|\,du\,dv=\iint_S 7\ du\,dv=7(14)=98.$

13. J(x,y) = 3 so $J(u,v)=\frac{1}{3}$. Then

$$\iint_R 1\ dV=\iint_S 1\cdot|J(u,v)|\,dV\iint_S \frac{1}{3}\ dV=\frac{1}{3}(15)=5.$$

15. The new uv domain is a triangle with vertices (0,), (1,0) and (0,1). $J(u,v)=\begin{vmatrix}2 & 1\\ 1 & 2\end{vmatrix}=3$.

$$\int_0^1\int_0^{1-u}[(2u+v)+3(u+2v)]\cdot 3\ dv\ du=6\ .$$

17. Using the substitution from Fig. 4, u=x+y, v=x-2y, so $J(x,y)=\begin{vmatrix}1 & 1\\ 1 & -2\end{vmatrix}=-3$ and $J(u,v)=-\frac{1}{3}$.

$$\int_3^6\int_0^6\left[\frac{u}{3}\right]\cdot\frac{1}{3}\ du\ dv=\int_3^6 2\ dv\ du=6$$ (Note: The substitution u=(x+y)/3 and v=(x–2y) from example leads to the same integral value since with this substitution u goes from 0 to 2 and v from 1 to 2.)

19. u=y–x v=2x+y so $J(x,y)=\begin{vmatrix}1 & 1\\ 1 & -2\end{vmatrix}=-3$ and $J(u,v)=-\frac{1}{3}$.

x=(v–u)/3 y=(2u+v)/3 so 3x+6y=3u+3v. $\int_6^9\int_0^2[3u+3v]\cdot\frac{1}{3}\ du\ dv=\int_6^9(2+2v)\ dv=51$

21. This is from Problem 8: x = u/v and y = uv

$$J(u,v)=\begin{vmatrix}\frac{1}{v} & -\frac{u}{v^2}\\ v & u\end{vmatrix}=\frac{2u}{v}$$ Then $u=\sqrt{xy},\ v=\sqrt{\frac{y}{x}}$ with u=1 to 2 and v=1 to 2.

$$\int_1^2\int_1^2[v+u]\cdot\left(\frac{2u}{v}\right)du\ dv=\int_1^2\int_1^2\left[2u+\frac{2u^2}{v}\right]du\ dv=\int_1^2 3+\frac{14}{3}\frac{1}{v}\ dv=3+\frac{14}{3}\ln(2)\ .$$

(Note: See the Two Transformations, Same Result page at the end of the problem answers.)

23. R is the region bounded by the ellipse. If x=au and y=bv then the uv shape is $u^2 + v^2 \le 1$, a circle of radius 1 and area π. $J(u,v) = \begin{vmatrix} a & 0 \\ 0 & b \end{vmatrix} = ab$ so $\iint_{\text{ellipse}} 1 \, dx \, dy = \iint_{\text{circle}} ab \, du \, dv = ab \cdot \iint_{\text{circle}} 1 \, du \, dv = ab\pi$

15.0 Introduction to Vector Calculus

This may seem like a strange title since we have already been doing calculus with vectors, but the topics in this chapter extend the main calculus ideas into fields of vectors such as the one illustrated in Fig. 1 (from **Modeling Waves and Currents Produced by Hurricanes Katrina, Rita, and Wilma** by Lie-Yauw Oey and Dong-Ping Wang) . First we will examine some examples of vector fields. Then on to line integrals (integrals along paths in 2D or 3D) that will enable us to calculate the work done moving an object along a path in a wind field or magnetic field. Finally we will consider surface integrals and then some important theorems that generalize the Fundamental Theorem of Calculus to vector fields.

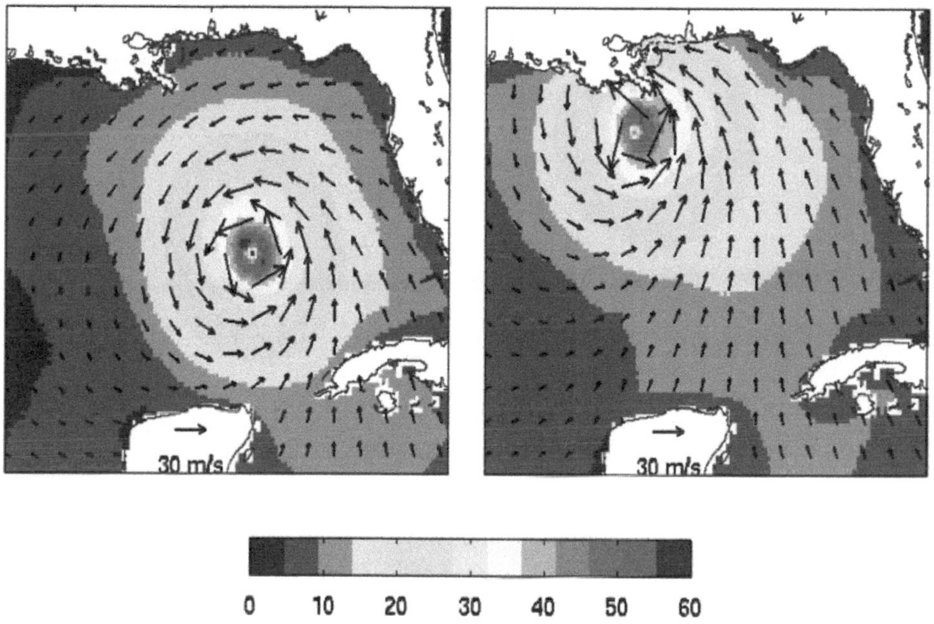

Fig. 1: Hurricane Katrina wind vectors (in m/s) on August 29, 2005

Vector fields contain a great deal of information and calculus can help us use that information. But even without calculus we can answer some questions about such fields.

Example 1: (a) Which way, clockwise or counterclockwise, where the winds blowing?

(b) Were the strongest winds greater in the left picture or in the right picture?

(c) Approximately what is the maximum wind speed in the right picture?

(d) In which direction was the eye of the hurricane traveling?

Solution: (a) Counterclockwise. (b) Right picture

(c) Using the "30 m/s" key at the bottom of each picture, the maximum wind speed was about 50-60 m/s (about 120 miles/hour) . (d) To the NNW.

Other common vector fields are water velocity vectors indicating currents (Fig. 2) and force fields indicating the strength and direction of attractive or repulsive forces either in 2 or 3 dimensions.

(Note: Excellent, animated wind velocity fields for various parts of the world are at http://www.wunderground.com/maps/us/WindSpeed.html)

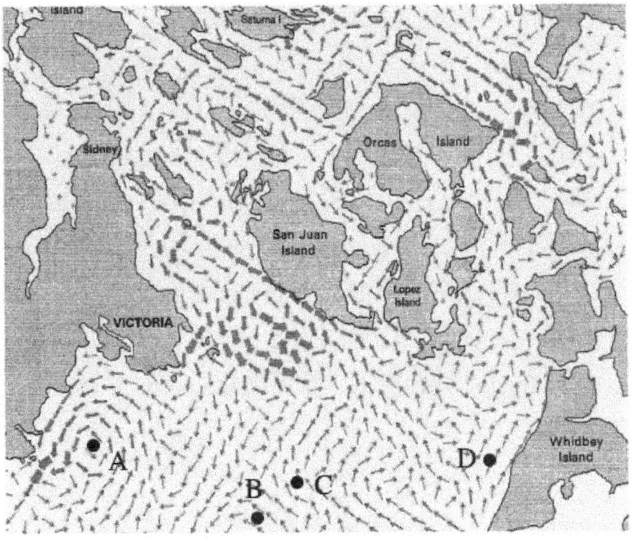

The biggest arrows represent current velocity over 2.5 mph, the smallest, under .25 mph. The arrows point the direction of flow.

Fig. 2: Currents in the San Juan Islands

Problems

Problems 1 to 4 refer to Fig. 2. This picture uses wider arrows to represent stronger current.

1. What is happening to the current at point A?

2. If we drop small pieces of wood into the water at points A and B, do they stay close or do they drift apart?

3. Is it easier to row from point D to San Juan Island or from San Juan Island to point D?

4. Is it easier to row from point C to point D or from point D to point C?

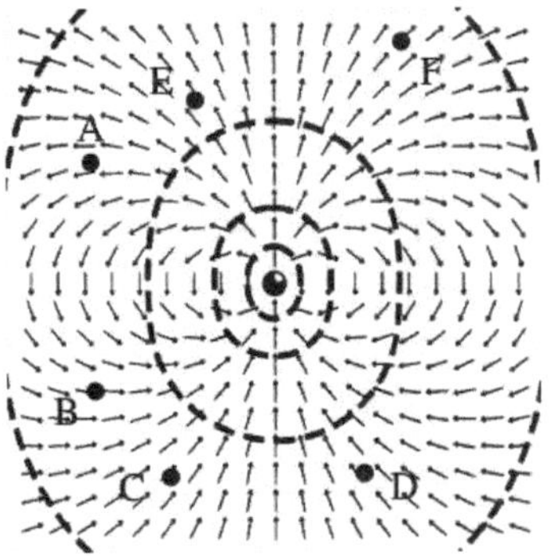

Fig. 3: Lines of Force for a bar magnet

Problems 5 to 8 refer to Fig. 3. The dotted curves each represent places of equal force. Moving from the center, the force at each dotted curve is 1/10 the force at the next closer dotted curve.

5. Is it easier to move from A to B or from B to A?

6. Is it easier to move from C to D or from D to C?

7. Is it easier to move from E to F or from F to E?

8. Imagine that this is a water current and that you dropped a cork into the water at A. Sketch the path of the cork.

15.1 Vector Fields

2D Vector Fields

> A **2D vector field** is a function **F** that assigns a 2D vector **F**(x,y) to each point (x,y) in the domain of the field. Since F(x,y) is a 2D vector we can write
>
> F(x,y) = P(x,y)**i** + Q(x,y) **j** = < P(x,y), Q(x,y) > or simply **F** = P **i** + Q **j**

Unlike our earlier work with vectors, these F(x,y) vectors have assigned locations. The vector field consists of an infinite number of vectors (one at each point) so we typically just graph enough of these vectors to make the pattern clear. The convention is to put the tail of the **F**(x,y) vector at the point (x,y).

Example 1: Plot the vectors **F**(x,y)=< y, -x> at the points (1,0), (1,1), (0,1), (-1,1), (-1,0) and (-1,-1).

Solution: **F**(1,0)=<0,-1>, **F**(1,1)=<-1,1>, **F**(0,1)=<1,0>, **F**(-1,1)=<1,1>, **F**(-1,0)=<0,1>, **F**(-1,-1)=<-1,1>. These are shown in Fig. 1.

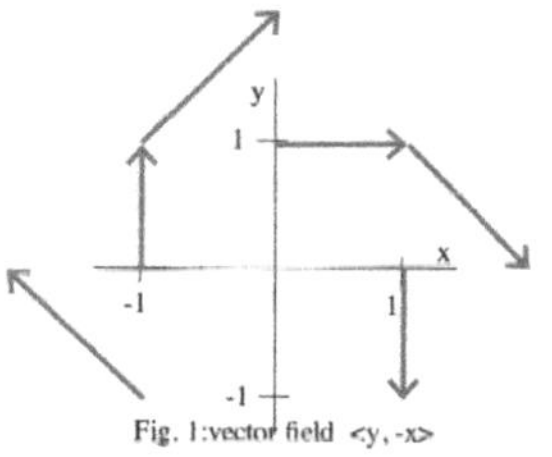

Fig. 1:vector field <y, -x>

It is very tedious to draw a vector field by hand, and some computer programs can do a very nice job. Fig. 2 is the same vector field **F**(x,y)=< y, -x> drawn by the program Maple with the command

with(plots): fieldplot([y, -x], x = -2 .. 2, y = -2 .. 2, arrows = THIN, color=red, thickness=2,grid=[11,11],title = "Vector Field <y, -x>", titlefont = ["ROMAN", 18]);

Maple automatically scaled the arrows to fit into the figure.

Fig. 2: Vector Field <y, -x>

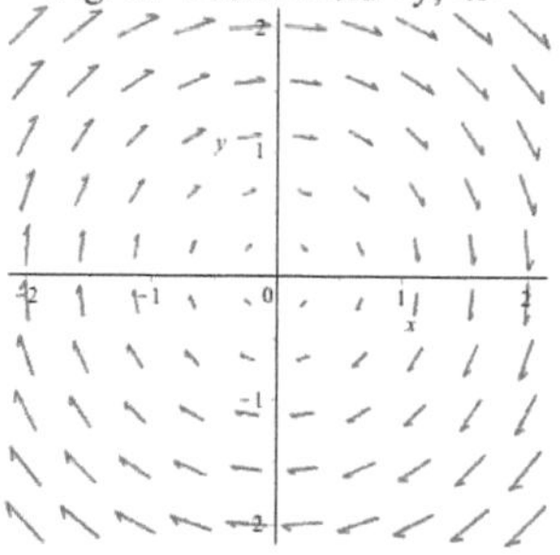

Practice 1: Plot the vectors **F**(x,y)=< x, y/2> at the same points as in Example 1.

A few vector fields are very common and you should be able to recognize them after plotting just a few representative vectors.

Radial Fields <x,y> and <-x,-y>. These are shown in Fig. 3a and 3b. These vectors all point toward the origin or all vectors point away from the origin (except at the origin).

Fig. 3a <x, y>

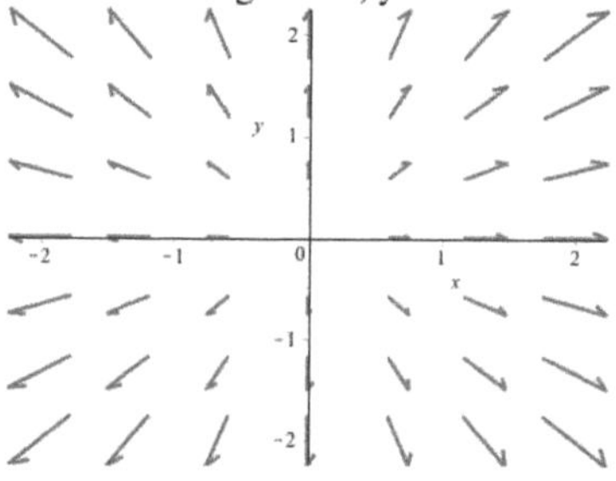

Fig. 3b <-x, -y>

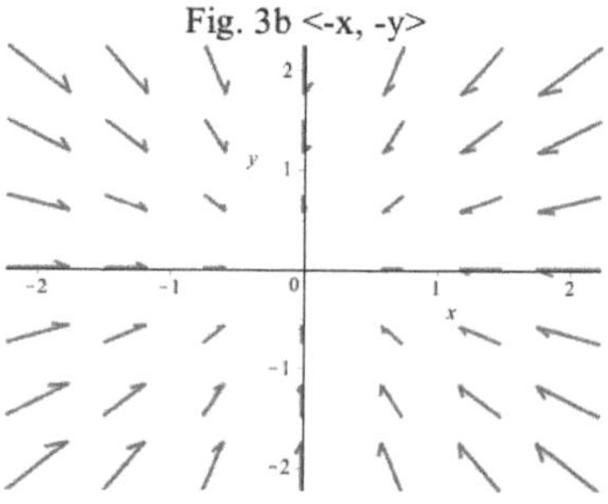

Rotational Fields: < -y, x> , < y, -x> and $\left\langle \frac{y}{\sqrt{x^2+y^2}}, \frac{-x}{\sqrt{x^2+y^2}} \right\rangle$. These are shown in Fig. 4.

The first and second of these fields have vectors that increase in magnitude the farther they are from the origin. The third field has vectors of constant magnitude 1.

Fig. 4a <-y, x>

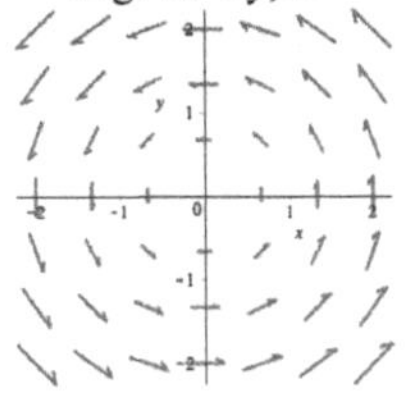

Fig. 4b <y, -x>

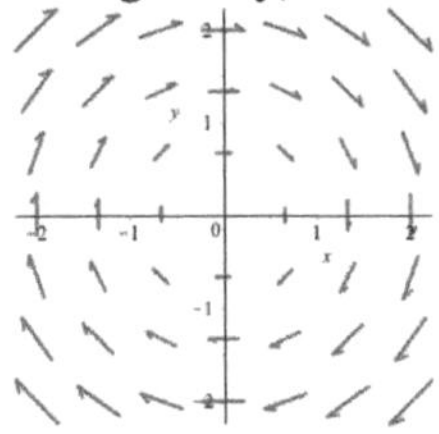

Fig. 4c <y/sqrt(x^2+y^2), -x/sqrt(x^2+y^2)>

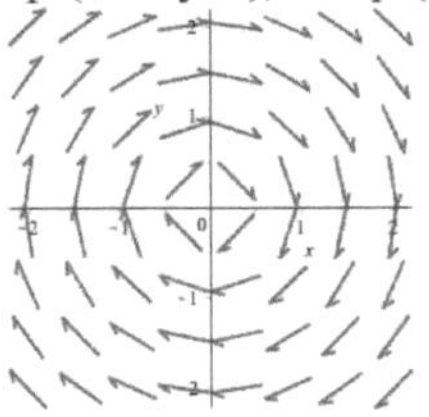

The chapter Appendix gives the Maple commands for creating 2D and 3D vector fields.

3D Vector Fields

A 3D vector field **F** assigns a 3D vector at each point in the 3D domain of **F**:

$F(x,y,z) = \langle P(x,y,z), Q(x,y,z), R(x,y,z) \rangle = P\mathbf{i} + Q\mathbf{j} + R\mathbf{k}$ where P, Q and R are scalar-valued functions.

Since we live in a 3D world, 3D vector (force) fields are very common in applications, but they are much more difficult to create by hand and to visualize on a 2D page. Fortunately, some software can do the work for us.

Radial Fields: F(x,y,z) = <x,y,z> and <-x,-y,-z>.

These are shown in Fig. 5a and 5b. In Fig. 5c all of the vectors of **F**(x,y,z) = <x,y,z> have been normalized to have the same length. The vectors in all three plots point toward or away from the origin (except at the origin).

Fig. 5a: <x,y,z>

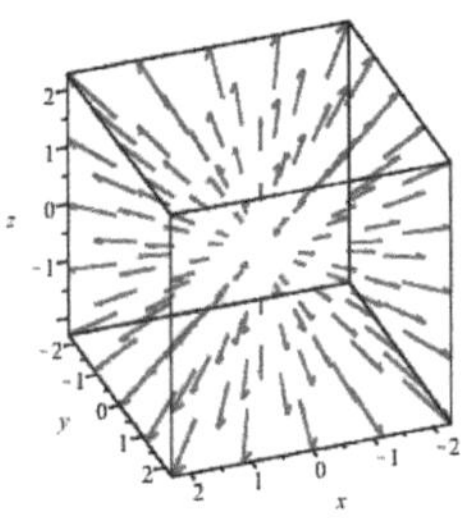

Fig. 5b: <-x,-y,-z>

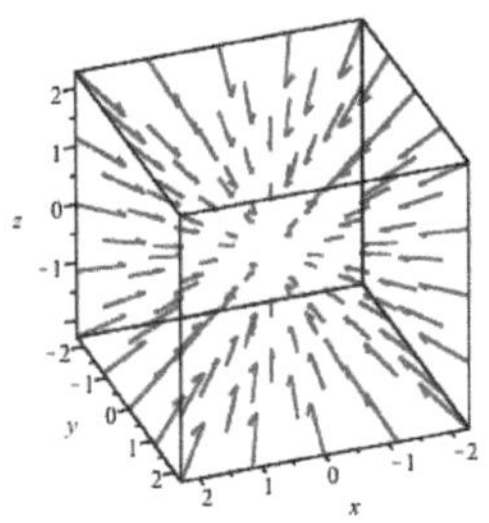

Fig. 5c:

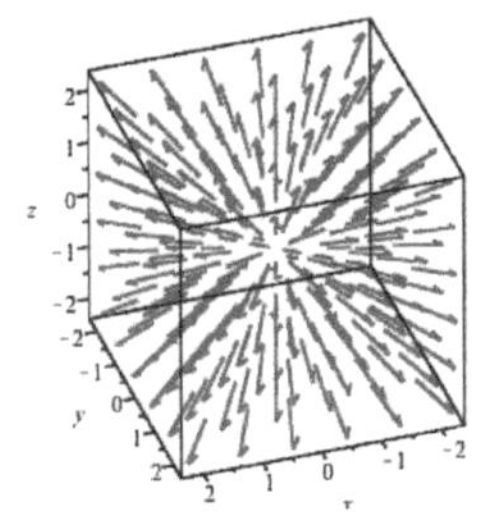

Rotational Fields: These are more difficult to visualize because the rotation can be around any of the axes, some other line, or simply around the origin. Three views of **F**(x,y,z) = <-y, x, 0> are shown in Fig. 6.

Fig. 6a <-y,x,0>

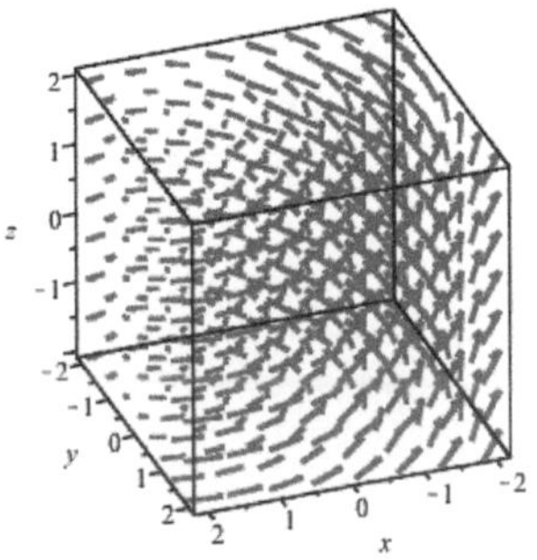

Fig. 6b <-y,x,0>

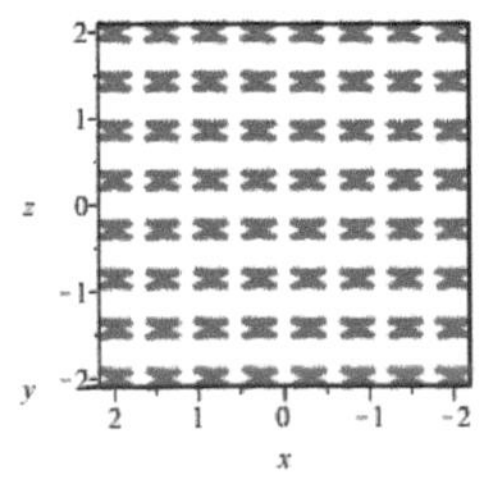

Fig. 6c <-y,x,0>

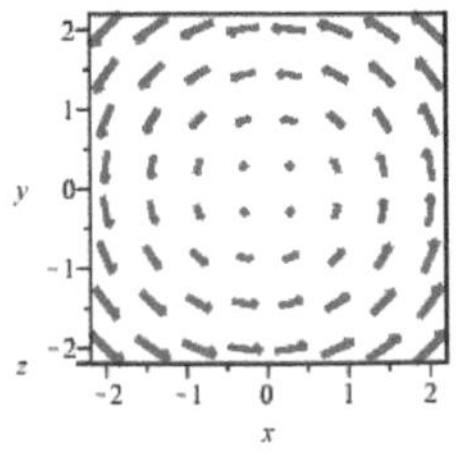

Fig. 7 shows the "swirl" field for $\mathbf{F}(x,y,z) = \langle y,z,x \rangle$, and Fig. 8 is the simple field $\mathbf{F}(x,y,z) = \langle 1,3,2 \rangle$ in which all of the vectors have the same magnitude and direction.

Fig. 7 < y, z, x >

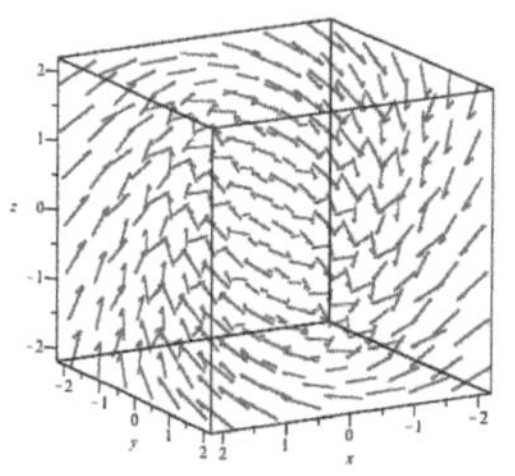

Fig. 8: <1,3,2>

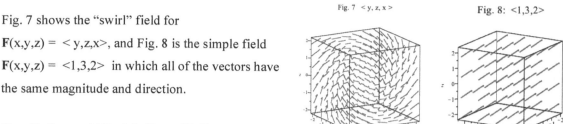

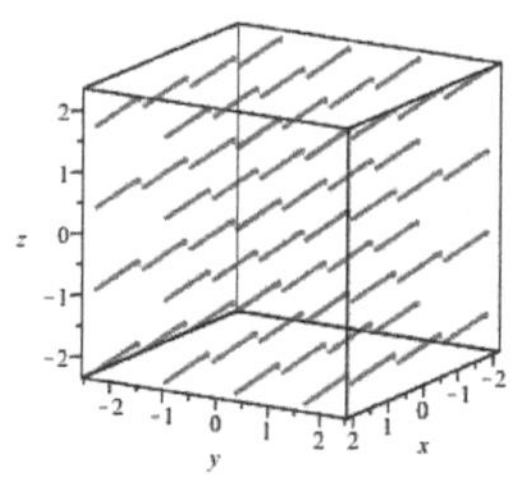

Gravitation and Electric Force Fields

If, as is the situation for gravitational and electric fields, the magnitude of the force is inversely proportional to the square of the distance between the objects, then $\mathbf{F}(x,y,z) = \dfrac{k}{x^2+y^2+z^2}$ where k is a positive or negative constant. The field will look like Fig. 5a or 5b.

Gravitational Field: Newton's Law of Gravitation says that k=GMm where m and M are the masses of the objects and G is the gravitational constant: $\mathbf{F} = \dfrac{GMm}{x^2+y^2+z^2}$. Suppose the object with mass M is located at the origin, and let $\mathbf{r} = \langle x,y,z \rangle$ be the position vector of the object of mass m. Then $|\mathbf{r}|^2 = x^2+y^2+z^2$ and the force is directed toward the origin so the direction of $\mathbf{r}$ is $-\dfrac{\mathbf{r}}{|\mathbf{r}|}$. Putting this together, the **gravitational field** is $\mathbf{F}(x,y,z) = \mathbf{F}(\mathbf{r}) = \dfrac{GMm}{|\mathbf{r}|^2}\left(-\dfrac{\mathbf{r}}{|\mathbf{r}|}\right) = \left(\dfrac{-GMm}{|\mathbf{r}|^3}\right)\mathbf{r}$. These vectors behave like those in Fig. 5b in which the vector at each point is directed toward the origin.

Electrical Field: Coulomb's Law says that the force **F** is inversely proportional to the square of the distance between the two charges. If an electric charge Q is located at the origin, and a charge q is located at (x,y,z) then $|\mathbf{F}| = \dfrac{eqQ}{|\mathbf{r}|^2}$ where $\mathbf{r} = \langle x,y,z \rangle$ and e is a constant. Finally, $\mathbf{F}(\mathbf{r}) = \left(\dfrac{eqQ}{|\mathbf{r}|^3}\right)\mathbf{r}$ with qQ<0 for unlike charges (attracting each other) and qQ>0 for like charges (repelling each other). If we consider the force per unit q of charge then we have the **electric field** $\mathbf{E}(\mathbf{r}) = \dfrac{1}{q}\mathbf{F}(\mathbf{r}) = \left(\dfrac{eQ}{|\mathbf{r}|^3}\right)\mathbf{r}$.

Gradient Fields

In section 13.5 the gradient vector, $\nabla f(x,y) = \left\langle \dfrac{\partial f}{\partial x}, \dfrac{\partial f}{\partial y} \right\rangle$, was introduced, and this very naturally assigns a vector at each (x,y) location where f has partial derivatives.

Example 2: For $z = f(x,y) = x^2 + y^2$, what is the gradient vector field for f?

Solution: $\frac{\partial f}{\partial x}=2x$ and $\frac{\partial f}{\partial y}=2y$ so the vector field is $\mathbf{F}=\langle 2x,2y\rangle$ .which looks similar to Fig. 3a.

Practice 2: $z=f(x,y)=xy$. Determine the gradient vector field for f(x,y) and sketch several vectors from this field.

Fig. 9: level curves and gradient vectors for z=x^2+y^2

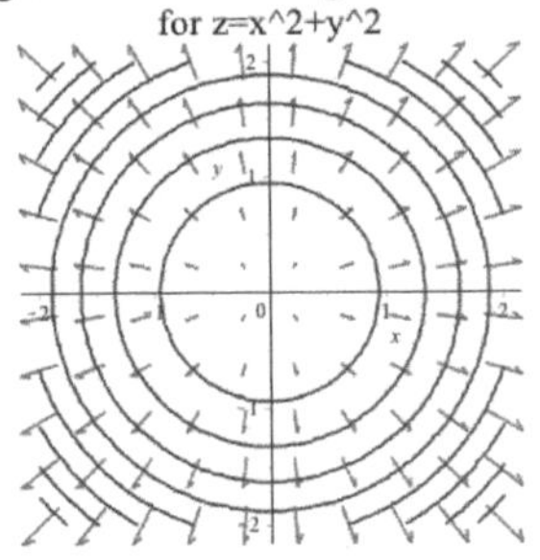

One of the important properties of the gradient vector $\nabla f(x,y)$ is that it is perpendicular to the level curve of f(x,y) at the point (x,y) and points "uphill." So if the level curves of z=f(x,y) are known, then it is easy to sketch the gradient vector field. Fig. 9 shows level curves and gradient vectors of the paraboloid $z=f(x,y)=x^2+y^2$.

Fig. 10a shows the surface $f(x,y)=\frac{-5x}{1+x^2+y^2}$. Fig. 10b shows level curves for f(x,y) as well as gradient vectors. The gradient vectors have been drawn to all have the same length to better show their directions and that they are perpendicular to the level curves.

Fig. 10a: z=-5*x/(1+x^2+y^2) Fig. 10b: level curves and gradients

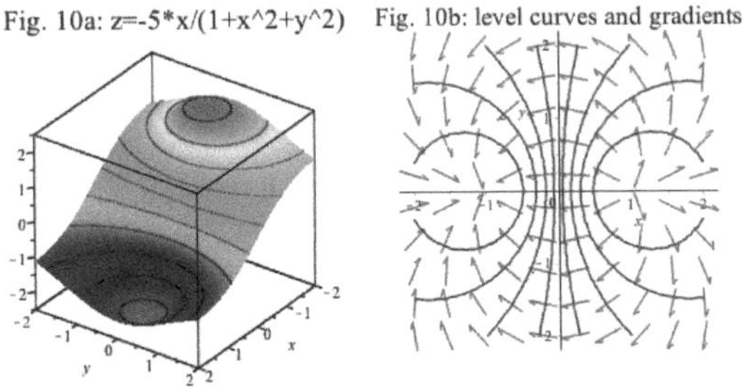

Practice 3: Fig. 11 shows the surface and the level curves for the saddle $f(x,y)=x^2-y^2$. Sketch the gradient field.

Fig. 11: z=x^2-y^2 level curves

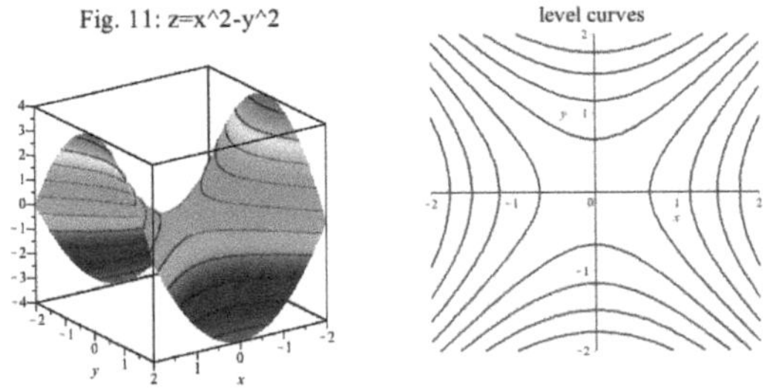

In a later section (give section number later) we will start with a vector field **F**(x,y) and determine if that field **F** is the gradient field of some function z=f(x,y). Such a field **F** is called **conservative field** and the function z=f(x,y) that gives rise to the field is called the **potential function** for **F**. Conservative fields have a number of important properties.

Problems

In problems 1 to 8, plot vectors from the given field **F**(x,y) at several locations (x,y) for integer values for x and y with $-2\le x\le 2$ and $-2\le y\le 2$.

1. $\mathbf{F}=\langle 2, 1\rangle$
2. $\mathbf{F}=\langle 1, x\rangle$
3. $\mathbf{F}=\langle x, x\rangle$
4. $\mathbf{F}=\langle y, y\rangle$
5. $\mathbf{F}=\langle x^2, y\rangle$
6. $\mathbf{F}=\langle 1, x\cdot y\rangle$
7. $\mathbf{F}=\langle x,-y\rangle$
8. $\mathbf{F}=\langle y,1-x\rangle$

In problems 9 to 11, match the vector field F to the 3D plot.

9. (a) $\mathbf{F} = \langle 0,0,1 \rangle$

(b) $\mathbf{F} = \langle -y,x,0 \rangle$

(c) $\mathbf{F} = \langle 1,1,0 \rangle$

9A

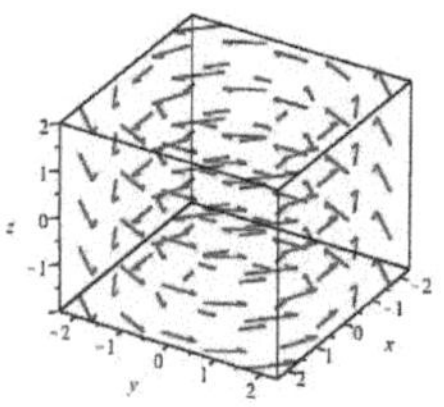

9B

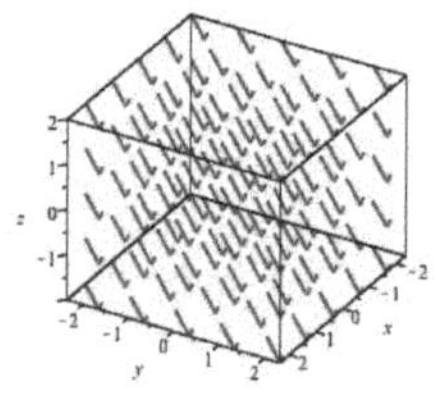

9C

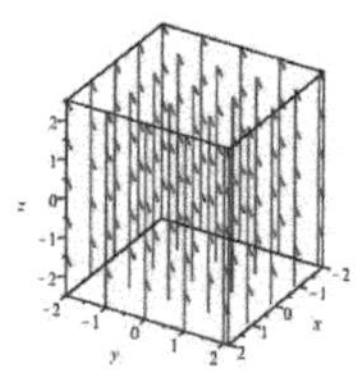

10. (a) $\mathbf{F} = \langle 0,z,-y \rangle$

(b) $\mathbf{F} = \langle -x,-y,-z \rangle / \sqrt{x^2+y^2+z^2}$

(c) $\mathbf{F} = \langle -x,-y,-z \rangle$

10A

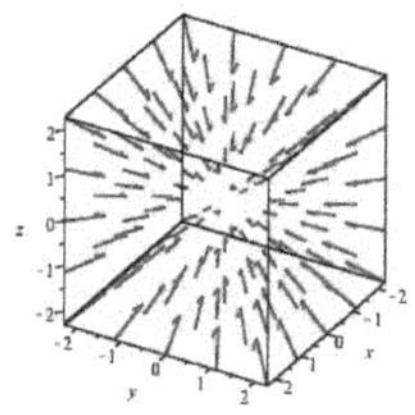

10B

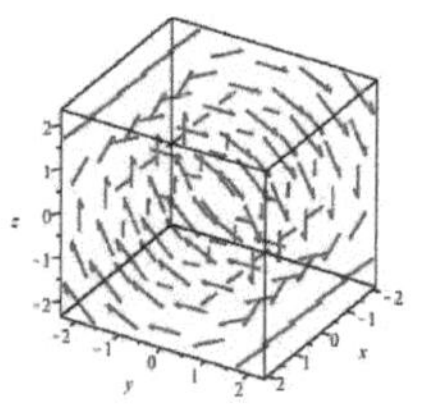

10C

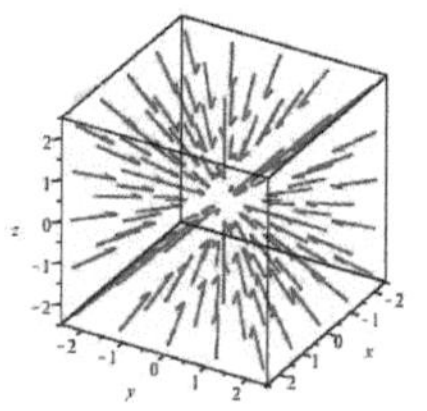

11. (a) $\mathbf{F} = \langle x,y,z \rangle$

(b) $\mathbf{F} = \langle -z,0,x \rangle$

(c) $\mathbf{F} = \langle 1,0,0 \rangle$

11A

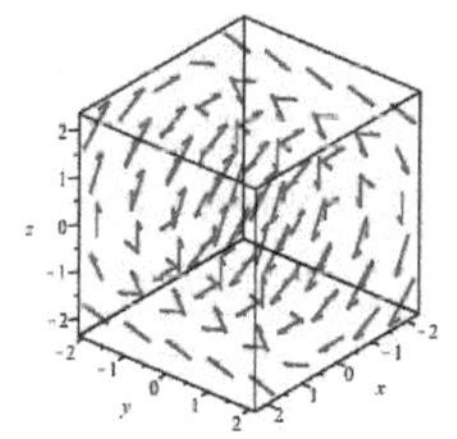

11B

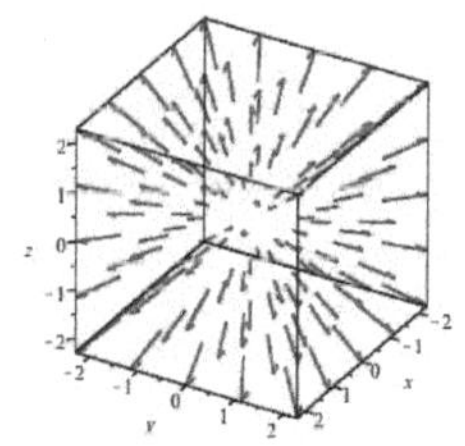

11C

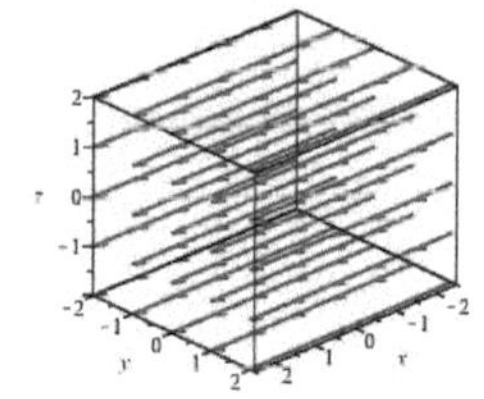

In problems 12 to 21, find the gradient field for the given functions.

12. $f(x,y) = x^2 + y^2$

13. $f(x,y) = x^2 - y^2$

14. $f(x,y) = x/y$

15. $f(x,y) = x \cdot y^2 - x^2 \cdot y$

16. $f(x,y) = x \cdot \sin(y) + e^{x \cdot y}$

17. $f(x,y,z) = 3x + 2y - z$

18. $f(x,y,z) = x \cdot y \cdot z$

19. $f(x,y,z) = \sqrt{x^2 + y^2 + z^2}$

20. $f(x,y,z) = x^3 \cdot y + 2y \cdot z + 5$

21. $f(x,y,z) = \sin(x \cdot y \cdot z)$

In problems 22 to 27, some level curves of z=f(x,y) are given. On the each figure sketch the gradient vector field for this function. (Do not put heads on the arrows -- just use short line segments.)

22.

Problem 22

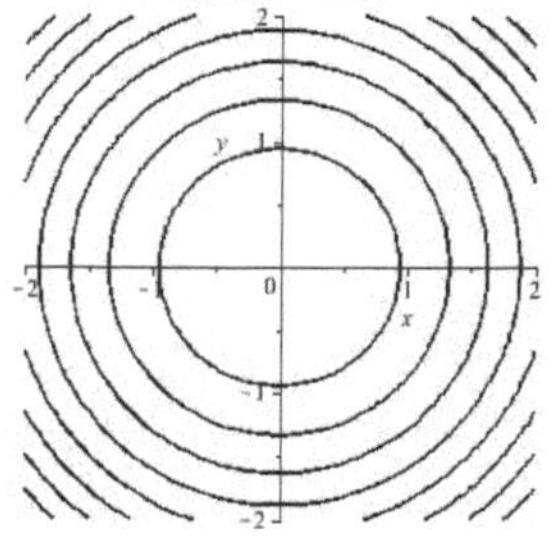

23.

Problem 23

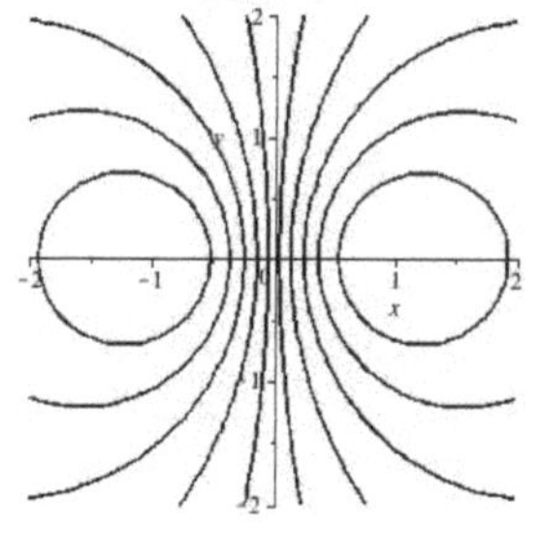

24.

Problem 24

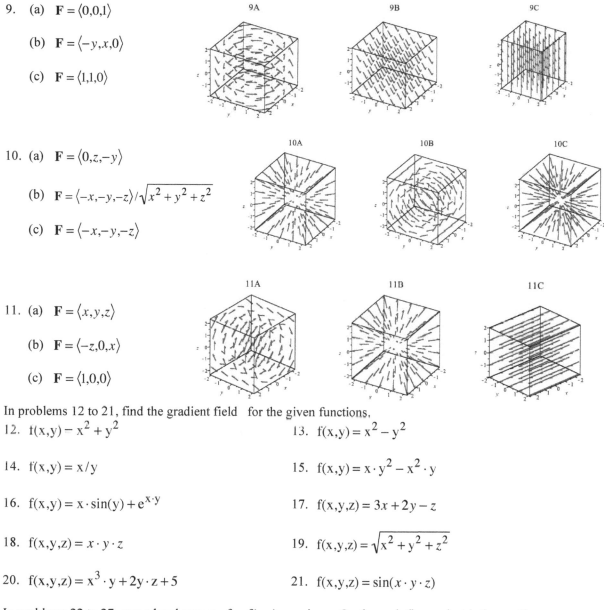

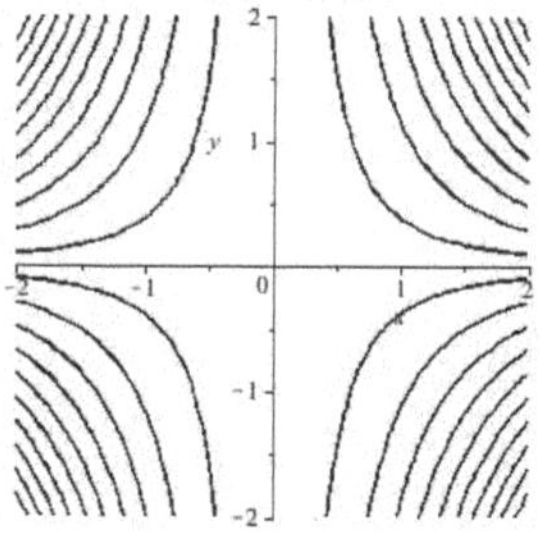

25.

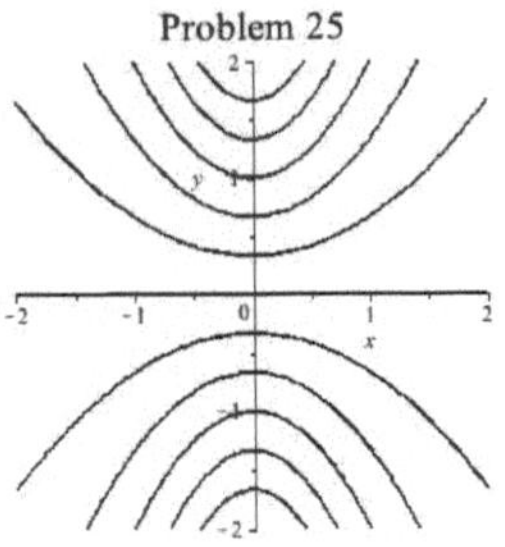

26.

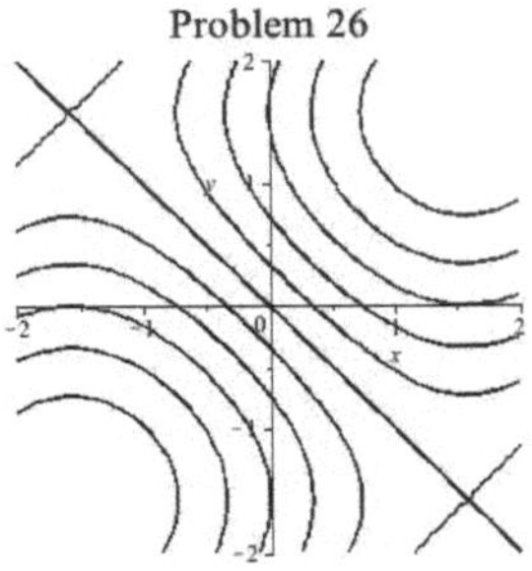

27.

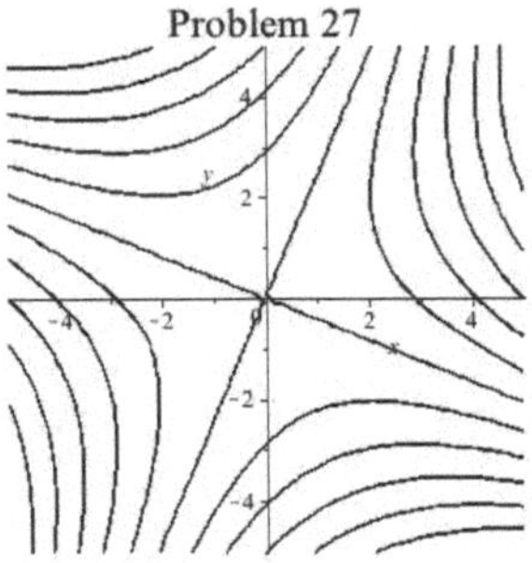

Practice Answers

Practice 1: See Fig. P1. Fig. P1 Maple shows the < x, y/2> field for -2≤x≤2 and -2≤y≤2.

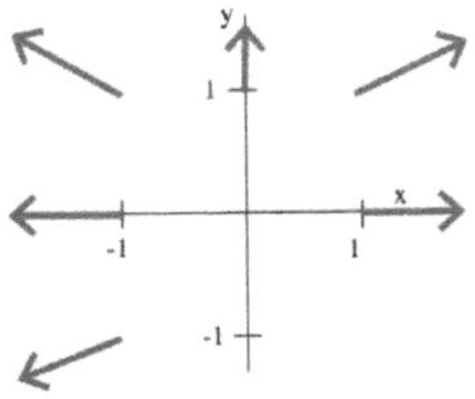

Fig. P1: vector field < x, y/2>

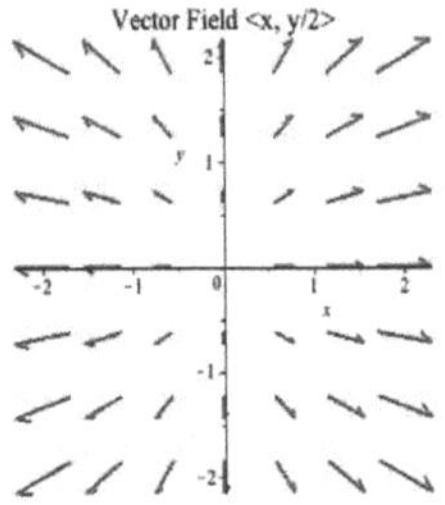

Fig. P1 Maple: < x, y/2 >

Practice 2: $z = f(x,y) = xy$. $\nabla f(x,y) = \langle y, x \rangle$.

The field is shown in Fig. P2.

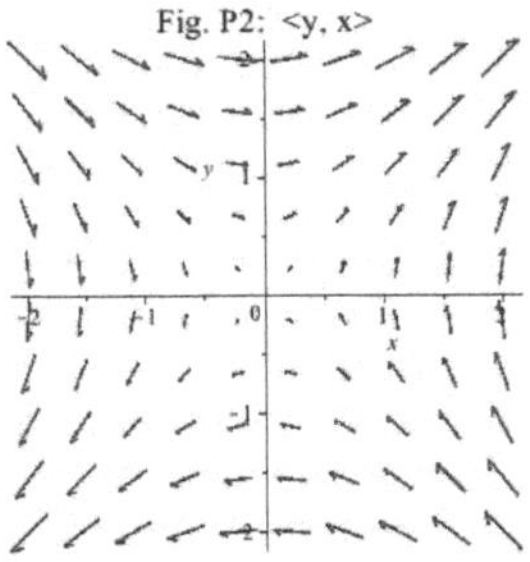

Practice 3: Fig. P3 shows gradient vectors for $f(x,y) = x^2 - y^2$. They have all been drawn the same length. The vectors closer to the origin should be shorter.

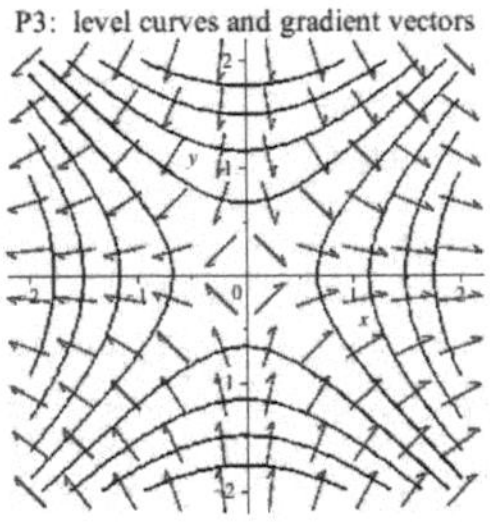

Appendix: Maple commands for plotting 2D and 3D vector fields

2D field <y,x> with(plots): fieldplot([y, x], x = -2..2, y = -2 .. 2);

This variation adds more details:

```
with(plots): fieldplot([y, x], x = -2..2, y = -2 .. 2, arrows = THIN, color = red, thickness=1,
grid=[10,10], scaling=constrained, title = "Fig. P2: <y, x>", titlefont = ["ROMAN", 18]);
```

3D field "swirl" <y,z,x> with(plots): fieldplot3d([y,z,x], x = -2 .. 2, y = -2 .. 2, z = -2 .. 2);

With more details:

```
with(plots): fieldplot3d([y,z,x], x = -2 .. 2, y = -2 .. 2, z = -2 .. 2, arrows = THIN, color=red,
thickness=1, grid=[8,8,8], title = "Fig. 6: <y,z,x>", titlefont = ["ROMAN", 18],orientation=[70,60,0]);
```

Note: Maple automatically centers each vector **F**(x,y) at the location (x,y).

Level curves and gradient vectors together

```
with(plots):
CP:=contourplot(x^2-y^2,x = -2..2, y = -2..2, color=blue, thickness=1, contours=8):
Gfld:=fieldplot([2*x,-2*y], x = -2..2, y = -2..2, arrows = THIN, color = red, thickness=1,grid=[10,10]):
display(Gfld,CP);
```

15.2 Del Operator and 2D Divergence and Curl

The divergence and curl of a vector field F describe two characteristics of the field at each point in the field. They will be extended into 3D in later sections, but they are more easily understood graphically and intuitively in 2D, so we start there.

Divergence: div **F**

Assume that the vector field F describes the flow of a liquid in 2D, perhaps water contained between two close-together parallel plates. The **divergence** is the rate per unit area that the water dissipates (**departs**, leaves) at the point P. A positive value for the divergence means that more water is leaving at P than is entering at P. But we can't really see a point so imagine a small circle C centered at P, and consider whether more water s leaving or entering this circle. Fig. 1(a) shows more water leaving than entering the circle around P so div **F**(P) > 0, Fig. 1(b) has more entering than leaving so div **F**(P) < 0, and Fig, 1(c) shows the same amount leaving as entering so div **F** (P) = 0.

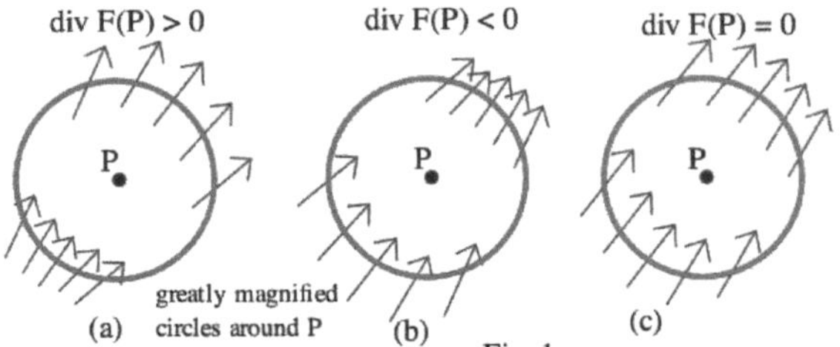

Fig. 1

Example 1: Estimate whether div **F**(P) is positive, negative or zero when P=A, B and C in Fig. 2.

Solution: (a) div F > 0 (b) div F = 0 (c) div F < 0

Practice 1: Estimate whether div **F**(P) is positive, negative or zero when P=A, B and C in Fig. 3.

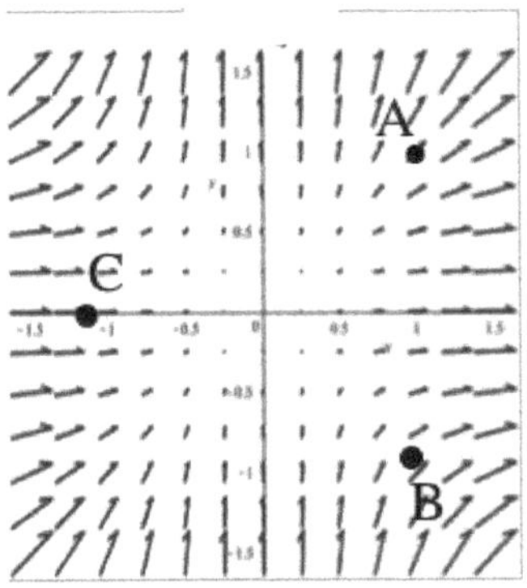

Fig. 2

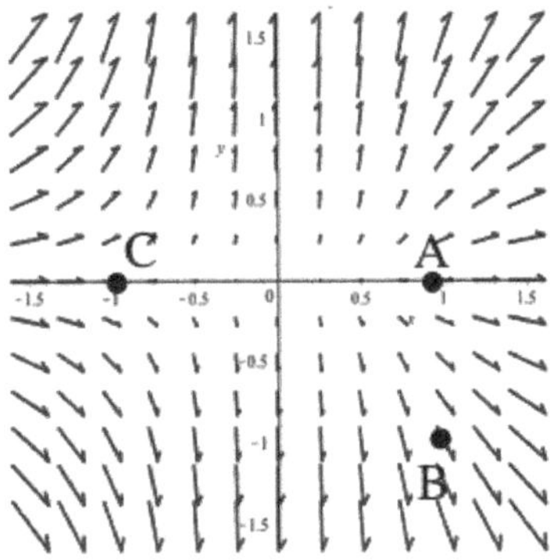

Fig. 3

It turns out that div **F**(P) is very easy to calculate.

> **Definition: Divergence** div F
>
> For a vector field $\mathbf{F}(x,y) = M\mathbf{i} + N\mathbf{j}$ with continuous partial derivatives,
>
> the divergence of F at Point P is $\text{div } \mathbf{F}(P) = \frac{\partial M}{\partial x} + \frac{\partial N}{\partial y}$.
>
> (In 3D with $\mathbf{F}(x,y,z) = M\mathbf{i} + N\mathbf{j} + P\mathbf{k}$, $\text{div } \mathbf{F}(P) = \frac{\partial M}{\partial x} + \frac{\partial N}{\partial y} + \frac{\partial P}{\partial z}$.)

We will justify this definition for divergence in section 15.10, but, for now, we will simply use it.

Example 2: The vector field in Fig. 2 is $\mathbf{F} = \left\langle x^2, y^2 \right\rangle$. Calculate div **F**(P) at P=A=(1,1), P=B=(1,–1) and P=C=(–1,0).

Solution: $\text{div } \mathbf{F} = \frac{\partial M}{\partial x} + \frac{\partial N}{\partial y} = 2x+2y$. (a) at A div **F** =4, at B div **F** = 0, at C div **F** = –2.

Practice 2: The vector field in Fig. 3 is $\mathbf{F} = \left\langle \mathbf{x^2},\ 2 \cdot y \right\rangle$. Calculate div **F**(P) at P=A=(1,0), P=B=(1,–1) and P=C=(–1,0).

For the radial field $\mathbf{F} = \langle x, y \rangle$, div $\mathbf{F} = 2$ at every point. For the radial field $\mathbf{F} = \langle -x, -y \rangle$, div $\mathbf{F} = -2$ at every point. And for the rotational field $\mathbf{F} = \langle -y, x \rangle$, div $\mathbf{F} = 0$ at every point. (Fig. 4)

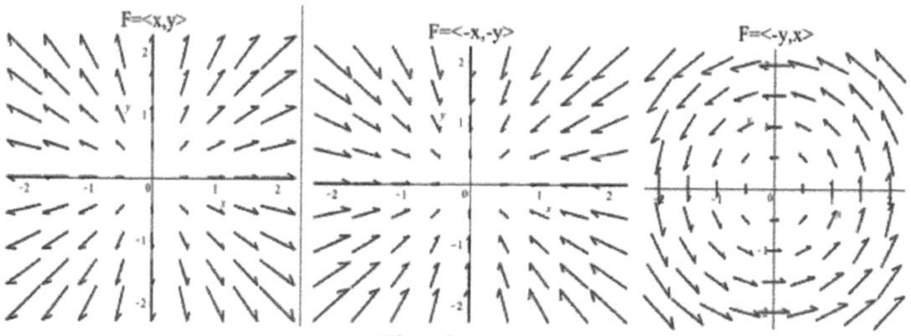

Fig. 4

Curl: curl F

The curl **F** (P) measures the rotation of the vector field **F** at the point P. Again picture a small circle centered at P, but this time imagine an axle at P and small paddles on the circle (Fig. 5). If the water vectors would rotate the little wheel counterclockwise (Fig. 6a) we say that curl **F** (P) > 0, if the rotation is clockwise (Fig. 6b) curl **F**(P) < 0, and if there is no rotation then curl **F** (P) = 0 (Fig. 6c)

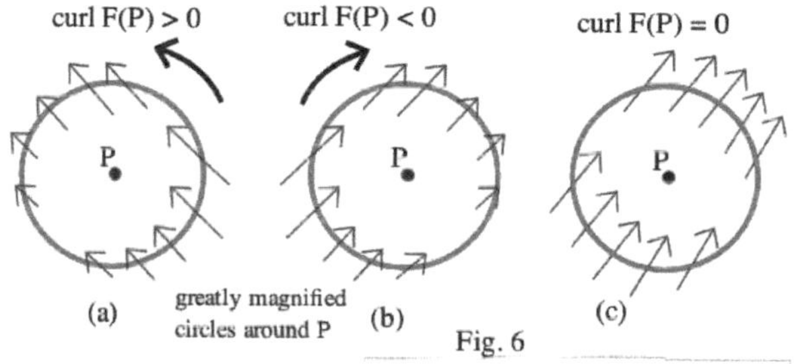

Fig. 6

Example 3: Estimate whether curl **F**(P) is positive, negative or zero when P=A, B and C in Fig. 7.

Solution: (a) at A curl **F** > 0, at B curl **F** < 0, at C curl **F** > 0

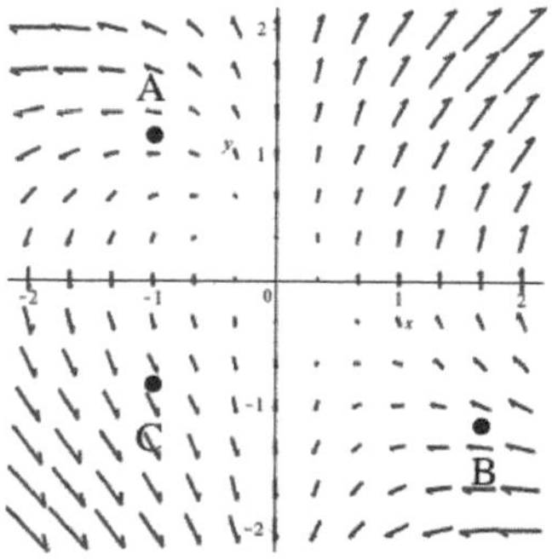

Fig. 7

> **Definition: Curl in 2D** curl F
>
> For a vector field $\mathbf{F}(x,y) = M\mathbf{i} + N\mathbf{j}$ with continuous partial derivatives,
>
> then the curl of F at Point P is $\text{curl } \mathbf{F}(P) = \frac{\partial N}{\partial x} - \frac{\partial M}{\partial y}$.

Note: Later we will define curl F to be a vector quantity in 3D (still measuring a rotation) as

$$\text{curl F} = \left(\frac{\partial P}{\partial y} - \frac{\partial N}{\partial z}\right)\mathbf{i} + \left(\frac{\partial M}{\partial z} - \frac{\partial P}{\partial x}\right)\mathbf{j} + \left(\frac{\partial N}{\partial x} - \frac{\partial M}{\partial y}\right)\mathbf{k}$$

For now we will just need and use the k component of this 3D vector. That will tell us about the rotation in the xy-plane,

Example 4: The vector field in Fig. 7 is $\mathbf{F} = \langle xy, x+y \rangle$ with A=(–1,1), B=(1.8,–1) and C=(–1,–1). Calculate curl **F**(P) at each point.

Solution: curl $\mathbf{F} = \frac{\partial N}{\partial x} - \frac{\partial M}{\partial y} = 1 - x$. At A curl **F** = 2, at B curl **F** = -0.8, at C curl **F** = 2.

Practice 3: Calculate curl **F** for the radial field $\mathbf{F} = \langle x, y \rangle$ and the rotational field $\mathbf{F} = \langle -y, x \rangle$.

If curl F=0 at all points in the field, then F is called **irrotational**.

The divergence and the curl measure completely different characteristics of the field at a point, and knowing the sign of one does not tell us anything about the sign of the other. For example, the curl of $\mathbf{F} = \langle x^2y + y, y^2 \rangle$ is curl $\mathbf{F} = \frac{\partial N}{\partial x} - \frac{\partial M}{\partial y} = (0) - (x^2 + 1) < 0$ at all points, but div $\mathbf{F} = \frac{\partial M}{\partial x} + \frac{\partial N}{\partial y} = 2xy + 2y$ which can be positive, negative or zero depending on the location (x,y).

Practice 4: Find a vector field **F** so that div **F** > 0 at all points but curl **F** can be positive, negative or zero depending on the location (x,y).

del operator $\nabla = \left\langle \frac{\partial}{\partial x}, \frac{\partial}{\partial y}, \frac{\partial}{\partial z} \right\rangle$

A mathematical operator is like a function but it typically operates on functions or other advanced objects. For example, differentiation $\frac{d}{dx}[\ ]$ and integration $\int [\ ]dx$ are operators that do things to whatever function is put into the brackets. Similarly, the del operator does things to functions and even vector fields. And the del notation is a compact way to represent complicated operations (and an easy way to remember them).

in 2D: $\nabla f = \left\langle \frac{\partial f}{\partial x}, \frac{\partial f}{\partial y} \right\rangle$ which we met earlier and called the 2D gradient of f

$\nabla \bullet \mathrm{F} = \left\langle \frac{\partial f}{\partial x}, \frac{\partial f}{\partial y} \right\rangle \bullet \langle M, N \rangle = \frac{\partial M}{\partial x} + \frac{\partial N}{\partial y}$ which is the 2D divergence, div **F**

$\nabla \mathbf{x} \mathrm{F} = \begin{vmatrix} \mathbf{i} & \mathbf{j} & \mathbf{k} \\ \frac{\partial}{\partial x} & \frac{\partial}{\partial y} & 0 \\ M & N & 0 \end{vmatrix} = \frac{\partial N}{\partial x} - \frac{\partial M}{\partial y}$ which is the 2D curl, curl **F**

In 3D: $\nabla f = \left\langle \frac{\partial f}{\partial x}, \frac{\partial f}{\partial x}, \frac{\partial f}{\partial x} \right\rangle$ the 3D gradient of f

$\nabla \bullet \mathrm{F} = \left\langle \frac{\partial f}{\partial x}, \frac{\partial f}{\partial y}, \frac{\partial f}{\partial z} \right\rangle \bullet \langle M, N, P \rangle = \frac{\partial M}{\partial x} + \frac{\partial N}{\partial y} + \frac{\partial P}{\partial z}$ the 3D divergence, div **F**

$$\nabla\mathbf{x}\mathrm{F} = \begin{vmatrix} \mathbf{i} & \mathbf{j} & \mathbf{k} \\ \dfrac{\partial}{\partial x} & \dfrac{\partial}{\partial y} & \dfrac{\partial}{\partial z} \\ \mathrm{M} & \mathrm{N} & \mathrm{P} \end{vmatrix} = \left(\frac{\partial P}{\partial y} - \frac{\partial N}{\partial y}\right)\mathbf{i} + \left(\frac{\partial M}{\partial z} - \frac{\partial P}{\partial x}\right)\mathbf{j} + \left(\frac{\partial N}{\partial x} - \frac{\partial M}{\partial y}\right)\mathbf{k} \quad \text{the 3D curl of F}$$

You should recognize the z-component of $\nabla\mathbf{x}\mathrm{F}$ as the 2D curl of F.

The del operator can be combined with itself to create new operations:

the Laplacian: $\nabla \bullet \nabla f = \left\langle \frac{\partial}{\partial \mathbf{x}}, \frac{\partial}{\partial \mathbf{y}}, \frac{\partial}{\partial \mathbf{z}} \right\rangle \bullet \left\langle \frac{\partial f}{\partial \mathbf{x}}, \frac{\partial f}{\partial \mathbf{y}}, \frac{\partial f}{\partial \mathbf{z}} \right\rangle = \frac{\partial^2 f}{\partial \mathbf{x}^2} + \frac{\partial^2 f}{\partial \mathbf{y}^2} + \frac{\partial^2 f}{\partial \mathbf{z}^2}$

is important in mathematical physics.

And the del operator can help illuminate connections between objects that seem unrelated:

$$\text{div curl F} = \nabla \bullet (\nabla\mathbf{x}\mathrm{F}) = \left\langle \frac{\partial}{\partial \mathbf{x}}, \frac{\partial}{\partial \mathbf{y}}, \frac{\partial}{\partial \mathbf{z}} \right\rangle \bullet \left(\frac{\partial P}{\partial y} - \frac{\partial N}{\partial y}\right)\mathbf{i} + \left(\frac{\partial M}{\partial z} - \frac{\partial P}{\partial x}\right)\mathbf{j} + \left(\frac{\partial N}{\partial x} - \frac{\partial M}{\partial y}\right)\mathbf{k} = \ \dots \ = 0$$

You can fill in the " ..." by taking the dot product and then recognizing that the various mixed partial derivatives are equal (e.g, $\frac{\partial^2 f}{\partial \mathbf{x}\partial \mathbf{y}} = \frac{\partial^2 f}{\partial \mathbf{y}\partial \mathbf{x}}$).

Wrap up

The justifications for the definitions of the divergence and curl of a vector field at a point will come later in this chapter, but the ideas and calculations in 2D are relatively easy and they will help with the ideas and calculations when we get to Green's Theorem.

Problems

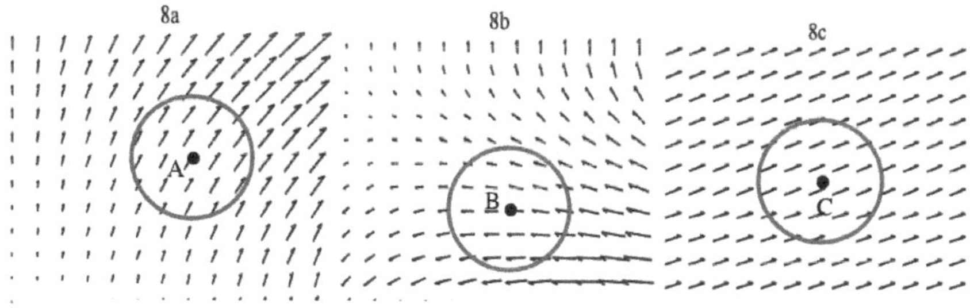

1. Estimate whether the div **F** is positive, negative or zero at A, B and C in Fig. 8.
2. Estimate whether the div **F** is positive, negative or zero at A, B and C in Fig. 9.

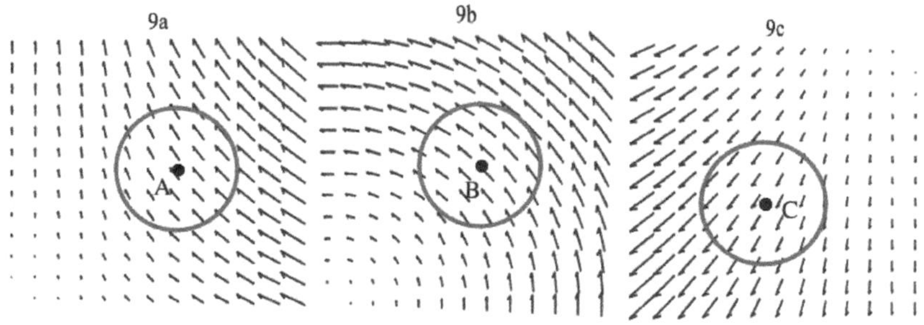

In problems 3 to 6, calculate div **F** at the given points.

3. $\mathbf{F} = \left\langle x^2 + 3y,\ 2y + x \right\rangle$ at A=(1, 3), B=(2, –1) and C=(–1, 3).
4. $\mathbf{F} = \left\langle x \cdot y^2,\ x^3 \cdot y + 3 \right\rangle$ at A=(1, 1), B=(2, –1) and C=(0, –2).
5. $\mathbf{F} = \langle 5x - 3y,\ x + 2y \rangle$ at A=(3, 2), B=(0, 3) and C=(1, 4).

6. $\mathbf{F} = \langle x^2 - y^2, x^2 + y^2 \rangle$ at A=(2, 3), B=(-2, 2) and C=(3, –4).

7. In Fig. 10 add additional vectors so div **F**(P) > 0.

8. In Fig. 10 add additional vectors so div **F**(Q) < 0.

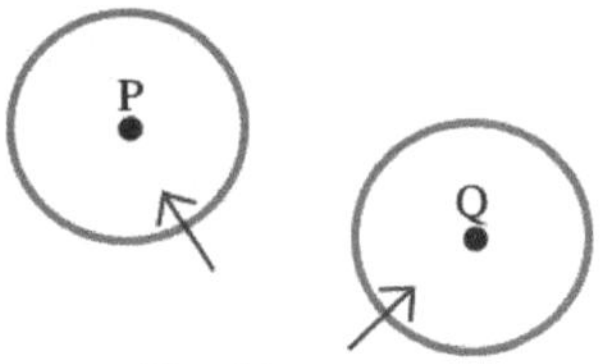

Fig. 10

9. Estimate whether the curl **F** is positive, negative or zero at A, B and C in Fig. 8.

10. Estimate whether the curl **F** is positive, negative or zero at A, B and C in Fig. 9.

In problems 11 to 14, calculate curl **F** at the given points.

11. $\mathbf{F} = \langle x^2 + 3y, 2y + x \rangle$ at A=(1, 3), B=(2, –1) and C=(–1, 3).

12. $\mathbf{F} = \langle x \cdot y^2, x^3 \cdot y + 3 \rangle$ at A=(1, 1), B=(2, –1) and C=(0, –2).

13. $\mathbf{F} = \langle 5x - 3y, x + 2y \rangle$ at A=(3, 2), B=(0, 3) and C=(1, 4).

14. $\mathbf{F} = \langle x^2 - y^2, x^2 + y^2 \rangle$ at A=(2, 3), B=(–2, 2) and C=(3, –4).

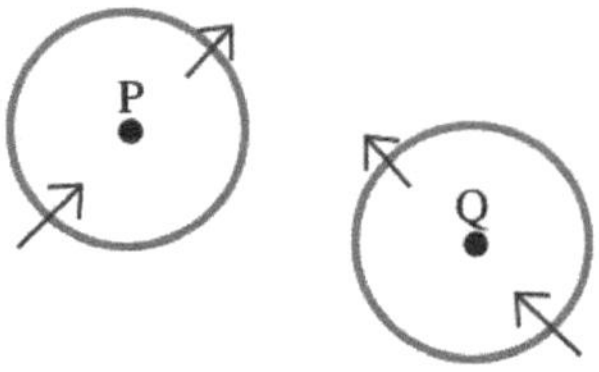

Fig. 11

15. In Fig. 11 add additional vectors so curl **F**(P) > 0.

16. In Fig. 11 add additional vectors so curl **F**(Q) < 0.

17. Show that the div and curl are always 0 for a constant field $\mathbf{F} = \langle a, b \rangle$.

18. Show that the div and curl are constant for a linear field $\mathbf{F} = \langle ax + by, cx + dy \rangle$.

19. If a rotational field is flowing counterclockwise, does that mean the curl at each point is positive. Justify your conclusion.

20. If the y-component of **F** is always 0, what can you conclude about the divergence?

21. If the y-component of **F** is always 0, what can you conclude about the curl?

Practice Answers

Practice 1: (a) div F > 0 (b) div F = 0 (c) div F < 0

Practice 2: div $\mathbf{F} = \frac{\partial M}{\partial x} + \frac{\partial N}{\partial y} = 2x + 2$. (a) at A div **F** = 4, at B div **F** = 4, at C div **F** = 0.

Practice 3: For $\mathbf{F} = \langle x, y \rangle$ curl $\mathbf{F} = \frac{\partial N}{\partial x} - \frac{\partial M}{\partial y} = 0$ at every point. No rotation anywhere.

For $\mathbf{F} = \langle -y, x \rangle$ curl $\mathbf{F} = \frac{\partial N}{\partial x} - \frac{\partial M}{\partial y} = (1) - (-1) = 2$ at every point. The rotation is counterclockwise at every point.

Practice 4: You need a field $\mathbf{F} = \langle M, N \rangle$ so that $\text{div } \mathbf{F} = \frac{\partial M}{\partial x} + \frac{\partial N}{\partial y} > 0$ everywhere but so $\text{curl } \mathbf{F} = \frac{\partial N}{\partial x} - \frac{\partial M}{\partial y}$ contains variables. $\mathbf{F} = \left\langle x^3 + x + y^2,\ y^3 \right\rangle$ works.

$\text{div } \mathbf{F} = \frac{\partial M}{\partial x} + \frac{\partial N}{\partial y} = 3x^2 + 1 + 3y^2 > 0$ and $\text{curl } \mathbf{F} = \frac{\partial N}{\partial x} - \frac{\partial M}{\partial y} = (0) - (2y)$.

15.3 Line Integrals

A curtain is hanging from a very bent rod (Fig. 1). If we have an equation z=f(x,y) or (x(t), y(t), z(t)) for the rod, how can we calculate the area of the curtain between the rod and the floor (the xy-plane). And if the rod has different densities, $\delta(x,y,z)$ at locations (x,y,z), along its length, how can we calculate the total mass of the rod?

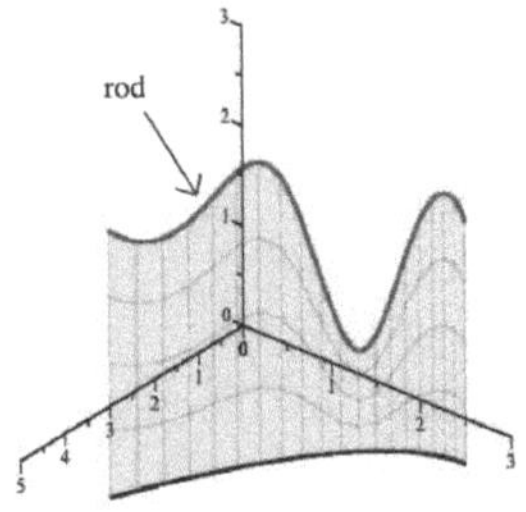

Fig. 1: Curtain on a bent rod

These are two silly questions, but their solutions illustrate the main idea of this section: an integral along a curve, a **line integral**. One of the main applications of line integrals is to determine the work done to move an object in a force field in two or three dimensions, and there are others.

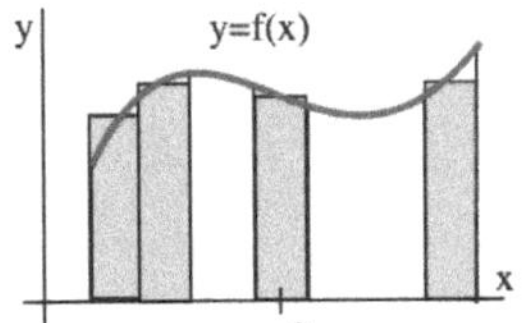

Fig. 2

Curtain area solution: In beginning calculus we created integral of f(x) on the interval a≤x≤b on the x-axis (Fig. 2) by partitioning the interval, picking a representative point x* in each subinterval, calculating the area of the little rectangle above the subinterval as $f(x^*)\cdot\Delta x$, forming the Riemann sum $\sum f(x^*)\cdot\Delta x$ of these little areas, and finally taking the limit as $\Delta x \to 0$ to get an integral: $\int_a^b f(x)\,dx$.

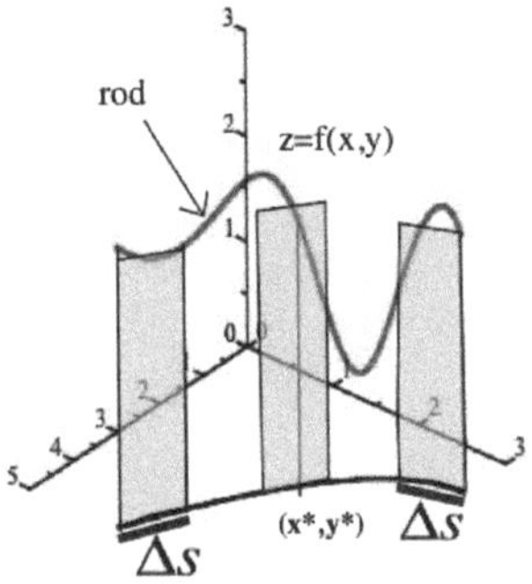

Fig. 3: Partitioned curtain

The same strategy works for the shower curtain, except here we partition the curve C in the xy-plane (Fig. 3), pick a representative point (x*, y*) in each subinterval, and find the area of each little rectangle as $f(x^*,y^*)\cdot\Delta s$ where Δs is the length of the subinterval. Then as $\Delta s \to 0$ we get that $\sum f(x^*,y^*)\cdot\Delta s \to \int f(x,y)\,ds$. If our curve in the xy-plane is parameterized by t, $\mathbf{r}(t)=\langle x(t),y(t)\rangle$ then z(t)=f(x(t), y(t)) and $\Delta s=\sqrt{(\Delta x)^2+(\Delta y)^2}$. The Riemann sum becomes

$$\sum f(x^*(t),y^*(t))\cdot\Delta s=\sum f(x^*(t),y^*(t))\cdot\frac{\sqrt{(\Delta x)^2+(\Delta y)^2}}{\Delta t}\cdot\Delta t=\sum z(t^*)\cdot\sqrt{\left(\frac{\Delta x}{\Delta t}\right)^2+\left(\frac{\Delta y}{\Delta t}\right)^2}\cdot\Delta t\ .$$

Taking the limit of this as $\Delta s \to 0$, we get that

$$\{\textbf{area between f(x,y) and xy-plane}\}=\int_C f\,ds=\int_{t=a}^{t=b} z(t)\cdot\sqrt{\left(\frac{dx}{dt}\right)^2+\left(\frac{dy}{dt}\right)^2}\cdot dt\ .$$

Example 1: Suppose the curve C is parameterized by $\mathbf{r}(t)=\langle x(t),y(t)\rangle$ with $x(t)=1+t^2$ and $y(t)=3-t$, and that $f(x,y)=2+\sin(3xy)=2+\sin(3(1+t^2)(3-t))$. This is the situation in Fig. 1. Write an integral in terms of t for the area between the curve C in the xy-plane and f(x,y) for t from 0 to 2. If the units of x and y are meters and the t units are seconds, then what are the units of the result.

Solution: area between f(x,y) and the xy - plane $= \int_0^2 \left(2+\sin(3(t^2+1)(3-t))\right)\sqrt{(2t)^2+(-1)^2}\ dt$

Unfortunately we can not evaluate this integral by hand, but a calculator can: area=10.754 m^2.

Practice 1: Represent the area between the curve C parameterized by $\mathbf{r}(t) = \langle 2\cos(t),\ 2\sin(t)\rangle$ f(r(t))=2+cos(5t) (Fig. 4) as a definite integral and evaluate the integral.

Fig. 4: <2cos(t), 2sin(t), 2+cos(5t)>

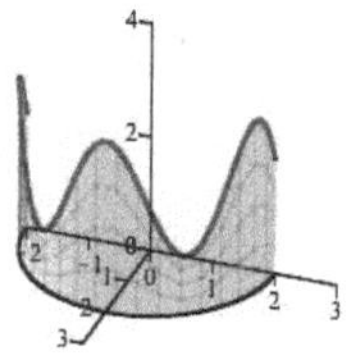

General case for line integral of a scalar function f over a path C : $\int_C f\ ds$

Suppose the path C is parameterized by $\mathbf{r}(t) = \langle x(t), y(t)\rangle$ where x(t) and y(t) are smooth (differentiable) functions of t, and that t varies from t=a to t=b. Then

$$\int_C f\,ds = \int_{t=a}^{b} f(\mathbf{r}(t))\cdot\left|\frac{d\mathbf{r}}{dt}\right|\ dt = \int_{t=a}^{b} f(\mathbf{r}(t))\cdot|\mathbf{r}'(t)|\ dt$$

Note: $ds = |\mathbf{r}'(t)|\ dt$ simply says that {change in position}={speed} ·{change in time}: distance =rate·time.

Example 2: C1 is the semicircular path from (2,0) to (-2,0) parameterized by $\mathbf{r}(t) = \langle 2\cos(t),\ 2\sin(t)\rangle$ for t= 0 to π, and f(x,y)= 5+2x+4y (Fig. 5). C2 is the straight line path from (2,0) to (-2,0) parameterized by the path $\mathbf{r}(t) = \langle 2 - 4t,\ 0\rangle$ for t=0 to t=1.

Evaluate $\int_{C1} f\ ds$ and $\int_{C2} f\ ds$.

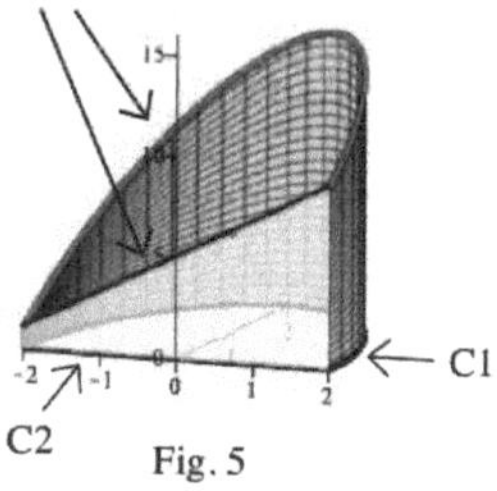

Fig. 5

Solution: Along C1 $|\mathbf{r}'(t)| = \sqrt{(-2\sin(t))^2+(2\cos(t))^2} = 2$ and $f(\mathbf{r}(t)) = 5 + 2\cdot 2\cdot\cos(t) + 4\cdot 2\cdot\sin(t)$ so

$$\int_{C1} f\,ds = \int_0^{\pi} (5+4\cdot\cos(t)+8\cdot\sin(t))(2)\ dt = 32+10\pi \approx 63.4 .$$

Along C2 $|\mathbf{r}'(t)| = \sqrt{(-4)^2+(0)^2} = 4$ and $f(\mathbf{r}(t)) = 5+2\cdot(2-4t)+4\cdot(0) = 9-8t$ so

$$\int_{C2} f\,ds = \int_0^1 (9-8t)(4)\ dt = 20 .$$

Even though C1 and C2 begin and end at the same points, the values of the line integrals are different.

Note: We can always parameterize the straight line from A to B by $P(t) = (1-t)\cdot A + t\cdot B$ for t=0 to t=1. Circles are typically parameterized using variations of $x = r\cdot\cos(t)$ and $y = r\cdot\sin(t)$ that take directions and shifts into account.

Units: In the previous example, suppose f(x,y) is the density (in kg/m) of the curve **r** at the location (x,y), that the location (x,y) is in meters, m, and that time is in seconds, s. Then the units of the integral are

$$\int_C f\,ds = \int_{t=a}^{b} f(\mathbf{r}(t))|\mathbf{r}'(t)|\,dt = \text{kg}$$

$$\left(\frac{\text{kg}}{\text{m}}\right)\left(\frac{\text{m}}{\text{s}}\right)(\text{s})$$

Practice 2: C is the straight line path from (0,1) to (4,2) and f(x,y)=2x+y. Evaluate $\int_C f\,ds$.

If the units of f are pounds, the x and y units are feet, and the t units are minutes, then what are the units of this line integral?

3D Mass of a Rod

Suppose a curve $C(t) = \left(t, t^2, 1+t^2\right)$ meters (Fig. 6) has linear density $\delta(x,y,z) = x - y + z - 1$ g/m. We can represent the mass of the curve from t=1 to t=3 as an integral in t by proceeding as before and starting with a partition of the curve into segments of lengths $\Delta s = \sqrt{(\Delta x)^2 + (\Delta y)^2 + (\Delta z)^2} = \sqrt{1 + (2t)^2 + (2t)^2} = \sqrt{1+8t^2}$.

The density at each location (x*, y*, z*) on the curve is $\delta(x,y,z) = x - y + z - 1 = (t) - (t^2) + (1+t^2) - 1 = t$ so the mass along each segment is $\{\text{segment mass}\} = \text{density} \cdot \text{length} = t \cdot \sqrt{1+8t^2}$. Finally, the total mass of the rod is $\int_{t=1}^{3} t \cdot \sqrt{1+8t^2}\,dt = \frac{1}{24}\left(1+8t^2\right)^{3/2} \Big|_1^3 = \frac{73}{24}\sqrt{73} - \frac{9}{8} \approx 24.86\ g$.

Fig. 6: <t, t^2, 1+t^2>

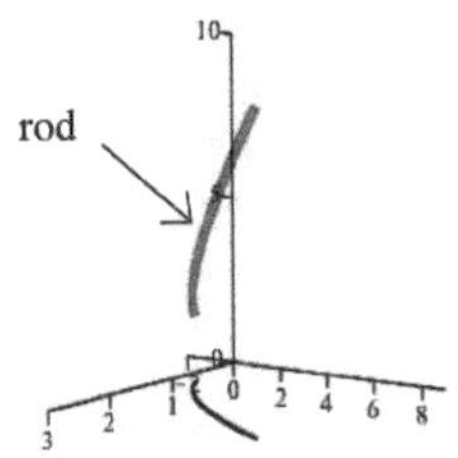

Generalizing this approach to any continuous function f(x,y,z) along a smooth curve C parameterized by $\mathbf{r}(t) = \langle x(t), y(t), z(t) \rangle$ from t=a to t=b, we again have

$$\int_C f\,ds = \int_{t=a}^{t=b} f(\mathbf{r}(t)) \cdot |\mathbf{r}'(t)| \cdot dt\ .$$

Practice 3: Represent the total mass of the curve $C(t) = \left(2t, t^2, t\right)$ that has density $\delta(x,y,z) = x + z$ from t=1 to t=4 as an integral and evaluate the integral.

Note: The various application formulas for first and second moments and centers of mass in section 14.6 also apply here, but we replace the triple integrals $\iiint_R$ with $\int_C$ and the dV with ds.

Work moving along a curve

Previously in this section the function f was a scalar-valued function, but very interesting situations arise when we move along a curve in a vector field.

In section 11.4 we saw that the elementary idea that "work=force times distance" could be extended to "work=(force in the direction of movement) · (displacement) = $\mathbf{F} \bullet \mathbf{D}$," the dot product of **F** and **D**, where **F** is a force vector and **D** is the displacement vector. This is exactly the idea we need to calculate the work moving an object along a curve in two or three dimensions.

Line integral of a vector-valued function F over a path C

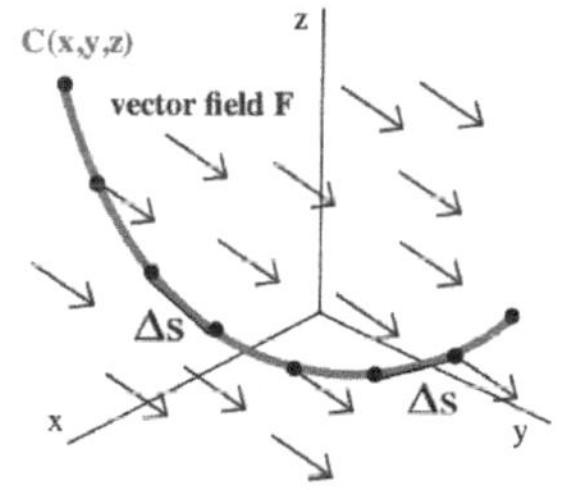

Fig. 7: Curve C and vector fiekd F

Work: Suppose C is a smooth curve in 3D parameterized by $\mathbf{r}(t) = \langle x(t), y(t), z(t) \rangle$ for $a \le t \le b$ and that **F**(x,y,z) is a 3D force vector field. If we partition C (Fig. 7) into small time increments, then at location P*=(x(t*), y(t*), z(t*) the displacement is tangent to the curve and has length Δs so the displacement is $\Delta s \cdot \mathbf{T}(t^*)$ where $\mathbf{T}(t^*)$ is the unit tangent vector at P* (Fig. 8). The work for **F** to move the object along that small Δs segment of the curve is $\mathbf{F} \bullet \mathbf{T} \cdot \Delta s$. Then the total work is approximately

$$\sum \mathbf{F}(x(t^*), y(t^*), z(t^*)) \bullet \mathbf{T}(x(t^*), y(t^*), z(t^*)) \cdot \Delta s$$

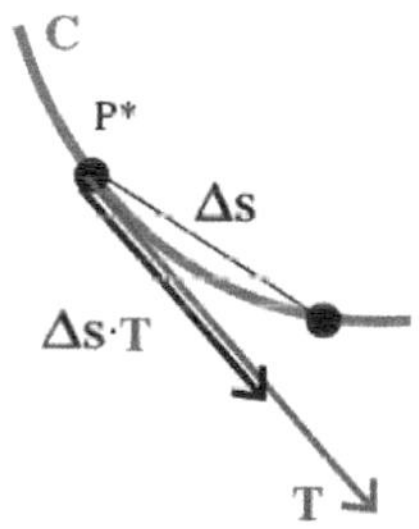

Fig. 8: work on one segment

As $\Delta s \to 0$ the approximations become better and better, and the work to move an object along C is defined to be

$$\text{work} = \int_C \mathbf{F} \bullet \mathbf{T}\ ds = \int_C \mathbf{F}(x(t), y(t), z(t)) \bullet \mathbf{T}(x(t), y(t), z(t))\ ds$$

But $\mathbf{T}(t) = \dfrac{\mathbf{r}'(t)}{|\mathbf{r}'(t)|}$ and $ds = |\mathbf{r}'(t)|\ dt$ so

$$\text{work} = \int_C \mathbf{F} \bullet \mathbf{T}\ ds = \int_{t=a}^{b} \mathbf{F}(\mathbf{r}(t)) \bullet \mathbf{r}'(t)\ dt$$

Note: The units for work are the units of **F** times the units of length, ds.

Fig. 9: C=(cos(t),sin(t),t) F=<1,1,z>

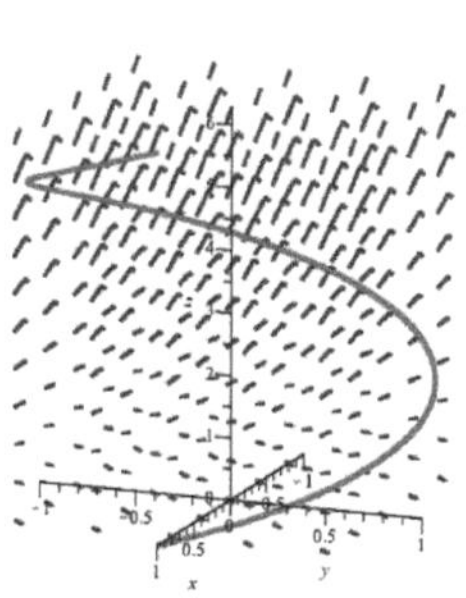

Example 2: Represent the work done in the field $\mathbf{F}(x,y,z) = \langle 1, 1, z \rangle$ to move an object along the helix $\mathbf{r}(t) = \langle \cos(t), \sin(t), t \rangle$ for t=0 to t=2π (Fig. 9) as an integral and then evaluate the integral.

Solution: $\mathbf{F}(\mathbf{r}(t)) \bullet \mathbf{r}'(t) = \langle 1, 1, z \rangle \bullet \langle -\sin(t), \cos(t), 1 \rangle = -\sin(t) + \cos(t) + t$ so $\text{work} = \int_{t=0}^{2\pi} (-\sin(t) + \cos(t) + t)\ dt = 2\pi^2$

Practice 4: Represent the work done in the constant field $\mathbf{F}(x,y,z) = \langle 3,2,1 \rangle$ to move an object along the curve $\mathbf{r}(t) = \langle t, t^2, t^3 \rangle$ for t=0 to t=2 as an integral and then evaluate the integral.

Definition
If C is a smooth curve given by the vector function $\mathbf{r}(t)$ for $a \le t \le b$, and $\mathbf{F}$ is a continuous vector field defined on C, then the **line integral of F along C** is $$\int_C \mathbf{F}(\mathbf{r}(t)) \bullet d\mathbf{r} = \int_{t=a}^{b} \mathbf{F}(\mathbf{r}(t)) \bullet \mathbf{r}'(t)\, dt = \int_C \mathbf{F} \bullet \mathbf{T}\, ds$$

Note: This pattern will occur often in future sections, and you should be familiar with all three notations. The middle integral is usually the easiest to use for computations.

Note: If $\mathbf{F} = \langle M,N,P \rangle$ and $\mathbf{T} = \left\langle \frac{dx}{ds}, \frac{dy}{ds}, \frac{dz}{ds} \right\rangle$ then $\mathbf{F} \bullet \mathbf{T} ds = M dx + N dy + P dz$. Units = (F units)(ds units).

You should also recognize what is happening geometrically (Fig. 10). If the angle θ between **F** and **r'** is 90° then $\mathbf{F} \bullet \mathbf{T} = |\mathbf{F}|\,|\mathbf{T}| \cos(\theta) = 0$, if the **F** and **r'** angle is acute ($-90^\circ < \theta < 90^\circ$) then $\mathbf{F} \bullet \mathbf{T} = |\mathbf{F}|\,|\mathbf{T}| \cos(\theta) > 0$, and if the **F** and **r'** angle is obtuse ($90^\circ < \theta < 270^\circ$) then $\mathbf{F} \bullet \mathbf{T} = |\mathbf{F}|\,|\mathbf{T}| \cos(\theta) < 0$.

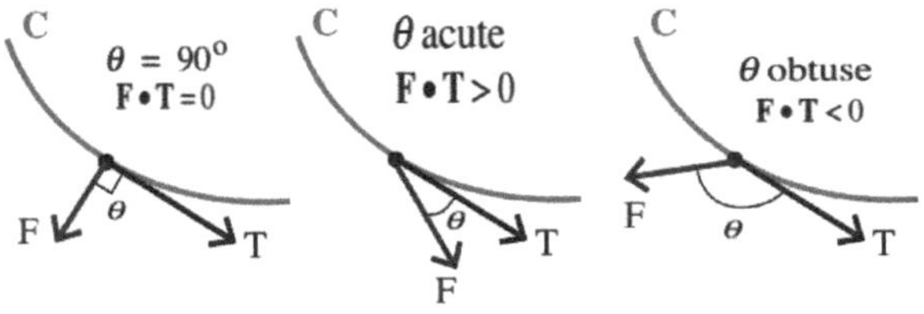

Fig. 10: Geometry of $\mathbf{F} \bullet \mathbf{T}$

Practice 5: C is the semicircle given by $\mathbf{r}(t) = \langle \cos(t), \sin(t) \rangle$ for $0 \le t \le \pi$, and four vector fields are $\mathbf{F1} = \langle 1,0 \rangle$, $\mathbf{F2} = \langle -1,0 \rangle$, $\mathbf{F3} = \langle 0,1 \rangle$ and $\mathbf{F4} = \langle x,y \rangle$. Use sketches of the path C and a few vectors from the field **F** to estimate whether each line integral $\int_C \mathbf{F}(\mathbf{r}(t)) \bullet d\mathbf{r}$ is positive, negative or zero.

Work in Gravitational, Electrical and Magnetic Fields: In all three of these situations the force between points is inversely proportional to the square of the distance between the points and acts along the straight line connecting the points. If one point is at the origin and the other is at (x,y,z) then the magnitude of **F** is

$$|\mathbf{F}| = \frac{k}{|\mathbf{r}|^2} = \frac{k}{(x^2+y^2+z^2)}$$ and the direction of **F** is $$\frac{\mathbf{r}}{|\mathbf{r}|} = \frac{\langle x,y,z \rangle}{(x^2+y^2+z^2)^{1/2}}$$ so $$\mathbf{F}(\mathbf{r}) = \frac{k\langle x,y,z \rangle}{(x^2+y^2+z^2)^{3/2}} = \frac{k\mathbf{r}}{|\mathbf{r}|^3}.$$

Example 3: Find the work to moving an object along the path from (1,1,1) in a straight line to (c,c,c) where c>1.

Solution: The line can be parameterized by $\mathbf{r}(t) = \langle t,t,t \rangle$ as t goes from 1 to c, and

$$\mathbf{F}(\mathbf{r}(t)) \bullet \mathbf{r}'(t) = \frac{k\langle t,t,t \rangle}{|\langle t,t,t \rangle|^3} \bullet \langle 1,1,1 \rangle = \frac{k3t}{(3t^2)^{3/2}} = \frac{k}{\sqrt{3}} \frac{1}{t^2}$$ so the work done is

$\text{work} = \int_{t=1}^{c} \mathbf{F}(\mathbf{r}(t)) \bullet \mathbf{r}'(t)\, dt = \frac{k}{\sqrt{3}} \int_{t=1}^{c} \frac{1}{t^2}\, dt = \frac{k}{\sqrt{3}} \left(-\frac{1}{t} \right) \Big|_{t=1}^{c} = \frac{k}{\sqrt{3}} \left(1 - \frac{1}{c} \right)$. As the object is moved farther and farther away, as $c \to \infty$, the work approaches the finite value $k/\sqrt{3}$.

Flow along a curve C in a vector field F

> **Flow:** If a smooth curve C is parameterized by r(t) in a continuous vector field F, then the **flow along C** from t=a to t=b is $\text{flow} = \int_C \mathbf{F} \bullet \mathbf{T}\, ds$.

F might represent the velocity field of a fluid in a region of space (water in a river channel, air in a wind tunnel), then the integral of $\mathbf{F} \bullet \mathbf{T}$ along a curve C in that space is the "flow" along that curve. In this case, if the units of F are m/sec, then the units for flow are m^2/sec.

If C is a closed loop, then $\int_C \mathbf{F} \bullet \mathbf{T}\, ds$ is called the **circulation** around C.

Flow is calculated in the same way as work.

Flux across a closed curve C in 2D

Just as flow measured the accumulation of a vector field along a curve C, the flux measures the accumulation as the vector field crosses perpendicular to the curve so we need the velocity of the field in the direction of the normal vector to the curve at each point. And then we want to accumulate all of those little values, an integral.

> $\{\text{Flux of F across closed curve C}\} = \int_C \mathbf{F} \bullet \mathbf{n}\, ds$ where **n** is the (outward) unit normal vector to C and the curve C is traversed exactly once in the counterclockwise direction.

Flow is the line integral of the scalar component $\mathbf{F} \bullet \mathbf{T}$ of $\mathbf{F}$ in the direction of the unit tangent vector to C.

Flux is the line integral of the scalar component $\mathbf{F} \bullet \mathbf{n}$ of $\mathbf{F}$ in the direction of the unit normal vector to C.

If $\mathbf{F}(x,y,z) = \langle \mathbf{M(x,y)}, \mathbf{N(x,y)} \rangle = \mathbf{Mi} + \mathbf{Nj}$ is a vector field in 2D and C is parameterized in the xy-plane in the counterclockwise direction by $\mathbf{r}(t) = \langle x(t), y(t) \rangle$ then the unit normal vector is $\mathbf{n} = \mathbf{T}\mathbf{x}\mathbf{k}$ where $\mathbf{T}$ is the unit tangent vector and $\mathbf{k} = \langle 0,0,1 \rangle$ (Fig. 11). Then

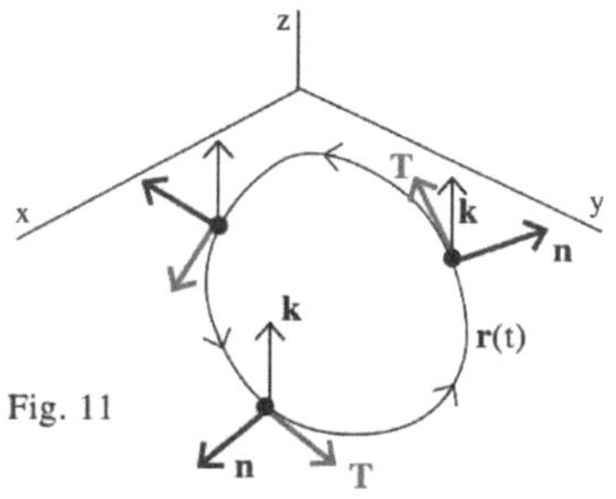

Fig. 11

$$\mathbf{n} = \mathbf{T}\mathbf{x}\mathbf{k} = \begin{vmatrix} \mathbf{i} & \mathbf{j} & \mathbf{k} \\ \frac{dx}{ds} & \frac{dy}{ds} & 0 \\ 0 & 0 & 1 \end{vmatrix} = \left\langle \frac{dy}{ds}, -\frac{dx}{ds}, 0 \right\rangle \text{ so}$$

Fig. 12: Flux across C

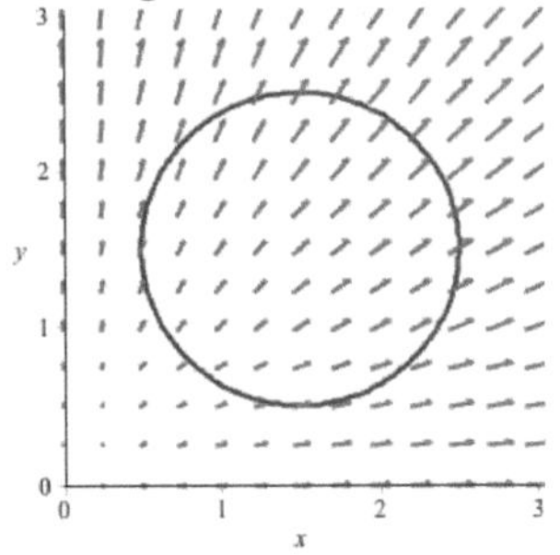

$$\mathbf{F} \bullet \mathbf{n} = \langle M, N \rangle \bullet \left\langle \frac{dy}{ds}, -\frac{dx}{ds} \right\rangle = M\frac{dy}{ds} - N\frac{dx}{ds} \text{ and}$$

$$\text{flux} = \int_C \mathbf{F} \bullet \mathbf{n}\, ds = \int_C \left(M\frac{dy}{ds} - N\frac{dx}{ds} \right) ds = \int_C M dy - N dx .$$

Example 4: Calculate the flux across the circle C parameterized by

$\mathbf{r}(t) = \langle 1.5 + \cos(t), 1.5 + \sin(t) \rangle$ for the vector field $\mathbf{F}(x,y) = \langle x,y \rangle = x\mathbf{i} + y\mathbf{j}$. (Fig. 12)

Solution: M=1.5+cos(t), N=1.5+sin(t), dx=-sin(t) dt, dy=cos(t) dt, $0 \le t \le 2\pi$. This path traverses the circle in the counterclockwise direction. Then

$$\text{flux} = \int_C M dy - N dx = \int_{t=0}^{2\pi} (1.5 + \cos(t))(\cos(t)\, dt) - (1.5 + \sin(t))(-\sin(t)\, dt)$$

$$= \int_{t=0}^{2\pi} \left(1.5\cos(t) + \cos^2(t) + 1.5\sin(t) + \sin^2(t)\right) dt = 2\pi$$

The net outward flow across the circle is positive – more is leaving than is entering the circular region.

Practice 6: What flux do you expect for this curve C if the field is reversed to become $\mathbf{F}(x,y) = \langle -x,-y \rangle$?

Calculate the flux over this C for the field.

If the simple closed curve C encloses a source of water, then the flux across C will be positive and equal to the rate of water input from the source. If C contains a sink (a drain), then the flux across C will be negative. If C contains both sources and sinks, then the flux across C will be the signed sum of the sources (counted as positive) and the sinks (counted as negative).

Wrap up

This section has examined line integrals and a variety of their applications, but there are really only two mathematical situations: when the function is scalar-valued and when the function is vector-valued.

If f is a **scalar-valued** function then:

$$\int_C \mathbf{f}\, ds = \int_{t=a}^{b} \mathbf{f}(\mathbf{r}(t)) \cdot \left|\frac{dr}{dt}\right| dt = \int_{t=a}^{b} \mathbf{f}(\mathbf{r}(t)) \cdot |\mathbf{r}'(t)|\, dt$$

$$\text{area} = \int_C f(\mathbf{r}(t))ds = \int_{t=a}^{b} f(\mathbf{r}(t)) \cdot |\mathbf{r}'(t)|\, dt$$

$$\text{mass} = \int_C \delta(\mathbf{r}(t))ds = \int_{t=a}^{b} \delta(\mathbf{r}(t)) \cdot |\mathbf{r}'(t)|\, dt$$

If $\mathbf{F}$ is a **vector-valued** function then:

$$\int_C \mathbf{F}\bullet\mathbf{T}\,ds = \int_C \mathbf{F}(\mathbf{r}(t))\bullet d\mathbf{r} = \int_{t=a}^{b} \mathbf{F}(\mathbf{r}(t))\bullet\mathbf{r}'(t)\,dt = \int_C (Mdx + Ndy + Pdz)$$

$$\text{work} = \int_C \mathbf{F}\bullet\mathbf{T}\;ds$$

$$\text{flow} = \int_C \mathbf{F}\bullet\mathbf{T}\;ds$$

$$\text{flux} = \int_C \mathbf{F}\bullet\mathbf{n}\,ds = \int_C (Mdy - Ndx) \quad \text{(C a simple, closed 2D curve oriented counterclockwise)}$$

Later in this chapter we will consider 3D electric and magnetic vector fields and their flows and fluxes.

Problems

1. Determine the area between the curve $\mathbf{r}(t)=\langle 2t+1, 3+t^2\rangle$ for t from 0 to 3 in the xy-plane and $f(x,y)=x^2-4y+11$.
2. Determine the area between the curve $\mathbf{r}(t)=\langle \sin(t), 1+\cos(t)\rangle$ for t from 0 to π in the xy-plane and $f(x,y)=2+xy$
3. Create an integral to calculate the area between $r(t)=\langle x(t),\ y(t),\ z(y)\rangle$ $(x,y,z\geq 0)$ and the xz-plane.
4. Create an integral to calculate the area between $r(t)=\langle x(t),\ y(t),\ z(y)\rangle$ $(x,y,z\geq 0)$ and the yz-plane.
5. A pipe is parameterized by $\mathbf{r}(t)=\langle 3\cos(t),\ 3\sin(t)\rangle$ for $0\leq t\leq \pi/2$ and the density of the pipe is $\delta(x,y)=1+x+2y$ at location (x,y). Find the total mass of the pipe.
6. Find the mass of the pipe in Problem 5 if $\delta(x,y)=1+3x$.
7. A pipe is parameterized by $\mathbf{r}(t)=\langle 1+t,\ 3t,\ 2+t\rangle$ for $0\leq t\leq 2$ and has density $\delta(x,y,z)=y+z$ at (x,y,z). Find the mass of the pipe.
8. Find the mass of the pipe in Problem 7 if $\delta(x,y,z)=x+2y+3z$.

In problems 9 to 14, evaluate $\int_C f\,ds$ for the given function f on the curve C.

9. $f(x,y)=2x+y$ on C given by $\mathbf{r}(t)=\langle 3t+2, 5-4t\rangle$ for $0\leq t\leq 2$.
10. $f(x,y)=x-2y$ on C given by $\mathbf{r}(t)=\langle 12t+1, 5t-4\rangle$ for $1\leq t\leq 4$.
11. $f(x,y)=x^2y+y$ on C given by $\mathbf{r}(t)=\langle 3, t^2+1\rangle$ for $0\leq t\leq 3$.
12. $f(x,y)=xy$ on C given by $\mathbf{r}(t)=\langle 4t^2, 3t^2+3\rangle$ for $0\leq t\leq 1$.
13. $f(x,y)=x+y$ on C given by $\mathbf{r}(t)=\langle 2-\sin(t), 1+\cos(t)\rangle$ for $0\leq t\leq \pi$.
14. $f(x,y)=x-2y$ on C given by $\mathbf{r}(t)=\langle 2+5\cos(t), 1+2\cos(t)\rangle$ for $0\leq t\leq 2\pi$.

15. In Fig. 13 is the work along each path A and B positive, negative or zero?
16. In Fig. 13 is the work along each path C and D positive, negative or zero?
17. In Fig. 14 is the flow along each path A and B positive, negative or zero?
18. In Fig. 14 is the flow along each path C and D positive, negative or zero?

Fig. 13

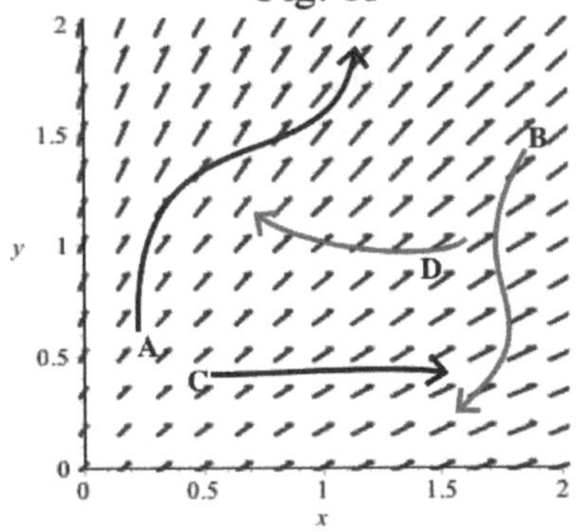

19. Calculate the work to move an object along the path $\mathbf{r}(t) = \langle 2+3t, 4t \rangle$ for $0 \le t \le 3$ in the field $\mathbf{F} = \langle x, x+y \rangle$.
20. Calculate the work to move an object along the path $\mathbf{r}(t) = \langle t^2, t \rangle$ for $0 \le t \le 2$ in the field $\mathbf{F} = \langle -y, x \rangle$
21. Calculate the work to move an object along the path $\mathbf{r}(t) = \langle \cos(t), t, \sin(t) \rangle$ for $0 \le t \le \pi$ in the field $\mathbf{F} = \langle 1,2,3 \rangle$.
22. Calculate the work to move an object along the path $\mathbf{r}(t) = \langle \cos(t), t, \sin(t) \rangle$ for $0 \le t \le \pi$ in the field $\mathbf{F} = \langle x,y,z \rangle$.

Fig. 14

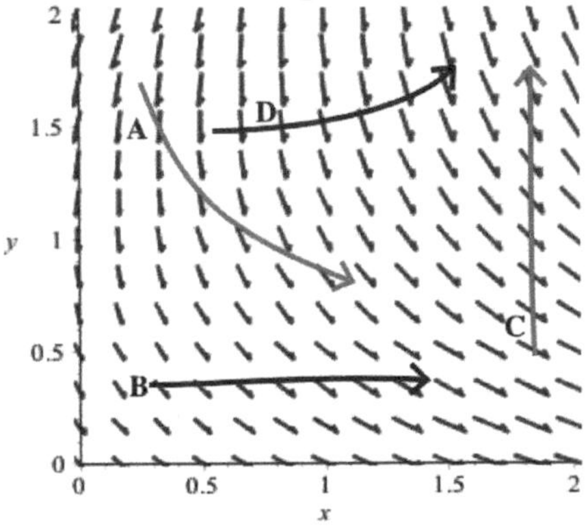

23. Calculate the flow along the path $\mathbf{r}(t) = \langle t, 4t, t^2 \rangle$ for $1 \le t \le 2$ in the field $\mathbf{F} = \langle z, 2y, x \rangle$.
24. Calculate the flow along the path $\mathbf{r}(t) = \langle t^2, 3+t \rangle$ for $0 \le t \le 5$ in the field $\mathbf{F} = \langle -y, 2x \rangle$.
25. Calculate the flux around the closed path $\mathbf{r}(t) = \langle \cos(t), 2 \cdot \sin(t) \rangle$ for $0 \le t \le 2\pi$ in the field $\mathbf{F} = \langle 2x, 1+2y \rangle$.
26. Calculate the flux around the closed path $\mathbf{r}(t) = \langle \cos(t), 2 \cdot \sin(t) \rangle$ for $0 \le t \le 2\pi$ in the field $\mathbf{F} = \langle 3x, 1+2y \rangle$.
27. If the circulation is 0 around a closed path in a vector field **F**, can the flux be positive, negative, zero? (Think about the unit circle path in a radial vector field.)
28. If the work along a path C in a vector field is positive, what can be said about the flow along that path?

In problems 29 to 34, assume that the orientation of the path is reversed. Is the original value changed or not?

29. What happens to the area? Why?
30. What happens to the mass? Why?
31. What happens to the work? Why?
32. What happens to the flow? Why?
33. What happens to the circulation? Why?

34. What happens to the flux? Why?

Practice Answers

Practice 1: $$\textbf{area} = \int_{t=-\pi/2}^{\pi/2} (2+\cos(5t))\cdot\sqrt{(-2\sin(t))^2+(2\cos(t))^2}\cdot dt = \int_{t=-\pi/2}^{\pi/2} ((2+\cos(5t))\cdot 2\cdot dt$$

$$= 2\left(2t+\frac{1}{5}\sin(5t)\right)\Big|_{-\pi/2}^{\pi/2} = 4\pi+\frac{4}{5}\approx 13.37$$

Practice 2: C is parameterized by $\mathbf{r}(t)=\langle(1-t)\cdot 0+t\cdot 4,(1-t)\cdot 1+t\cdot 2\rangle=\langle 4t,1+t\rangle$ for t=0 to t=1. Then $|\mathbf{r}'(t)|=\sqrt{17}$ and $f(\mathbf{r}(t))=2(4t)+(1+t)=9t+1$ so

$\int_C f\,ds = \int_{t=0}^{1} f(\mathbf{r}(t))\left|\frac{dr}{dt}\right| dt = \int_{t=0}^{1} (9t+1)(\sqrt{17})\,dt = \frac{11}{2}\sqrt{17}$. The units are $(\$)\left(\frac{\text{feet}}{\text{min}}\right)(\text{min}) = \$\cdot\text{feet}$

Practice 3: total mass $= \int_{t=1}^{4} 3t\cdot\sqrt{5+4t^2}\,dt = \frac{1}{4}\left(5+4t^2\right)^{3/2}\Big|_1^4 = \frac{69}{4}\sqrt{73}-\frac{27}{4}\approx 136.54$

Practice 4: $\mathbf{F}(\mathbf{r}(t))\bullet\mathbf{r}'(t)=\langle 3,2,1\rangle\bullet\langle 1,\ 2t,\ 3t^2\rangle = 3+4t+3t^2$ and work $= \int_{t=0}^{2}(3+4t+3t^2)\,dt = 22$

Practice 5: C and the vector fields are shown in Fig. P5. The line integral is negative for F1, positive for F2, and 0 for F3 (by symmetry) and F4.

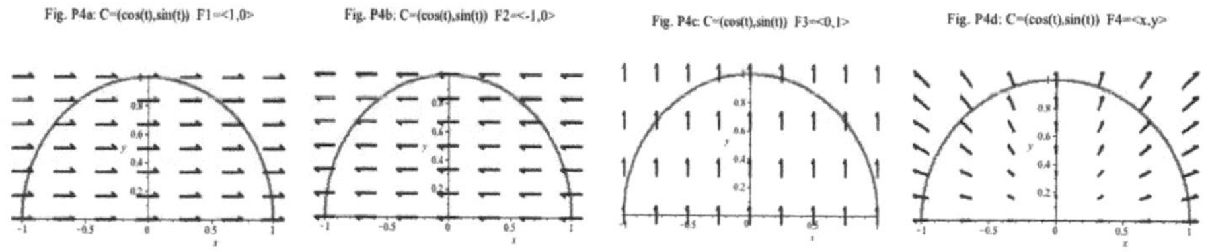

Fig. P5

Practice 6: M=-(1.5+cos(t)), N=-(1.5+sin(t)), dx=-sin(t) dt, dy=cos(t) dt, $0\le t\le 2\pi$.

$$\text{flux} = \int_C M\,dy - N\,dx = \int_{t=0}^{2\pi} -(1.5+\cos(t))(\cos(t)\,dt)+(1.5+\sin(t))(-\sin(t)\,dt)$$

$= -\int_{t=0}^{2\pi}\left(1.5\cos(t)+\cos^2(t)+1.5\sin(t)+\sin^2(t)\right)dt = -2\pi$, the opposite of the Example 4 value.

15.4 The Fundamental Theorem of Line Integrals and Potential Functions

Section 15.2 introduced the line integral along a curve in a vector field F, $\int_C \mathbf{F} \bullet \mathbf{T} \, ds$, discussed some applications of these integrals and showed how to calculate them. But something curious is going on.

Let $\mathbf{F}(x,y) = \langle x, y + x \rangle$. If we take three different paths from (0,0) to (2,4) in this vector field we get three different values. Along the curve C_1 given by $\mathbf{r}_1(t) = \langle 2t, 4t \rangle$

$$\int_{C_1} \mathbf{F} \bullet \mathbf{T} \, ds = \int_{C_1} \mathbf{F}(\mathbf{r}_1(t)) \bullet \mathbf{r}'_1(t) \, dt = \int_{t=0}^{1} \langle 2t, 4t + 2t \rangle \bullet \langle 2, 4 \rangle \, dt = \int_{t=0}^{1} 28t \, dt = 14 .$$

But along the curve C_2 given by $\mathbf{r}_2(t) = \langle 2t, 4t^2 \rangle$, $\int_{C_2} \mathbf{F} \bullet \mathbf{T} \, ds = 46/3$.

And along the curve C_3 given by $\mathbf{r}_3(t) = \langle 2t, 4t^5 \rangle$, $\int_{C_3} \mathbf{F} \bullet \mathbf{T} \, ds = 50/3$.

In this vector field the work to move an object from (0,0) to (2,4) depends on the path of the object.

However, if we change the vector field slightly, to $\mathbf{F}(x,y) = \langle x, y \rangle$, and calculate the work along the same three paths from (0,0) to (2, 4), the results are always the same:

$$\int_{C_1} \mathbf{F} \bullet \mathbf{T} \, ds = \int_{t=0}^{1} \langle 2t, 4t \rangle \bullet \langle 2, 4 \rangle \, dt = \int_{t=0}^{1} 20t \, dt = 10 ,$$

$$\int_{C_2} \mathbf{F} \bullet \mathbf{T} \, ds = \int_{t=0}^{1} \langle 2t, 4t^2 \rangle \bullet \langle 2, 8t \rangle \, dt = \int_{t=0}^{1} 32t^3 + 4t \, dt = 10 ,$$

and $$\int_{C_3} \mathbf{F} \bullet \mathbf{T} \, ds = \int_{t=0}^{1} \langle 2t, 4t^5 \rangle \bullet \langle 2, 20t^4 \rangle \, dt = \int_{t=0}^{1} 80t^9 + 4t \, dt = 10 .$$

And the result will be 10 no matter what other smooth paths we take from (0, 0) to (2, 4) in this field.

This field $\mathbf{F}(x,y) = \langle x, y \rangle$ is a special type of vector field called a **gradient field** or a **potential field** or a **conservative field**, and it has the wonderful property that the value of the line integral does not depend on the path. This property is called **path independence**.

Definition: $\int_C \mathbf{F} \bullet \mathbf{T} \, ds$ **is independent of path** if $\int_{C_1} \mathbf{F} \bullet \mathbf{T} \, ds = \int_{C_2} \mathbf{F} \bullet \mathbf{T} \, ds$ for any two paths C_1 and C_2 with the same initial and ending points.

Conservative Fields

But what are the conservative vector fields and how can we determine if a given field is conservative?

Definitions: Conservative Field and Potential Function

A vector field **F** is called a **conservative** (or gradient, or potential) **field** on a region R in 2D or 3D if there is a scalar function f on R so that $\nabla \mathbf{f} = \mathbf{F}$.

This scalar function f is called the **potential function** for the conservative field **F** .

Example 1: Each of these scalar-valued functions generates a conservative vector field. Determine the vector field

$$\mathbf{F} \text{ for } f_1(x,y) = xy \ , \ f_2(x,y) = \sqrt{x^2 + y^2} \text{ and } f_3(x,y,z) = x^2 y + 2yz \ .$$

Solution: All that is needed is the gradient of each function: $\mathbf{F}_1(x,y) = \langle y, x \rangle$,

$$\mathbf{F}_2(x,y) = \left\langle \frac{x}{\sqrt{x^2+y^2}}, \frac{y}{\sqrt{x^2+y^2}} \right\rangle \text{ and } \mathbf{F}_3(x,y,z) = \left\langle 2xy, x^2 + 2z, 2y \right\rangle.$$

Each of these is a conservative vector field since each $\mathbf{F} = \nabla \mathbf{f}$ for a differentiable function f.

Practice 1: Determine the conservative vector fields generated by $f_1(x,y) = 3x - 2y$, $f_2(x,y) = \sin(xy)$ and

$$f_3(x,y,z) = \frac{k}{x^2 + y^2 + z^2} \ .$$

Fundamental Theorem of Line Integrals

If there is a scalar function f on an open, connected region R in 2D or 3D so that $\nabla \mathbf{f} = \mathbf{F}$ (**F** is a conservative field), and C is any piecewise smooth curve in R from point A (when t=a) to point B (when t=b)

then $$\int_C \mathbf{F} \bullet \mathbf{T} \, ds = \int_{t=a}^{b} \mathbf{F}(\mathbf{r}(t)) \bullet \mathbf{r}'(t) \, dt = \int_{t=a}^{b} \mathbf{F} \bullet d\mathbf{r} = f(B) - f(A)$$

and the line integral does not depend on the path C (path independence).

This is considered one of the four fundamental theorems of vector calculus (with Green's and Stokes's and the Divergence theorems). The meanings of "open," and "connected" are given in the chapter Appendix.

Proof: Suppose $\mathbf{F}(x,y,z) = \nabla \mathbf{f}(x,y,z) = \mathbf{f_x}(x,y,z)\mathbf{i} + \mathbf{f_y}(x,y,z)\mathbf{j} + \mathbf{f_z}(x,y,z)\mathbf{k}$ for some smooth scalar-valued function f. Then

$$\mathbf{F} \bullet \mathbf{r}' = \nabla \mathbf{f} \bullet \mathbf{r}' = \left\langle \frac{\partial \mathbf{f}}{\partial \mathbf{x}}, \frac{\partial \mathbf{f}}{\partial \mathbf{y}}, \frac{\partial \mathbf{f}}{\partial \mathbf{z}} \right\rangle \bullet \left\langle \frac{dx}{dt}, \frac{dy}{dt}, \frac{dz}{dt} \right\rangle = \frac{\partial \mathbf{f}}{\partial \mathbf{x}} \cdot \frac{dx}{dt} + \frac{\partial \mathbf{f}}{\partial \mathbf{y}} \cdot \frac{dy}{dt} + \frac{\partial \mathbf{f}}{\partial \mathbf{x}} \cdot \frac{dz}{dt} = \frac{df}{dt} \quad \text{(by the Chain Rule)}$$

so $\int_{t=a}^{b} \mathbf{F}(\mathbf{r}(t)) \bullet \mathbf{r}'(t)\, dt = \int_{t=a}^{b} \frac{df(\mathbf{r}(t))}{dt}\, dt = f(\mathbf{r}(b)) - f(\mathbf{r}(a)) = f(B) - f(A)$.

Example 2: $\mathbf{F}(x,y) = \langle y, x \rangle$. Evaluate $\int_C \mathbf{F} \bullet \mathbf{T}\, ds$ for a curve C that starts at A=(1,2) and ends at B=(4,3).

Solution: If we recognize (from Example 1) that $\mathbf{F} = \nabla \mathbf{f}$ for f(x,y)=xy then by the Fundamental Theorem of Line Integrals, $\int_C \mathbf{F} \bullet \mathbf{T}\, ds = f(B) - f(A) = f(4,3) - f(1,2) = 12 - 2 = 10$

along any smooth path C from A to B. You might want to check this value by using the path $\mathbf{r}(t) = \langle 1 + 3t, 2 + t \rangle$ for t from 0 to 1 and explicitly calculating $\int_{t=0}^{1} \mathbf{F}(\mathbf{r}(t)) \bullet \mathbf{r}'(t)\, dt$.

If we know a scalar function f whose gradient is the vector function $\mathbf{F}$, then a difficult calculus problem becomes an easy arithmetic problem. You should recognize the similarity of this calculation with the way integrals were evaluated in beginning calculus. In beginning calculus if f was an antiderivative of F (Df=F), then $\int_a^b F(x)\, dx = f(b) - f(a)$. Here if f is a potential function of $\mathbf{F}$ ($\nabla \mathbf{f} = \mathbf{F}$), then $\int_{t=a}^{b} \mathbf{F} \bullet d\mathbf{r} = f(B) - f(A)$.

Practice 2: $\mathbf{F}(x,y,z) = \langle 2xy, x^2 + 2z, 2y \rangle$. Evaluate $\int_C \mathbf{F} \bullet \mathbf{T}\, ds$ for the curve C that starts at A=(1,0,4) and ends at B=(4,3,2). (Suggestion: Look at the answers in Example 1.)

Example 3: Fig. 1 shows the level curves for a smooth function z=f(x,y) and the vector field $\mathbf{F} = \nabla \mathbf{f}$. Calculate the amount of work done moving an object from point A to point B along the two given paths.

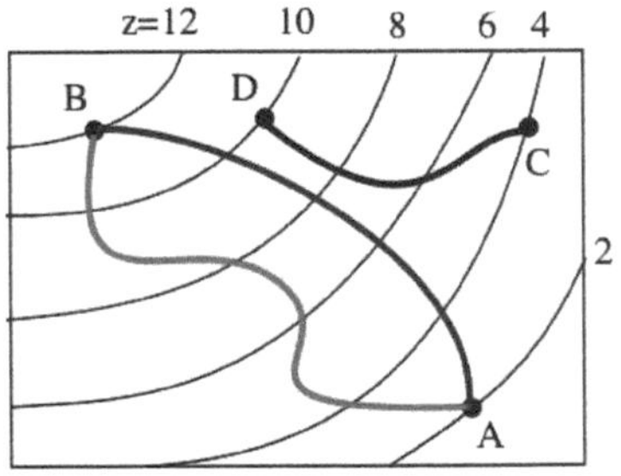

Fig. 1: Level curves of z=f(x,y)

Solution: work = $\int_C \mathbf{F} \bullet \mathbf{T}\, ds = f(B) - f(A) = 10$ along each path.

Practice 3: Using the same figure, calculate the amount of work done moving an object from point C to point D along the given curve. What is the total work to go from C to D and then back to C?

Theorem: $\int_C \mathbf{F} \bullet \mathbf{T}\, ds$ is **independent of path** on open, simply-connected region R

if and only if $\int_C \mathbf{F} \bullet \mathbf{T}\, ds = 0$ for every closed path C in R.

Finding potential functions

Finding potential functions is a lot like finding antiderivatives but now we need to match two antiderivatives. If $\mathbf{F} = \langle M(x,y),\ N(x,y) \rangle$, then the potential function f must satisfy both $f_x = \frac{\partial f}{\partial x} = M$ and $f_y = \frac{\partial f}{\partial y} = N$. so typically two antiderivatives are needed, one with respect to x and one with respect to y.

Example 4: Find a potential function f for $\mathbf{F} = \langle 4y,\ 4x + 5 \rangle$.

Solution: $\frac{\partial f}{\partial x} = 4y$ so (taking an antiderivative with respect to x) f(x,y)=4xy + g(y) (since $\frac{\partial g(y)}{\partial x} = 0$).

But $\frac{\partial}{\partial y}(4xy + g(y)) = 4x + g'(y)$ must equal N=4x+5, so $g'(y) = 5$ and $g(y) = 5y$ (typically the "+k" is omitted). Then $f(x,y) = 4xy + 5y$. A quick check shows that $f_x = 4y$ and $f_y = 4x + 5$ so $\nabla \mathbf{f} = \mathbf{F}$. .

Note: We could have first taken the antiderivative of N=4x+5 with respect to y.

Practice 4: Find a potential functions f for $\mathbf{F} = \left\langle y \cdot e^{xy} + 2,\ x \cdot e^{xy} + 3 \right\rangle$ and $\mathbf{F} = \left\langle 2xy + \cos(x),\ x^2 \right\rangle$.

But the vector field $\mathbf{F}(x,y) = \langle x,\ y + x \rangle$ at the beginning of this section was not path independent so it is not a conservative field and does not have a potential function. There is a very easy way to determine if a vector field has a potential function and is conservative.

Theorem: Test for a conservative field

If R is an open, connected, and simply-connected region then

2D: $\mathbf{F} = \langle M(x,y),\ N(x,y) \rangle$ is a conservative field on R if and only if $\frac{\partial M}{\partial y} = \frac{\partial N}{\partial x}$.

3D: $\mathbf{F} = \langle M(x,y),\ N(x,y),\ P(x,y) \rangle$ is a conservative field on R if and only if

$$\frac{\partial M}{\partial y} = \frac{\partial N}{\partial x}, \quad \frac{\partial M}{\partial z} = \frac{\partial P}{\partial x} \quad \text{and} \quad \frac{\partial N}{\partial z} = \frac{\partial P}{\partial y}.$$

Proof: These follow from Clairaut's Theorem which says that mixed partial derivatives are equal: $f_{xy} = f_{yx}$.

If has a potential function f, then $f_{xy} = \frac{\partial}{\partial y}\left(\frac{\partial f}{\partial x}\right) = \frac{\partial M}{\partial y}$ and $f_{yx} = \frac{\partial}{\partial x}\left(\frac{\partial f}{\partial y}\right) = \frac{\partial N}{\partial y}$ so $\frac{\partial M}{\partial y} = \frac{\partial N}{\partial x}$.

The proof for the 3D case follows from Clairaut's Theorem that $f_{xy} = f_{yx}$, $f_{xz} = f_{zx}$ and $f_{yz} = f_{zy}$.

The proof in 2D that $M_y = N_x$ implies that $\mathbf{F} = \langle M,\ N \rangle$ is conservative requires Green's Theorem which appears later.

For the vector field $\mathbf{F}(x,y) = \langle x, y+x \rangle$, $\frac{\partial M}{\partial y} = 0$ and $\frac{\partial N}{\partial x} = 1$ so the field does not have a potential function.

Practice 5: Which of these fields are conservative: $\mathbf{F_1} = \langle 2xy^2+3, 2x^2y \rangle$, $\mathbf{F_2} = \langle -y, x \rangle$, $\mathbf{F_3} = \langle yz, xz, xy+2x \rangle$ and $\mathbf{F_4} = \langle yz, xz, xy+2 \rangle$?

Potential or Gradient fields are also called Conservative fields because the total energy, kinetic plus potential, is constant at each point in the field, energy is conserved. The derivation of this result is shown in an Appendix after the Practice Answers.

Summary of results

If R is an open, connected, simply-connected region, then the following are equivalent:

(1) $\mathbf{F}$ is a conservative field.

(2) There is a potential function f so that $\mathbf{F} = \nabla f$ . .

(3) $\int_C \mathbf{F} \bullet \mathbf{T}\, ds = \int_{t=a}^{b} \mathbf{F} \bullet \mathbf{dr} = f(B) - f(A)$ for all points A and B in R and for all smooth curves C.

(4) $\int_C \mathbf{F} \bullet \mathbf{T}\, ds = \int_{t=a}^{b} \mathbf{F} \bullet \mathbf{dr} = 0$ for all simple, closed, smooth curves in R.

Problems

In problems 1 to 6, determine the conservative field generated by the given potential function.

1. $f(x,y) = x^2 + 3y^2$
2. $f(x,y) = 3x^2 - 4y^2$
3. $f(x,y) = \sin(3x+2y)$
4. $f(x,y) = x^2y^3 + xy^2$
5. $f(x,y) = \ln(2x+5y) + e^y$
6. $f(x,y) = \tan(x) - \sec(y)$

In problems 7 to 12, determine the work to move an object in the given field from point A to point B.

7. $\mathbf{F} = \langle 2x, 2y \rangle$, $A = (1,2)$, $B = (5,1)$
8. $\mathbf{F} = \langle y, x \rangle$, $A = (1,3)$, $B = (3,5)$
9. $\mathbf{F} = \langle x, x \rangle$, $A = (0,2)$, $B = (3,6)$
10. $\mathbf{F} = \langle 3x^2y, 3x^2 \rangle$, $A = (1,0)$, $B = (3,1)$
11. $\mathbf{F} = \langle yz, xz, xy \rangle$, $A = (1,0,0)$, $B = (4,2,1)$
12. $\mathbf{F} = \langle -x,-y,-z \rangle$, $A = (0,0,0)$, $B = (2,4,6)$

In problems 13 to 16, determine the work to move the object along the given paths from A to B and from C to D. Each figure shows the level curves of z=f(x,y) for some smooth function f.

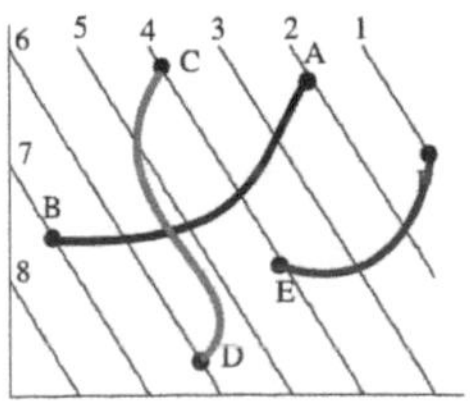

Problem 13

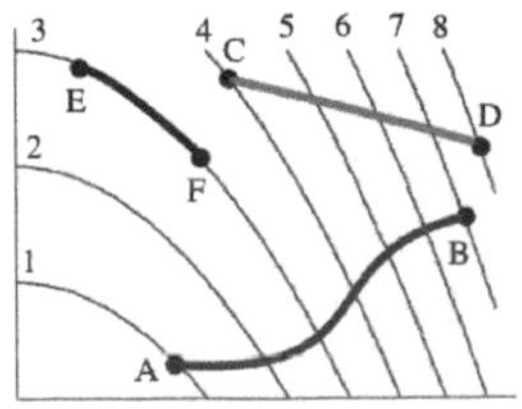

Problem 14

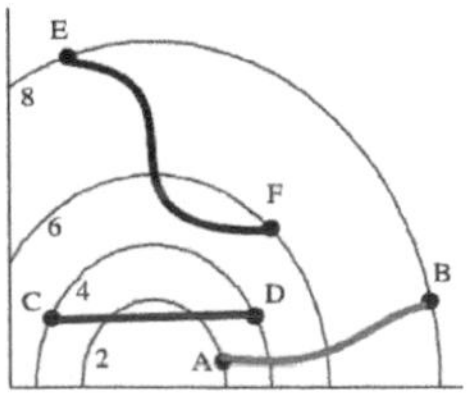

Problem 15

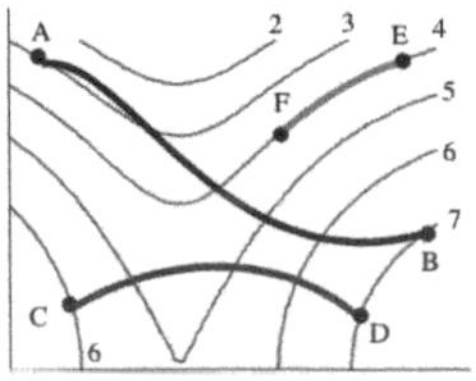

Problem 16

In problems 17 to 24, find a potential function for the given vector field F or show that F does not have a potential function.

17. $\langle 3x^2 + 4,\ 6 \rangle$

18. $\left\langle \frac{2}{2x+3y},\ \frac{3}{2x+3y} + 3y^2 \right\rangle$

19. $\langle y \cdot \cos(xy) + 3x^2y,\ x \cdot \cos(xy) + x^3 \rangle$

20. $\langle xy^3 + 3x^2y,\ 3xy^2 + x^2 \rangle$

21. $\langle \sin(y),\ x \cdot \cos(y) \rangle$

22. $\langle 2y + 3z,\ 2x,\ 3x \rangle$

23. $\langle yz \cdot \cos(xyz),\ xz \cdot \cos(xyz) + z,\ xy \cdot \cos(xyz) \rangle$

24. $\langle y \cdot \cos(xy),\ x \cdot \cos(xy) + z,\ y \rangle$

25. (a) For f(x,y)=arctan(y/x) show that $\nabla f = \left\langle \frac{-y}{x^2+y^2},\ \frac{x}{x^2+y^2} \right\rangle = \langle M, N \rangle$.

(b) Show that $M_y = N_x$.

(c) Show that for the closed circle $\mathbf{r}(t) = (\cos(t), \sin(t))$ for $0 \le t \le 2\pi$, $\int_C \mathbf{F} \bullet \mathbf{T}\ ds = \int_{t=a}^{b} \mathbf{F} \bullet \mathbf{dr} = 2\pi$.

(d) Why do these three results not violate the equivalent statements in the Summary?

Practice Answers

Practice 1: $\mathbf{F}_1(x,y) = \langle 3,-2 \rangle$, $\mathbf{F}_2(x,y) = \langle y \cdot \cos(xy),\ x \cdot \cos(xy) \rangle$ and

$$\mathbf{F}_3(x,y,z) = \left\langle \frac{-2kx}{(x^2+y^2+z^2)^2},\ \frac{-2ky}{(x^2+y^2+z^2)^2},\ \frac{-2kz}{(x^2+y^2+z^2)^2} \right\rangle = \frac{2k}{(x^2+y^2+z^2)^2}\langle -x,\ -y,\ -z \rangle$$

Note: $\mathbf{F}_3$ is a radial field with $|\mathbf{F}_3| = |2k|$ and it always points toward the origin.

Practice 2: From Example 1, $\mathbf{F} = \nabla \mathbf{f}$ for $f(x,y) = x^2y + 2yz$ so $\int_C \mathbf{F} \bullet \mathbf{T}\ ds = f(4,3,2) - f(1,0,4) = 72$.

Practice 3: work $= \int_C \mathbf{F} \bullet \mathbf{T}\ ds = f(D) - f(C) = 6$ along the given path, and, in fact, along every piecewise smooth path from C to D.

The work to go from C to D and then back to C is 0: the work from C to D is f(D)-f(C), and the work from D to C is f(C)-f(D) so the total work is {f(D)-f(C)}+{f(C)-f(D)}=0.

Practice 4: $\nabla(e^{xy} + 22 + 3y) = \langle y \cdot e^{xy} + 2,\ x \cdot e^{xy} + 3 \rangle$ and $\nabla(x^2y + \sin(x)) = \langle 2xy + \cos(x),\ x^2 \rangle$

Practice 5: $\mathbf{F}_1$ and $\mathbf{F}_4$ are conservative. $\mathbf{F}_2$ and $\mathbf{F}_3$ are not conservative.

Appendix: A bit of vocabulary about Curves and Regions

simple, closed | simple, not closed | not simple, closed | not simple, not closed

Fig. A1: Curves

A curve C parameterized by r(t) for $a \le t \le b$ is **simple** if

$\mathbf{r}(t_1) \ne \mathbf{r}(t_2)$ for $a < t_1 < t_2 < b$. (Fig. A1)

A curve C parameterized by r(t) for $a \le t \le b$ is **closed** if $\mathbf{r}(a) = \mathbf{r}(b)$.

A region R is **open** if every point in the R is inside a circle that lies completely in R.

R= $\{(x,y):\ x^2 + y^2 < 1\}$ is **open**.

R= $\{(x,y):\ x^2 + y^2 \le 1\}$ is **not open** since every circle around a boundary point contains points not in R.

A region R is **connected** if any two points in R can be joined by a continuous curve that lies in R.

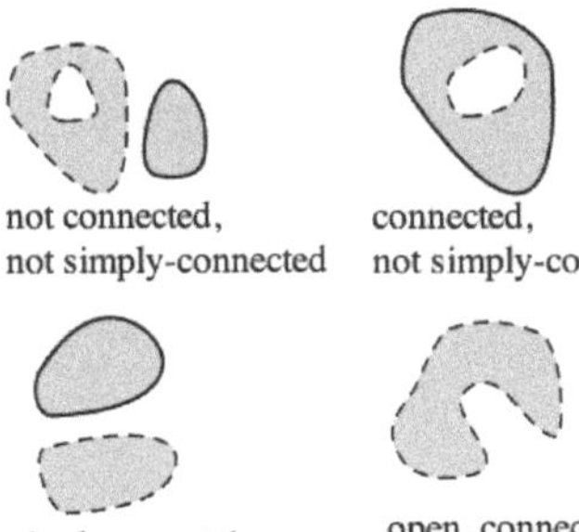

not connected, not simply-connected | connected, not simply-connected

simply-connected, not connected | open, connected, simply-connected

Fig. A3: Regions

A region R is **simply-connected** if every closed path in R can be contracted to a path without leaving R. A region with any holes (even just missing a single point) is not simply-connected.

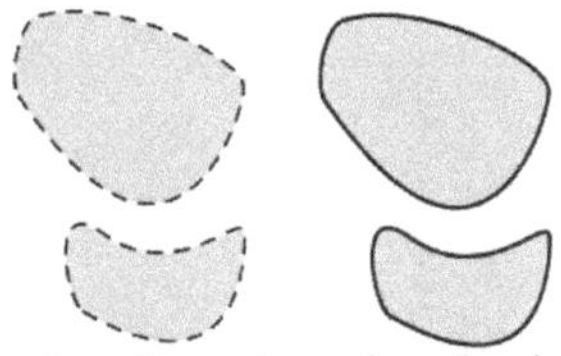

boundary not part of R | boundary is part of R

Open | Not Open

Fig. A2: Open and not open

Appendix Problems

For problems A1-A4, give examples of curves with the specified properties.

A1. closed, not simple

A2. not closed, not simple

A3. closed, simple

A4. not closed, simple

For problems A5-A10, give examples of regions with the specified properties. Give an example of a region that is

A5. open, connected and not simply-connected.

A6. open, simply-connected and not connected.

A7. connected, simply-connected, and not open.

A8. open, not connected and not simply-connected.

A9. connected, not open and not simply-connected.

A10. simply-connected, not open and not connected.

Appendix: Conservation of Energy

We need Newton's Second Law, $\mathbf{F} = m\mathbf{a}$ where $\mathbf{a}$ is the acceleration, so $\mathbf{F}(\mathbf{r}(t)) = m \cdot \mathbf{a}(t) = m \cdot \mathbf{r}''(t)$.

We also need that for any vector $\mathbf{w}$,

$$\frac{d}{dt}(\mathbf{w} \bullet \mathbf{w}) = \mathbf{w} \bullet \frac{d\mathbf{w}}{dt} + \frac{d\mathbf{w}}{dt} \bullet \mathbf{w} = 2\frac{d\mathbf{w}}{dt} \bullet \mathbf{w} \text{ so } \frac{d\mathbf{w}}{dt} \bullet \mathbf{w} = \frac{1}{2}\frac{d}{dt}(\mathbf{w} \bullet \mathbf{w}),$$

so if $\mathbf{w} = \mathbf{r}'$ then $\mathbf{r}'' \bullet \mathbf{r}' = \frac{d\mathbf{r}'}{dt} \bullet \mathbf{r}' = \frac{1}{2}\frac{d}{dt}(\mathbf{r}' \bullet \mathbf{r}')$.

Kinetic energy: work to move from A to B

$$\text{work} = \int_C \mathbf{F} \bullet d\mathbf{r} = \int_{t=a}^{b} \mathbf{F}(\mathbf{r}(t)) \bullet \mathbf{r}'(t)\, dt = \int_{t=a}^{b} m \cdot \mathbf{r}''(t) \bullet \mathbf{r}'(t)\, dt$$

$$= \int_{t=a}^{b} m \cdot \frac{1}{2}\frac{d}{dt}(\mathbf{r}' \bullet \mathbf{r}')\, dt = \frac{m}{2}\mathbf{r}' \bullet \mathbf{r}' \Big|_{t=a}^{b} = \frac{m}{2}|\mathbf{r}'(t)|^2 \Big|_{t=a}^{b} = \frac{m}{2}|\mathbf{v}(t)|^2 \Big|_{t=a}^{b} = \frac{m}{2}|\mathbf{v}(b)|^2 - \frac{m}{2}|\mathbf{v}(a)|^2$$

But kinetic energy $K = \frac{1}{2}mv^2$ so work $= K(B) - K(A)$.

Potential energy: work to move from A to B

If f is a conservative field so $\mathbf{F} = -\nabla f$ for a potential function f, then

$$\text{work} = \int_C \mathbf{F} \bullet d\mathbf{r} = \int_C -\nabla f \bullet d\mathbf{r} = (-f(\mathbf{r}(t))) \Big|_{t=a}^{b} = -f(\mathbf{r}(b)) + f(\mathbf{r}(a)) = f(A) - f(B)$$

Putting both type of energy together, $K(B)-K(A) = \text{work} = f(A) - f(B)$ so $K(B)+f(B)=K)A)+f(A)$.

The total energy, kinetic plus potential, at A is the same as at B for any two points A and B in R. In a conservative field the total energy is conserved.

Gravitational, electrical and magnetic fields are conservative fields.

15.5 Theorems of Green, Stokes and Gauss: An Introduction

The final sections of this text deal with the last three fundamental theorems of calculus, the theorems of Green, Stokes and Gauss. Each of these theorems extends the ideas of our earlier fundamental theorem of calculus to situations for vector-valued functions, and each has important applications to fields of physics and even to Maxwell's equations for magnetism and electricity. These theorems are technically sophisticated and difficult to prove, but the main ideas behind them are remarkably geometric and straightforward. The goal of this section is to approach these theorems geometrically and to illustrate why the ideas behind them are "easy and natural." The theorems will be clearly and precisely presented in the following sections and partial proofs will be given; the presentation in this section can help you understand what the theorems are saying and perhaps help you to remember them.

Introduction to Green's Theorem

Calculus deals with infinite collections of points, but sometimes a finite situation can give insight into the infinite.

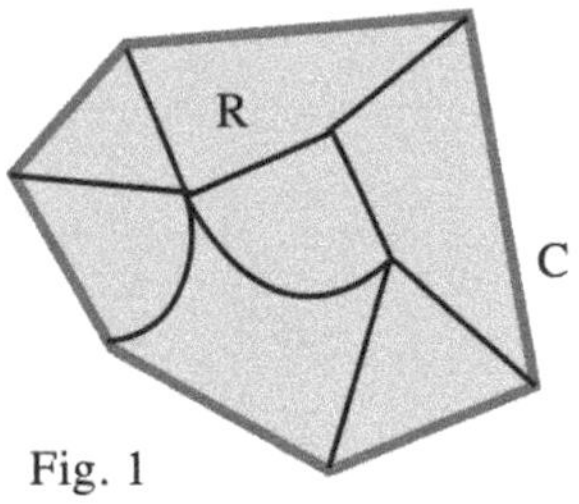

Fig. 1

In the following, R is a simple, simply-connected region consisting of a finite collection of cells. The boundary of R is a simple closed curve C (C consists of only the exterior edges of R). (Fig. 1)

Version 1 Green's Theorem: Divergence and Flux

Suppose water flows through the region R. Let's attach in-out flows to the edges of each cell. If we define the divergence of a cell to be the net outward flow of the cell then we can calculate the net outward flow of the collection of cells along the boundary of the collection – let's call this the flux across C.

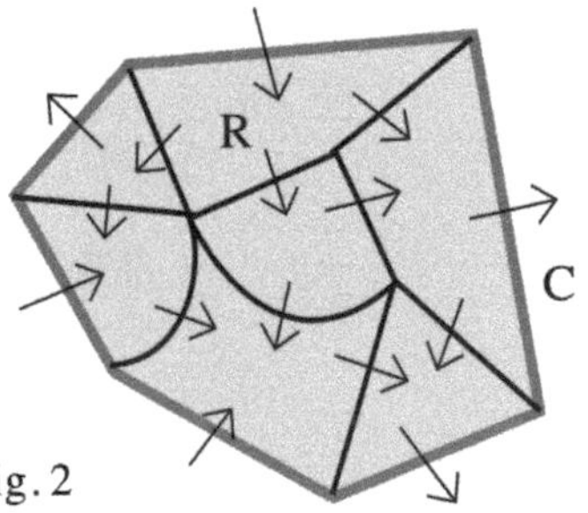

Fig. 2

We can calculate this flux in Fig. 2 in two ways. One way is to go around the boundary and add up the outward flows (counted as positive) and the inward flows (counted as negative). But if we add up the divergences for each cell in Fig. 2 we get the same net outward flow for the collection, the flux across C. This will always be true since for each inside boundary between cells, the outward flow from one cell becomes the inward flow into the next cell (Fig. 3) so the sum of those two flows will be zero, and that is the case for every shared edge inside the collection. Then the sum of all of the individual cell divergences is equal to just the sum of the flows on the outside edges (Fig. 4). This can be stated as

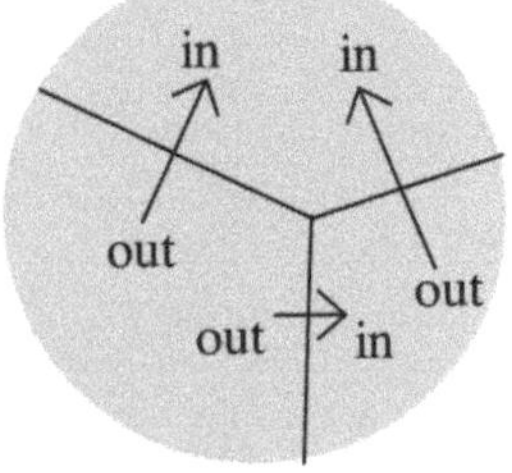

Fig. 3

In words: The sum of all of the individual cell divergences is equal to the sum of the flows on the outside edges (Fig. 4).

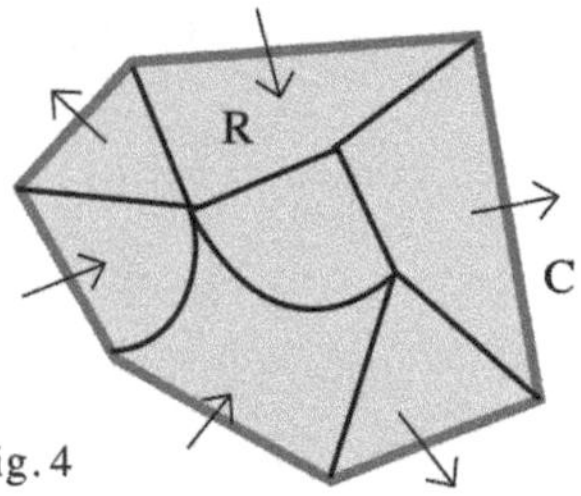

Fig. 4

Finite version (divergence-flux form Green's Theorem):

$$\sum_R \{\text{divergence of cell}\} = \sum_C \{\text{flow across each outside edge}\} = \{\text{flux across C}\}.$$

Integral version (divergence-flux form Green's Theorem): For $\mathbf{F} = \langle M(x,y), N(x,y)\rangle$

$$\iint_R \text{div}\,\mathbf{F}\ dA = \oint_{\mathbf{C}} \mathbf{F} \bullet \mathbf{n}\ ds = \text{flux across C}$$

Version 2 Green's Theorem: Curl and Circulation

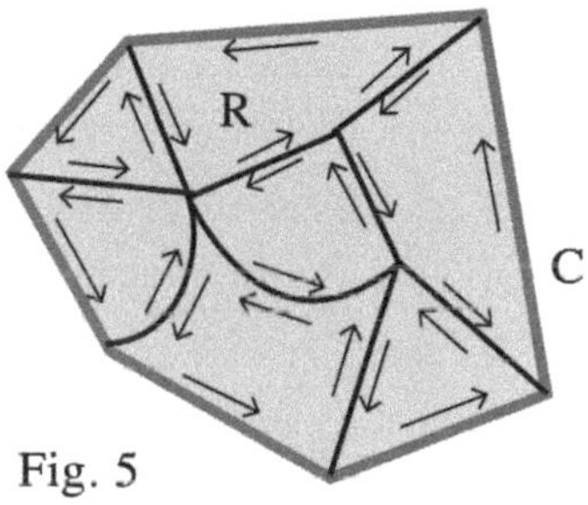

Fig. 5

Instead of looking at flow across edges of the cells, consider flow along each edge of a cell as in Fig. 5, and define the curl of each cell to be the sum of the flows around the edges of the cell counting flows in the counterclockwise direction to be positive and flows in the clockwise direction to be negative. On each inside shared edge (Fig. 6) the flow gets counted once as positive and once as negative so the sum of those two flows is 0. But this happens along every inside edge. If we add all of the curls together, the only flows that are not cancelled out in this way are the flows along the exterior edges of the collection, the flows along C (Fig. 7). This total flow around the boundary C of the collection is called the circulation.

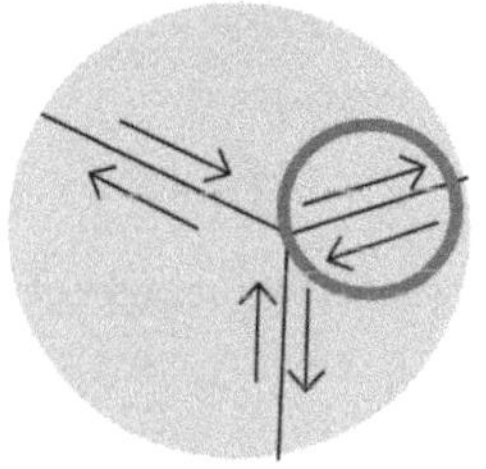
Fig. 6

In words: The circulation around the boundary C equals the sum of the circulations (curls) on the cells of R.

Finite version (Green's Theorem): $\sum_C \{\text{flow along outside edge}\} = \sum_R \text{curls dA}$

Integral version (curl-circulation form Green's Theorem): For $\mathbf{F} = \langle M(x,y), N(x,y)\rangle$

$$\oint_{\mathbf{C}} \mathbf{F} \bullet \mathbf{T}\ ds = \iint_R \text{curl}\,\mathbf{F}\ dA = \iint_R \left(\frac{\partial N}{\partial x} - \frac{\partial M}{\partial y}\right) dA$$

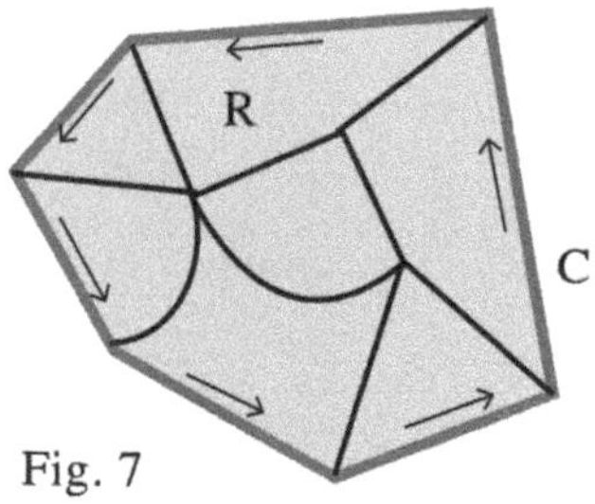

Fig. 7

If we could just take limits as all of the cells got smaller and smaller (it is not so easy) we would have both versions Green's Theorem which is discussed in Section 15.5:

$$\iint_R \left(\frac{\partial M}{\partial x} + \frac{\partial N}{\partial y}\right) = \oint_{\mathbf{C}} \mathbf{F} \bullet \mathbf{n}\ d = \int_C M\ dy - N\ dx = \text{flux}$$

$$\iint_R \left(\frac{\partial N}{\partial x} - \frac{\partial M}{\partial y}\right) = \oint_{\mathbf{C}} \mathbf{F} \bullet \mathbf{T}\ ds = \int_C M\ dx + N\ dy = \text{circulation}$$

Problems

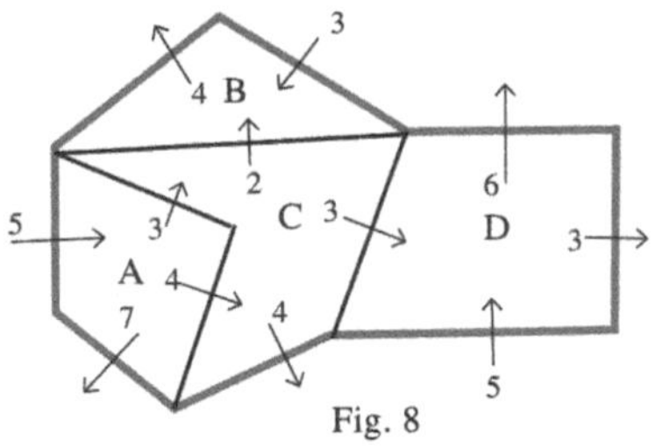

Fig. 8

1. For Fig. 8 verify that the sum of the divergences of all of the cells is equal to the flux across the boundary of the region.

2. For Fig. 9 verify that the sum of the curls of all of the cells is equal to the circulation around the boundary of the region.

Answer 1: A divergence = 9, B div = –1, C div = 2, D div = 1, so $\sum \text{div} = 11$.

Flux across the boundary = (7)+(4)+(–5)+(3)+(6)+(–3)+(4)+(–5)=11.

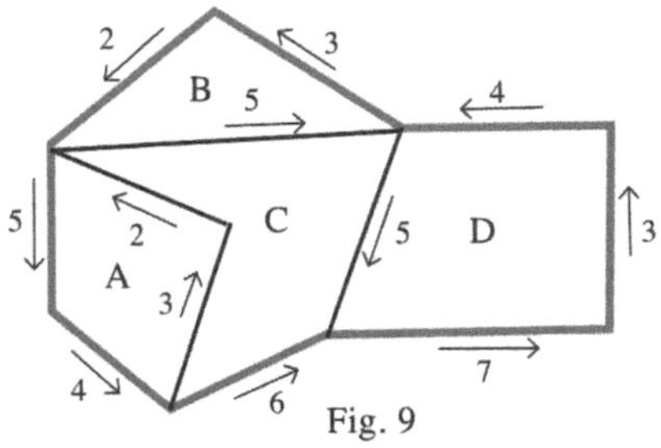

Fig. 9

Answer 2: A curl = 14, B curl = 10,

C curl =(6)+(–5)+(–5)+(–2)+(–3)= –9 , D curl = 19, so $\sum \text{curl} = 34$.

Circulation around the boundary = 34.

The following theorems of Stokes and Gauss extend Green's Theorem to higher dimensions.

Stokes' Theorem: Curl and Circulation

In Green's Theorem R was a planar region with boundary curve C (Fig. 10). Now suppose that the region R is a soap film and the boundary C is a rigid wire. If we gently blow on R to create an oriented, smooth surface with the same boundary C then each cell in the region R becomes a cell on

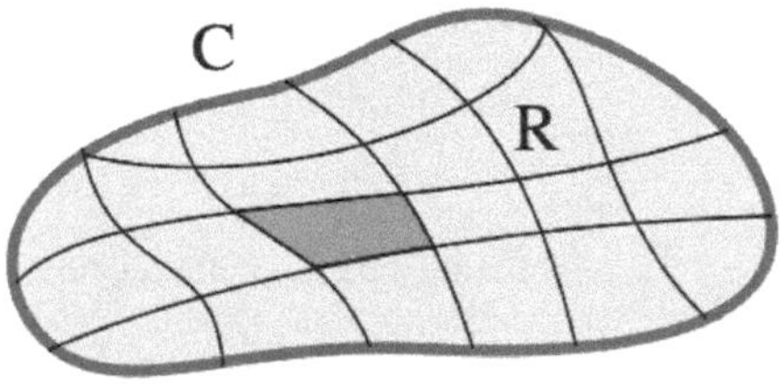

Fig. 10: Flat region R with boundary curve C

the surface S. Just like in Green's Theorem, on each inside shared edge (Fig. 11) the flow gets counted once as positive and once as negative so the sum of those two flows is 0. But this happens along every inside edge. If we add all of the curls together, the only flows that are not cancelled out in this way are the flows along the exterior edges of the collection, the flows along the boundary C. This total flow around the boundary C is called the circulation.

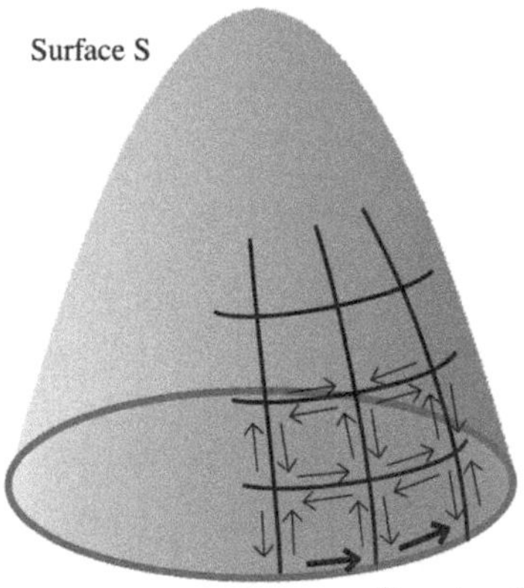

Fig. 11

In words: The circulation around the boundary C equals the sum of the circulations (curls) on the cells of surface S.

Finite version (Stokes' Theorem):

$$\sum_{C} \{\text{flow along outside edge}\} = \sum_{R} \text{curls } dA$$

Integral version (Stokes' Theorem): $$\oint_C \mathbf{F} \bullet \mathrm{T}\, ds = \oint_C \mathbf{F} \bullet \mathbf{dr} = \iint_S \text{curl } \mathbf{F} \bullet \mathbf{n}\, dS$$

Gauss/Divergence Theorem: Flux and Divergence

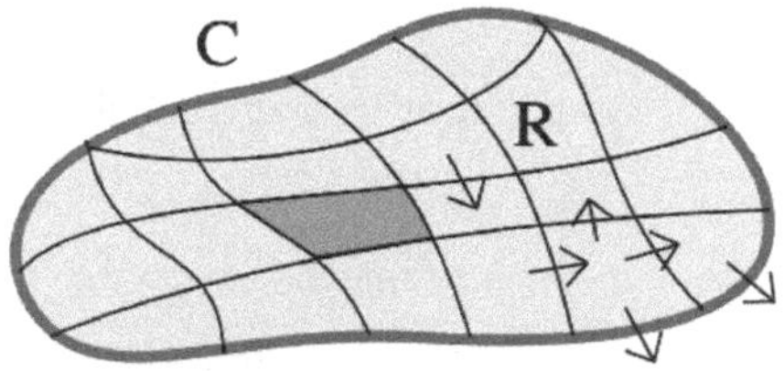

Fig. 12: Flat region R with boundary curve C

In Green's Theorem R was a planar region with boundary curve C, and the sum of the internal cell divergences was equal to the flux across the boundary C. Now suppose that instead of a region R in 2D there is a solid region E in 3D, a volume, and that the boundary (skin) of E is a surface S. Also imagine that E is partitioned into little 3D cells, and that each of these internal cells has a divergence, a net outward flow.

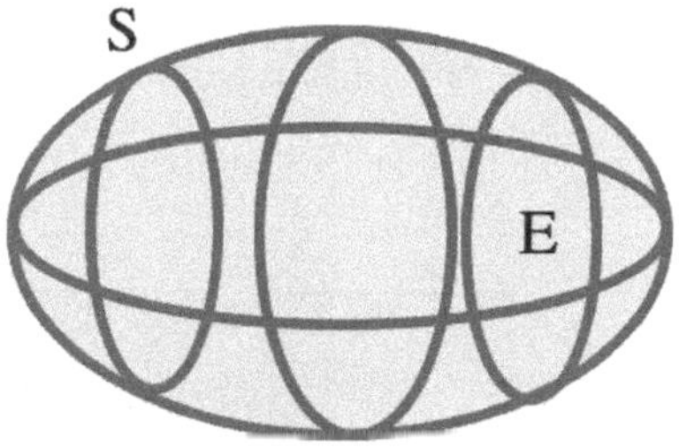

Fig. 13: E is the 3D region enclosed by the 2D boundary (skin) S

Just like in Green's Theorem, on each inside shared cell face the flow gets counted once as positive and once as negative so the sum of those two flows is 0. But this happens along every inside cell face. If we add all of the flows (divergences) for each cell together, the only flows that are not cancelled out in this way are the flows across the exterior faces of the collection, the flows across the boundary surface S. This total flow across the boundary S of the solid E is called the flux across S.

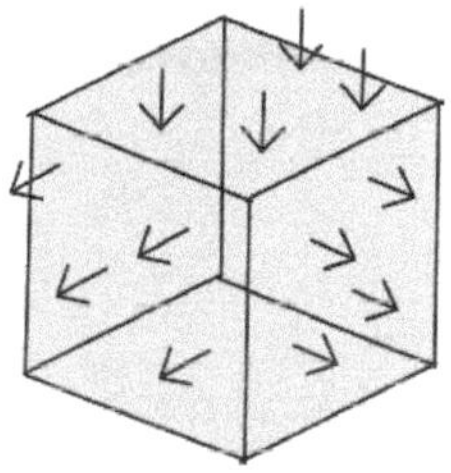
Fig. 14: cell in E

In words: The flux across the boundary S equals the sum of the divergences on the cells of solid E.

Finite version (Gauss/Divergence Theorem):

$$\text{flux across S} = \sum_{S} \mathrm{F} \bullet \mathbf{n}\, dA = \sum_{E} \text{divs } dV$$

Integral Version: $$\text{flux across S} = \iint_{S} \mathrm{F} \bullet \mathbf{n}\, dA = \iiint_{E} \text{div } \mathbf{F}\ dV$$

Wrap up

In the following sections these theorems will be more carefully presented and partially proved, and we will actually do calculations using them. These are the final big three theorems of calculus, and they are both beautiful and very useful.

15.6 Green's Theorem

Green's Theorem makes statements about the equivalence between what happens on the boundary of a 2D region with what is happening on the inside of the region. It says, in two ways, that a single integral around the boundary is equal to a double integral over the region enclosed by the boundary. In this way Green's Theorem is similar to the Fundamental Theorem of Calculus which relates the area above an interval to the values of the antiderivatives at the boundary (endpoints) of the interval.

As with some other theorems in mathematics. Green's Theorem allows us to trade one calculation for another one that may be easier. And it shows connections between ideas that do not seem to be related.

Green's Theorem:

If C is a simple, closed, and piecewise smooth curve (Fig. 1) and $\mathbf{F}(x,y) = \langle M(x,y), N(x,y)\rangle = M\mathbf{i} + N\mathbf{j}$ where M and N have continuous first partial derivatives in the region R enclosed by C,

then $\displaystyle \oint_C \mathbf{F}\bullet\mathbf{T}\, ds = \oint_C M\,dx + N\,dy = \iint_R \left(\frac{\partial N}{\partial x} - \frac{\partial M}{\partial y}\right) dA$ Circulation-Curl Form

and $\displaystyle \oint_C \mathbf{F}\bullet\mathbf{n}\, ds = \oint_C M\,dy - N\,dx = \iint_R \left(\frac{\partial M}{\partial x} + \frac{\partial N}{\partial y}\right) dA$. Flux-Divergence Form

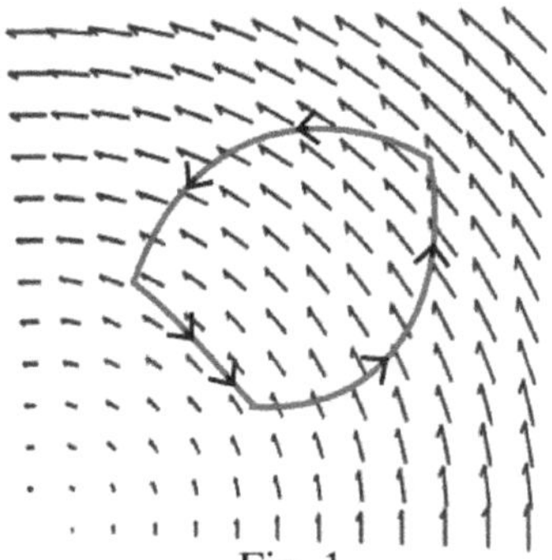
Fig. 1

Notation: The little circle on the integral sign indicates that C is a closed path parameterized in the counterclockwise direction..

The first conclusion says that the circulation is the accumulation of the interior curl. The second says that the flux is the accumulation of the interior divergence. It may help you make these connections by remembering that both circulation and curl deal with rotation while both flux and divergence deal with dissipation.

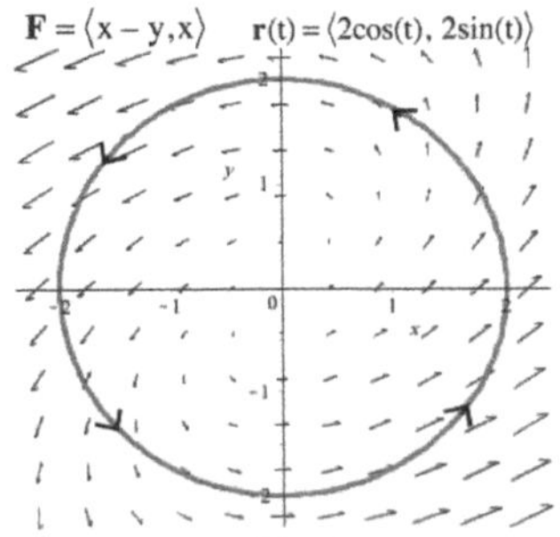

Fig. 2

A proof for simple regions is given in the Appendix of this section.

Example 1: For $\mathbf{F}=\langle x-y, x\rangle$ and $\mathbf{r}(t) = \langle 2\cdot\cos(t), 2\cdot\sin(t)\rangle$ (Fig. 2),

evaluate $\displaystyle\int_C (M dx + N dy)$ and $\displaystyle\iint_R \left(\frac{\partial N}{\partial x} - \frac{\partial M}{\partial y}\right) dA$.

Solution: First we need to write everything in terms of t:

$M = 2\cdot\cos(t) - 2\cdot\sin(t),\ \ N = 2\cdot\cos(t),\ \ dx = -2\sin(t)\ dt,\ \ dy = 2\cdot\cos(t)\ dt$. Then

$$\int_C (Mdx + Ndy) = \int_{t=0}^{2\pi} (2\cdot\cos(t) - 2\cdot\sin(t))(-2\sin(t)) + (2\cdot\cos(t))(2\cdot\cos(t))\ dt = \int_{t=0}^{2\pi} 4 - 4\cos(t)\cdot\sin(t)\ dt = 8\pi .$$

The circulation is 8π .

$\frac{\partial N}{\partial x} = 1,\ \frac{\partial M}{\partial y} = -1$ so $\frac{\partial N}{\partial x} - \frac{\partial M}{\partial y} = 2$ and $\iint_R (2)\ dA = 2(\text{area of circle of radius } 2) = 8\pi$.

As promised by the first conclusion of Green's Theorem, these two values are equal, and in this case the double integral was much easier to evaluate.

Practice 1: For $\mathbf{F} = \langle x - y,\ x\rangle$ and $\mathbf{r}(t) = \langle 2\cdot\cos(t),\ 2\cdot\sin(t)\rangle$, evaluate $\int_C (Mdy - Ndx)$ and $\iint_R \left(\frac{\partial M}{\partial x} + \frac{\partial N}{\partial y}\right) dA$.

Example 2: Evaluate $\int_C Mdy - Ndx$ and $\iint_R \left(\frac{\partial M}{\partial x} + \frac{\partial N}{\partial y}\right) dA$ for $\mathbf{F} = \langle -y, x\rangle$ and the triangular region R bounded by the x-axis, the line x=2 and the line y=x (Fig. 3).

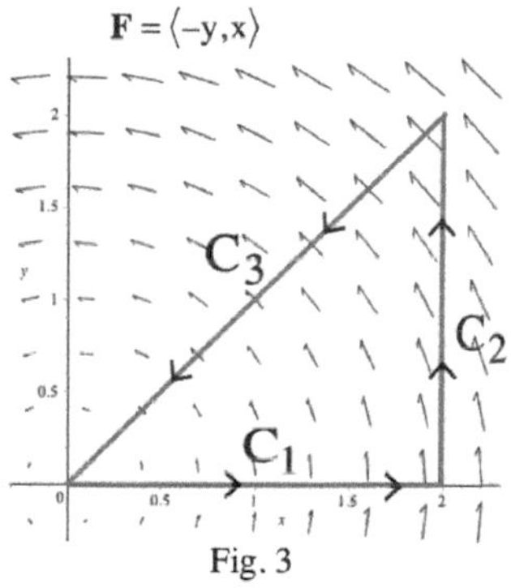

Fig. 3

Solution: $\frac{\partial M}{\partial x} = 0$ and $\frac{\partial N}{\partial y} = 0$ so we immediately have $\iint_R \left(\frac{\partial M}{\partial x} + \frac{\partial N}{\partial y}\right) dA = 0$.

The flux is 0.

The boundary of R consists of 3 line segments:

$C_1 = \langle 2t, 0\rangle$, $C_2 = \langle 2, 2t\rangle$, and $C_3 = \langle 2-2t, 2-2t\rangle$ with $0 \le t \le 1$. We need that choice for C_3 in order for the orientation to be counterclockwise.

On C_1, $\int_{t=0}^{1} Mdy - Ndx = \int_{t=0}^{1} (-0)(2) - (2t)(2)\ dt = \int_{t=0}^{1} -4t\ dt = -2$.

On C_2, $\int_{t=0}^{1} Mdy - Ndx = \int_{t=0}^{1} (-2t)(2) - (2)(0)\ dt = \int_{t=0}^{1} -4t\ dt = -2$.

On C_3, $\int_{t=0}^{1} Mdy - Ndx = \int_{t=0}^{1} (2t-2)(-2) - (2-2t)(-2)\ dt = \int_{t=0}^{1} 8 - 8t\ dt = 4$.

So $\int_{C_1} + \int_{C_2} + \int_{C_3} = (-2) + (-2) + (4) = 0$. Certainly the double integral was easier.

Practice 2: Evaluate $\int_C (Mdx + Ndy)$ and $\iint_R \left(\frac{\partial N}{\partial x} - \frac{\partial M}{\partial y}\right) dA$ for $\mathbf{F} = \langle -y, x\rangle$ and the triangular region R bounded by the x-axis, the line x=2 and the line y=x .

Green's Theorem can also be used to evaluate line integrals.

Example 3: Evaluate $\oint_C x^2y\,dy - y^2\,dx$ where C is the boundary of the rectangle

$R = \{(x,y):\ 0 \le x \le 2,\ 0 \le y \le 1\}$ oriented counterclockwise.

Solution: If we can match the form of the line integral with one of the forms of Green's Theorem then we can evaluate one double integral instead of the four line integrals around R.

If we use the Flux-Divergence form $\oint_C M dy - N dx = \iint_R \left(\frac{\partial M}{\partial x} + \frac{\partial N}{\partial y} \right) dA$ then we need

$M = x^2y$ and $N = y^2$ so $\oint_C x^2y\,dy - y^2\,dx = \iint_R (2xy + 2y)\,dA = \int_0^1 \int_0^2 (2xy + 2y)\,dx\,dy = \int_0^1 8y\,dy = 4$.

Practice 3: Use the Circulation-Curl form of Green's Theorem to evaluate the same line integral on the same region R.

In the previous examples we always traded a line integral for a double integral, but sometimes the opposite trade is useful.

Using Green's Theorem to Find Area

If the boundary of R is a simple closed curve C then area of $R = \iint_R 1\,dA$. If we can find M and N so that $\frac{\partial M}{\partial x} + \frac{\partial N}{\partial y} = 1$ then we can use $\iint_R 1\,dA = \iint_R \left(\frac{\partial M}{\partial x} + \frac{\partial N}{\partial y} \right) dA = \oint_C M dy - N dx$. Putting

$M = \frac{x}{2}$ and $N = \frac{y}{2}$ works so $\iint_R 1\,dA = \iint_R \left(\frac{\partial M}{\partial x} + \frac{\partial N}{\partial y} \right) dA = \oint_C M dy - N dx$.

> If the boundary of R is a simple closed curve C, then area of $R = \frac{1}{2} \oint_C x\,dy - y\,dx$.

Example 4: Use this result to determine the area of the elliptical region $\frac{x^2}{4} + \frac{y^2}{25} \le 1$.

Solution: The boundary of this region can be parameterized by $\mathbf{r}(t) = \langle 2\cos(t), 5\sin(t) \rangle$.

Then $\text{area} = \frac{1}{2} \oint_C (2\cos(t))(5\cos(t)) - (5\sin(t))(-2\sin(t))\,dt = \frac{1}{2} \int_0^{2\pi} 10\,dt = 10\pi$.

Practice 4: Use the same method to determine the area of the general elliptical region $\frac{x^2}{a^2} + \frac{y^2}{b^2} \le 1$.

Green's Theorem in More General Regions

The proof of Green's Theorem in the Appendix is valid for simple regions R (Fig. 4) in the plane for which any line parallel to an axis cuts the region R in at most 2 points or along an edge of R.

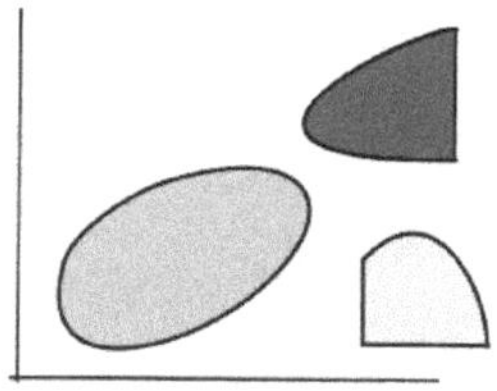

Fig. 4: Some simple regions

But Green's Theorem is true in much more complex regions if they can be decomposed into a union of simple regions.

If the region R is "bent," (Fig. 5) there are usually a finite number of cuts parallel to an axis so that R is the union of the some simple regions. Since the counterclockwise orientation on each simple region moves along each cut once in each direction (Fig. 6) the sum of those integral pieces is 0 and we are left with the integral around the original boundary of R.

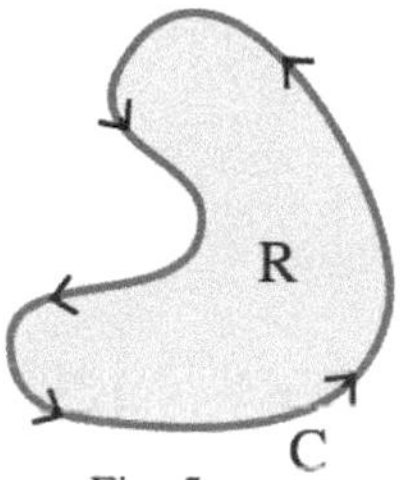

Fig. 5

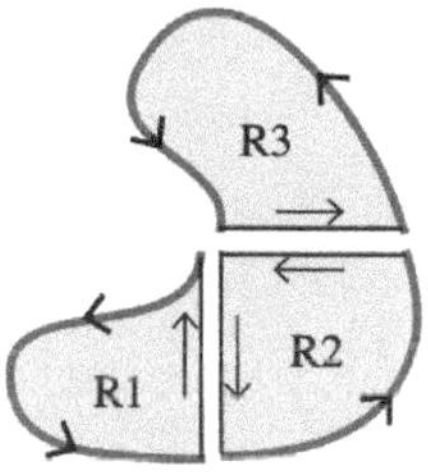

Fig. 6

Practice 5: Decompose the region in Fig. 7 into several simple regions. Indicate the direction(s) of the paths along each cut.

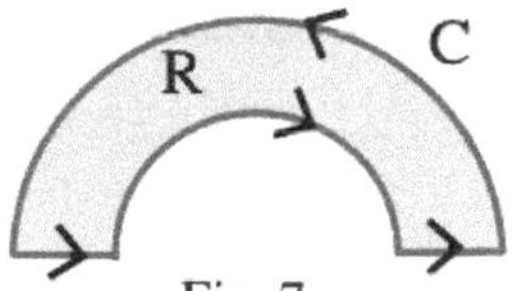

Fig. 7

Similarly, if R contains a finite number of holes (Fig. 8), then we can again create a single boundary for R by adding paths that connect to the holes. Then the integral along this new path will be the sum of the counterclockwise integrals around the outer boundary of R minus the sum of the counterclockwise integrals around the holes. The integrals along the added paths sum to 0 since they are traveled once in each direction. Remember, for a counterclockwise orientation of the curve C the region is always on our left hand side as we walk along C.

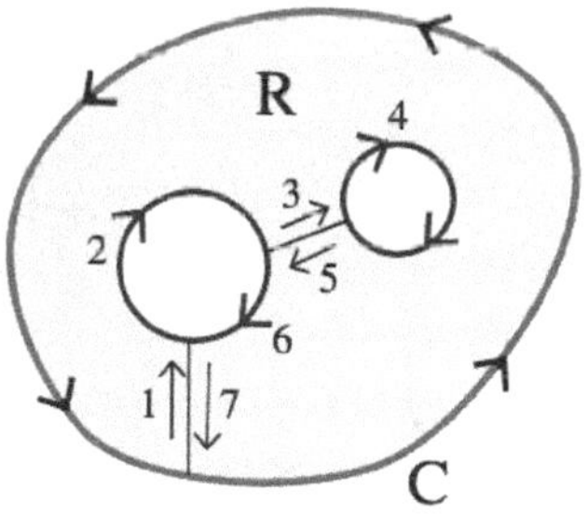

Fig. 8

Practice 6: Decompose the region in Fig. 9 into several simple regions. Indicate the directions and order of the paths along each cut.

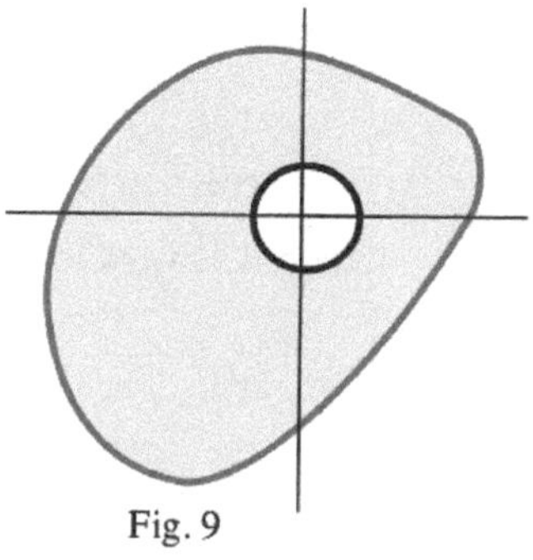

Fig. 9

Example 5: An interesting situation. $\mathbf{F}(x,y) = \left\langle \frac{-y}{x^2+y^2}, \frac{x}{x^2+y^2} \right\rangle$.

Let C_1 be any simple closed curve that does not enclose the origin and let C_2 be a simple closed curve that does enclose the origin (Fig. 10). Use the Circulation-Curl form of Green's Theorem to calculate the circulations around C_1 and C_2.

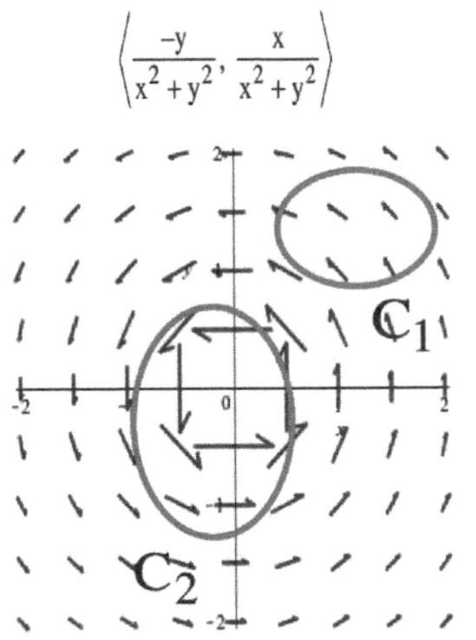

Fig. 10

Solution: On C_1 circulation

$$\oint_C \mathbf{F} \bullet \mathbf{T}\, ds = \oint_C M\,dx + N\,dy = \iint_R \left(\frac{\partial N}{\partial x} - \frac{\partial M}{\partial y} \right) dA.$$

$$\frac{\partial M}{\partial y} = \frac{(x^2+y^2)(-1) - (-y)(2y)}{(x^2+y^2)^2} = \frac{y^2-x^2}{(x^2+y^2)^2}. \quad \frac{\partial N}{\partial x} = \frac{(x^2+y^2)(1) - (x)(2x)}{(x^2+y^2)^2} = \frac{y^2-x^2}{(x^2+y^2)^2} \text{ so } \frac{\partial N}{\partial x} - \frac{\partial M}{\partial y} = 0$$

$$\text{so circulation} = \oint_C M\,dx + N\,dy = \iint_R \left(\frac{\partial N}{\partial x} - \frac{\partial M}{\partial y} \right) dA = \iint_R 0\, dA = 0 \text{ on } C_1.$$

Lets begin the "encloses the origin" by looking at the particular circle C that encloses the origin: $\mathbf{r}(t) = \langle h\cdot\cos(t),\, h\cdot\sin(t)\rangle$ for a positive value of h. In this case we can work with the line integral for circulation directly by putting everything in terms of t:

$$\text{circulation} = \oint_C M\,dx + N\,dy = \int_0^{2\pi} \left(\frac{-h\cdot\sin(t)}{h^2} \right)(-h\cdot\sin(t)) + \left(\frac{h\cdot\cos(t)}{h^2} \right)(h\cdot\cos(t))\, dt = \int_0^{2\pi} 1\, dt = 2\pi.$$

If C_2 is any simple closed curve that encloses the origin, we can take h small enough that the circle C is inside C_2. Then the region R bounded by D= the union of C_2 counterclockwise and C clockwise (Fig 10) does not contain the origin so $0 = \int_D = \int_{C_2} + \int_C = \int_{C_2} + (-2\pi)$

and $\{\text{circulation around } C_2\} = \oint_{C_2} M\,dx + N\,dy = 2\pi$.

For this vector field the circulation is 0 for any simple closed curve that does not surround the origin, and the circulation is always 2π for any simple. closed, positively-oriented curve that does surround the origin.

Problems

In problems 1 to 6 evaluate the line integral directly and by using Green's Theorem where C is positively oriented.

1. $\int_C x^2y\,dx + 3y\,dy$ when C is the rectangle $0 \le x \le 2$, $0 \le y \le 1$.

2. $\int_C xy^2\,dx + 5xy\,dy$ when C is the square $0 \le x \le 2$, $0 \le y \le 2$.

3. $\int_C 3xy\,dx + 2x^2\,dy$ when C is the triangle with vertices (0,0), (1,0) and (1,2).

4. $\int_C x^2\,dx + xy\,dy$ when C is the triangle with vertices (0,0), (0,2) and (2,2).

5. $\int_C ax\,dx + by\,dy$ when C is the circle $x^2 + y^2 = r^2$.

6. $\int_C ay\,dx + bx\,dy$ when C is the circle $x^2 + y^2 = r^2$.

In problems 7 to 10 use Green's Theorem to find the counterclockwise circulation and the outward flux for the field **F** and the curve C.

7. $\mathbf{F} = \langle x+2y, y-x \rangle$, C is the square $0 \le x \le 2$, $0 \le y \le 2$.

8. $\mathbf{F} = \langle 3x+2y, 4y-5x \rangle$, C is the rectangle $0 \le x \le 3$, $0 \le y \le 1$.

9. $\mathbf{F} = \langle x^2+y^2, x^2-y^2 \rangle$, C is the triangle with vertices (0,0), (0,2) and (2,2).

10. $\mathbf{F} = \langle x^2y, 3x+y^2 \rangle$, C is the triangle bounded by the lines x=0, y=1 and x=2y.

In problems 11 and 12 use Green's Theorem to find the area enclosed by the curve C.

11. C is given by $\mathbf{r}(t) = \langle t, t^2 \rangle$ as t goes from –2 to 2 and by $\mathbf{r}(t) = \langle t, 8-t^2 \rangle$ as t goes from 2 to –2.

12. C is the curve bounded by the x-axis and the cycloid
$\mathbf{r}(t) = \langle A(t-\sin(t)), A(1-\cos(t)) \rangle$.

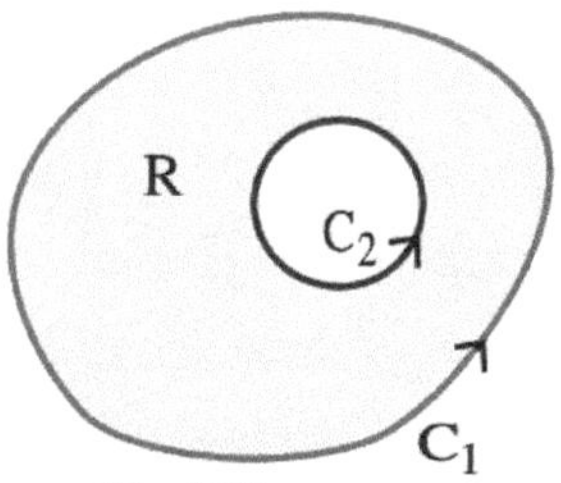

Fig. P13

13. $\mathbf{F} = \langle M, N \rangle$ and $\frac{\partial N}{\partial x} - \frac{\partial M}{\partial y} = 5$ on region R in Fig. P13. The area of R is 100, and $\int_{C_2} \mathbf{F} \bullet d\mathbf{r} = 20$. Use Green's Theorem to determine $\int_{C_1} \mathbf{F} \bullet d\mathbf{r}$.

14. $\mathbf{F} = \langle M,N \rangle$ and $\frac{\partial N}{\partial x} - \frac{\partial M}{\partial y} = 7$ on region R (inside C_1, outside C_2) in Fig.

P14. If $\int_{C_2} \mathbf{F} \bullet \mathbf{dr} = 3\pi$, use Green's Theorem to determine $\int_{C_1} \mathbf{F} \bullet \mathbf{dr}$.

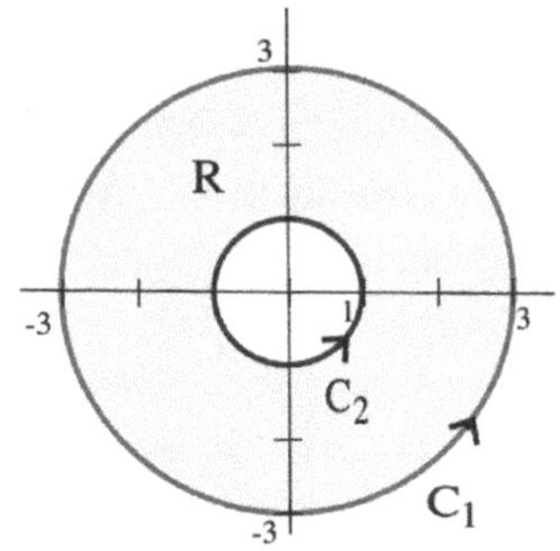

Fig. P14

15. $\mathbf{F} = \langle M,N \rangle$ and $\frac{\partial N}{\partial x} - \frac{\partial M}{\partial y} = 9$ on region R in Fig. P15,

$\int_{C_2} \mathbf{F} \bullet \mathbf{dr} = 3\pi$ and $\int_{C_3} \mathbf{F} \bullet \mathbf{dr} = 4\pi$. Use Green's

Theorem to determine $\int_{C_1} \mathbf{F} \bullet \mathbf{dr}$.

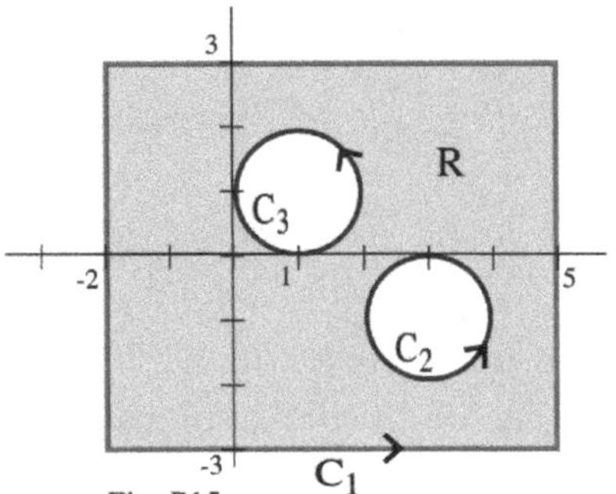

Fig. P15

16. $\mathbf{F} = \langle M,N \rangle$ and $\frac{\partial N}{\partial x} - \frac{\partial M}{\partial y} = 5$ on region R in Fig. P16, and

$\int_{C_2} \mathbf{F} \bullet \mathbf{dr} = 2\pi$. . Use Green's Theorem to determine $\int_{C_1} \mathbf{F} \bullet \mathbf{dr}$.

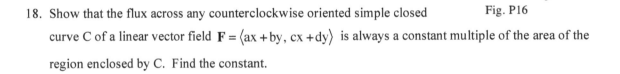

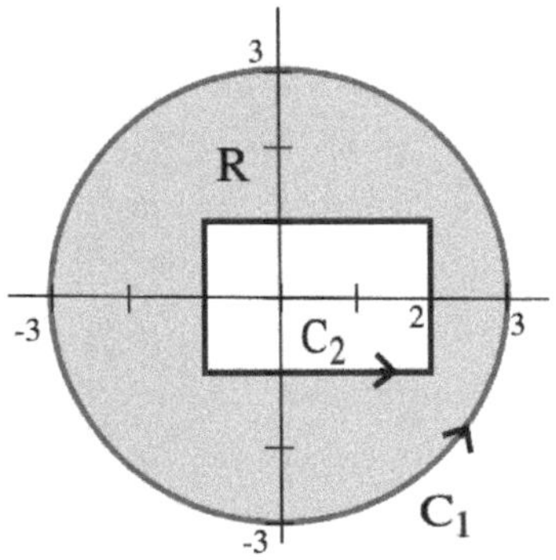

Fig. P16

17. Show that the circulation and flux of a constant field $\mathbf{F} = \langle a,b \rangle$ are 0 over every simply connected region R.

18. Show that the flux across any counterclockwise oriented simple closed curve C of a linear vector field $\mathbf{F} = \langle ax + by,\ cx + dy \rangle$ is always a constant multiple of the area of the region enclosed by C. Find the constant.

19. Show that the circulation around any counterclockwise oriented simple closed curve C of a linear vector field $\mathbf{F} = \langle ax + by,\ cx + dy \rangle$ is always a constant multiple of the area of the region enclosed by C. Find the constant.

Practice Answers

Practice 1: $M = 2\cdot\cos(t) - 2\cdot\sin(t)$, $N = 2\cdot\cos(t)$, $dx = -2\sin(t)\,dt$, $dy = 2\cdot\cos(t)\,dt$, $\frac{\partial M}{\partial x} = 1$, $\frac{\partial N}{\partial y} = 0$

$$\int_C (Mdy - Ndx) = \int_{t=0}^{2\pi} ((2\cdot\cos(t) - 2\cdot\sin(t))(2\cdot\cos(t)) - (2\cdot\cos(t)(-2\sin(t))\,dt = \int_{t=0}^{2\pi} 4\cdot\cos^2(t)\,dt = 4\pi .$$

$$\iint_R \left(\frac{\partial M}{\partial x} + \frac{\partial N}{\partial y}\right) dA = \iint_R (1+0)\,dA = \iint_R (1)\,dA = 1(\text{area of circle of radius 2}) = 4\pi.$$

As promised by the second conclusion of Green's Theorem, these two values are equal.

Practice 2: $\frac{\partial N}{\partial x} = 1$ and $\frac{\partial M}{\partial y} = -1$ so $\iint_R \left(\frac{\partial N}{\partial x} - \frac{\partial M}{\partial y}\right) dA = \iint_R (2)\,dA = 2(\text{triangle area}) = 2(2) = 4$.

The circulation is 4.

$\mathbf{F} = \langle -y, x\rangle$. $C_1 = \langle 2t, 0\rangle$, $C_2 = \langle 2, 2t\rangle$, and $C_3 = \langle 2-2t, 2-2t\rangle$ with $0 \le t \le 1$.

On C_1, $\int_{t=0}^{1} Mdx + Ndy = \int_{t=0}^{1} (-0)(2) + (2t)(0)\,dt = \int_{t=0}^{1} 0\,dt = 0$.

On C_2, $\int_{t=0}^{1} Mdx + Ndy = \int_{t=0}^{1} (-2t)(0) + (2)(2)\,dt = \int_{t=0}^{1} 4\,dt = 4$.

On C_3, $\int_{t=0}^{1} Mdx + Ndy = \int_{t=0}^{1} (2t-2)(-2) + (2-2t)(-2)\,dt = \int_{t=0}^{1} 0\,dt = 0$.

$\int_{C_1} + \int_{C_2} + \int_{C_3} = (0) + (4) + (0) = 4$. Again the double integral was easier.

The flows along C_1 and C_3 make sense in terms of Fig. 3. Along C_1 and C_3 the vector field is perpendicular to the boundary so there is no flow.

Practice 3: $\oint_C x^2y\,dy - y^2\,dx = \oint_C (Mdx + Ndy) = \iint_R \left(\frac{\partial N}{\partial x} - \frac{\partial M}{\partial y}\right) dA$ so $M = -y^2$ and $N = x^2y$.

$$\oint_C x^2y\,dy - y^2\,dx = \iint_R (2xy + 2y)\,dA = \int_0^1\int_0^2 (2xy+2y)\,dx\,dy = \int_0^1 8y\,dy = 4 .$$

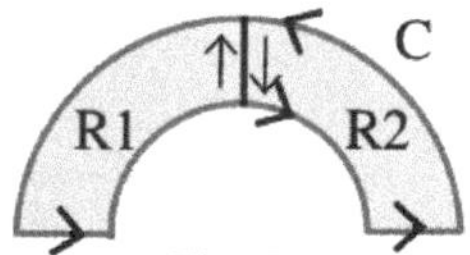

Fig. 7

Practice 4: Take $\mathbf{r}(t) = \langle a\cdot\cos(t), b\cdot\sin(t)\rangle$. Then

$$\text{area} = \frac{1}{2}\oint_C (a\cdot\cos(t))(b\cdot\cos(t)) - (b\cdot\sin(t))(-a\cdot\sin(t))\,dt = \frac{1}{2}\int_0^{2\pi} ab\,dt = ab\pi .$$

Practice 5: See Fig. 7

Practice 6: Fig. 9 shows one solution.

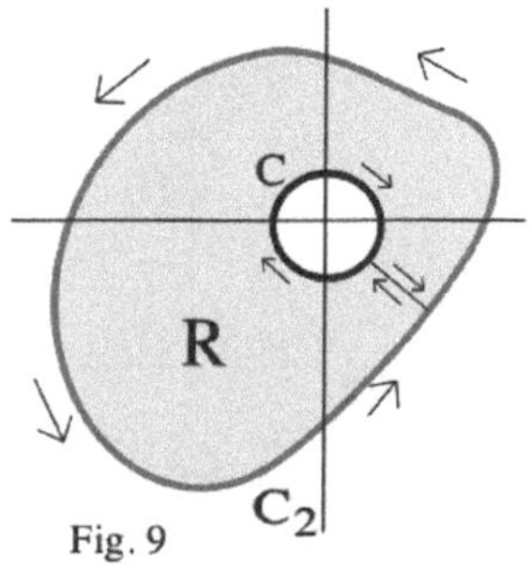

Fig. 9

Appendix A: Proof of Green's Theorem for Simple Regions

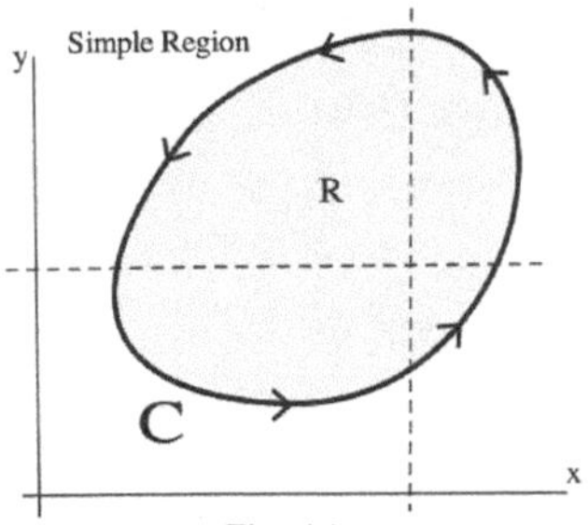

Fig. A1

A simple region R is one in which lines parallel to an axis intersect the boundary of R in at most two places (Fig. A1).

Label Fig. A2 so $f_1(x) \le y \le f_2(x)$ for $a \le x \le b$. C_1 is the curve $y = f_1(x)$ oriented counterclockwise as x goes from a to b , and C_2 is $y = f_2(x)$ which has a counterclockwise orientation as x goes from b to a. The closed curve C is the union of C_1 and C_2. With labeling we will compute

$\oint_C M\,dx$ and $\iint_R \frac{\partial M}{\partial y}\,dA$.

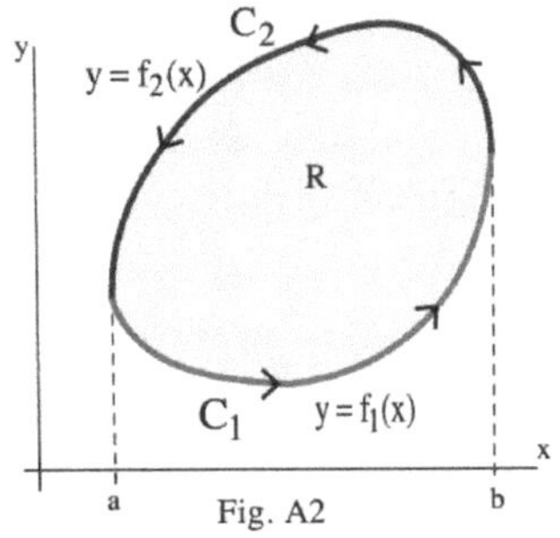

Fig. A2

$$\oint_C M\,dx = \oint_{C_1} M\,dx + \oint_{C_2} M\,dx = \int_a^b M(x,f_1)\,dx + \int_b^a M(x,f_2)\,dx$$

$$= \int_a^b M(x,f_1)\,dx - \int_a^b M(x,f_2)\,dx = \int_a^b M(x,f_1)\,dx - M(x,f_2)\,dx.$$

$$\iint_R \frac{\partial M}{\partial y}\,dA = \int_a^b \int_{f_1}^{f_2} \frac{\partial M}{\partial y}\,dy\,dx = \int_a^b M(x,f_2) - M(x,f_1)\,dx \quad \text{so} \quad \boxed{\oint_C M\,dx = -\iint_R \frac{\partial M}{\partial y}\,dA} \quad (1)$$

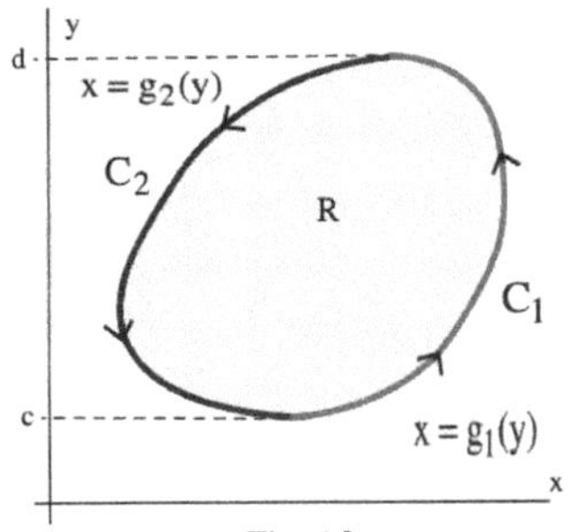

Fig. A3

To get the other part of the result we need, re-label the region R as in Fig. A3 so $x = g_1(y)$ for $c \le y \le d$. C_1 is the curve $x = g_1(x)$ oriented counterclockwise as y goes from c to d, and C_2 is $x = g_2(y)$ which has counterclockwise orientation as y goes from d to c. With this labeling we will compute $\oint_C N\,dy$ and $\iint_R \frac{\partial N}{\partial x}\,dA$.

$$\oint_C N\,dy = \oint_{C_1} N\,dy + \oint_{C_2} N\,dy = \int_c^d N(g_2,y)\,dy + \int_d^c N(g_1,y)\,dy$$

$$= \int_c^d N(g_2,y)\,dy - \int_c^d N(g_1,y)\,dy = \int_c^d N(g_2,y) - N(g_1,y)\,dy$$

$$\iint_R \frac{\partial N}{\partial x}\,dA = \int_c^d \int_{g_1}^{g_2} \frac{\partial N}{\partial x}\,dx\,dy = \int_c^d N(g_2,y) - N(g_1,y)\,dy \quad \text{so} \quad \boxed{\oint_C N\,dy = \iint_R \frac{\partial N}{\partial x}\,dA} \quad (2)$$

Adding result (1) and result (2)), we have $\boxed{\oint_C M\,dx + N\,dy = \iint_R \frac{\partial N}{\partial x} - \frac{\partial M}{\partial y}\,dA}$, the Circulation-Curl form of Green's Theorem.

Using a similar approach, you can show that $-\oint_C N\,dx = \iint_R \frac{\partial N}{\partial y}\,dA$ and $\oint_C M\,dy = \iint_R \frac{\partial M}{\partial x}\,dA$ so

$\oint_C M\,dy - N\,dx = \iint_R \frac{\partial M}{\partial x} + \frac{\partial N}{\partial y}\,dA$, the Flux-Divergence form of Green's Theorem.

On a Modified Simple Region

Modified Simple Region

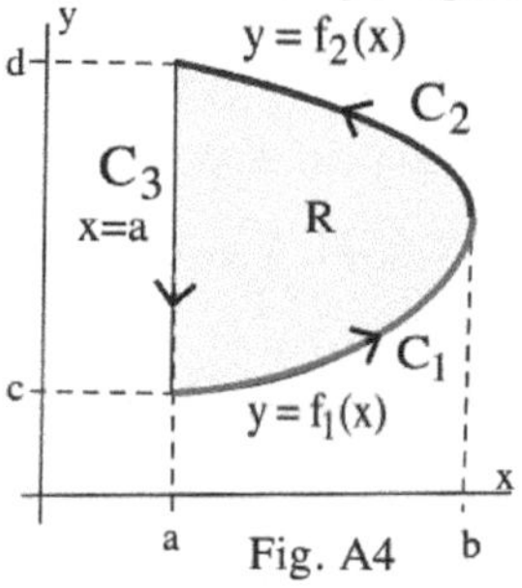

Fig. A4

Suppose the region R has one edge that if parallel to the y-axis (Fig. A4) labeled so $f_1(x) \le y \le f_2(x)$ for a≤x≤b. C_1 is the curve $y = f_1(x)$ oriented counterclockwise as x goes from a to b , and C_2 is $y = f_2(x)$ which has a counterclockwise orientation as x goes from b to a. C_3 is oriented counterclockwise with x=a as y goes from d to .The closed curve C is the union of C_1, C_2 and C_3.

$\oint_C M\,dx$ and $\iint_R \frac{\partial M}{\partial y}\,dA$.

$$\oint_C M\,dx = \oint_{C_1} M\,dx + \oint_{C_2} M\,dx + \oint_{C_3} M\,dx = \int_a^b M(x,f_1)\,dx + \int_b^a M(x,f_2)\,dx + \int_a^a M(x,f_2)\,dx$$

$$= \int_a^b M(x,f_1)\,dx + \int_b^a M(x,f_2)\,dx = \int_a^b M(x,f_1)\,dx - \int_a^b M(x,f_2)\,dx = \int_a^b M(x,f_1)\,dx - M(x,f_2)\,dx.$$

$$\iint_R \frac{\partial M}{\partial y}\,dA = \int_a^b \int_{f_1}^{f_2} \frac{\partial M}{\partial y}\,dy\,dx = \int_a^b M(x,f_2) - M(x,f_1)\,dx \quad \text{so} \quad \oint_C M\,dx = -\iint_R \frac{\partial M}{\partial y}\,dA.$$

To get the other part of the result we need, label the region R as in Fig. A5 so C_1 is the curve $x = g(y)$ oriented counterclockwise as y goes from c to d, and C_2 is $x = a$ which has counterclockwise orientation as y goes from d to c. With this labeling we will compute $\oint_C N\,dy$ and $\iint_R \frac{\partial N}{\partial x}\,dA$.

$$\oint_C N\,dy = \oint_{C_1} N\,dy + \oint_{C_2} N\,dy = \int_c^d N(g,y)\,dy + \int_d^c N(a,y)\,dy$$

$$= \int_c^d N(g,y)\,dy - \int_c^d N(a,y)\,dy = \int_c^d N(g,y) - N(a,y)\,dy$$

$$\iint_R \frac{\partial N}{\partial x}\,dA = \int_c^d \int_a^g \frac{\partial N}{\partial x}\,dx\,dy = \int_c^d N(g,y) - N(a,y)\,dy \quad \text{so}$$

Modified Simple Region

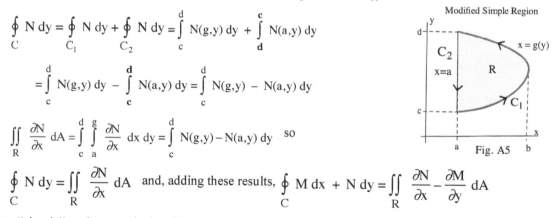

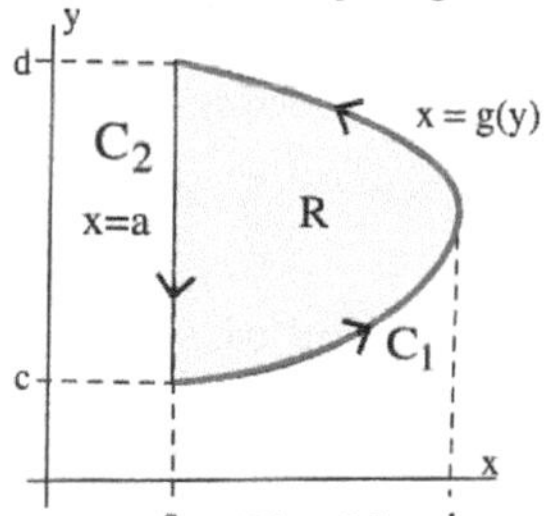

Fig. A5

$$\oint_C N\,dy = \iint_R \frac{\partial N}{\partial x}\,dA \quad \text{and, adding these results,} \quad \oint_C M\,dx + N\,dy = \iint_R \frac{\partial N}{\partial x} - \frac{\partial M}{\partial y}\,dA$$

Other "simple" regions can be handled in similar ways.

The proof for general regions is difficult and is not included here.

If the region R lies on a plane in 3D (R is flat), then after a rotation of axes R can be made to lie in a new x'y'-plane and Green's Theorem applies.

15.7 Divergence and Curl in 3D

The divergence and curl were introduced in Section 15.2 for a 2D vector field F since it is easier in 2D to visualize what they measure and because we only needed the 2D versions for Green's Theorem in Section 15.5. Here we extend those definitions to a 3D vector field.

Divergence: div **F** in 3D

Assume that the vector field F describes the flow of a liquid in 3D. The **divergence** is the rate per unit volume that the water dissipates (**departs**, leaves) at the point P. A positive value for the divergence means that more water is leaving at P than is entering at P. But we can't really see a point so imagine a small sphere C centered at P, and consider whether more water s leaving or entering this sphere. Fig. 1(a) shows more water leaving than entering the sphere around P so div **F**(P) > 0, Fig. 1(b) has more entering than leaving so div **F**(P) < 0, and Fig, 1(c) shows the same amount leaving as entering so div **F** (P) = 0. This is much more difficult to visualize in 3D than in 2D, but the calculations are not any harder.

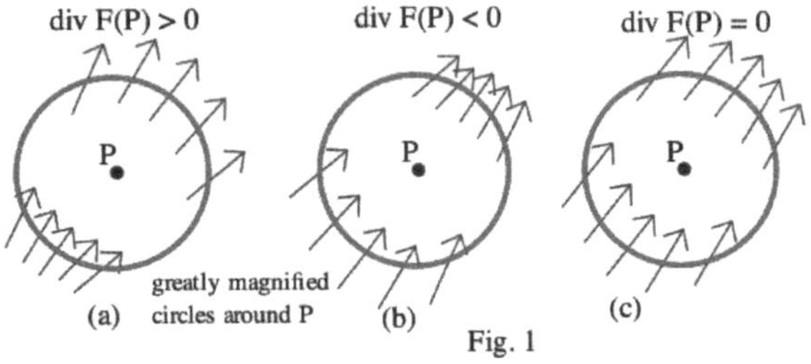

Fig. 1

Fig. 2: < x, y, z >

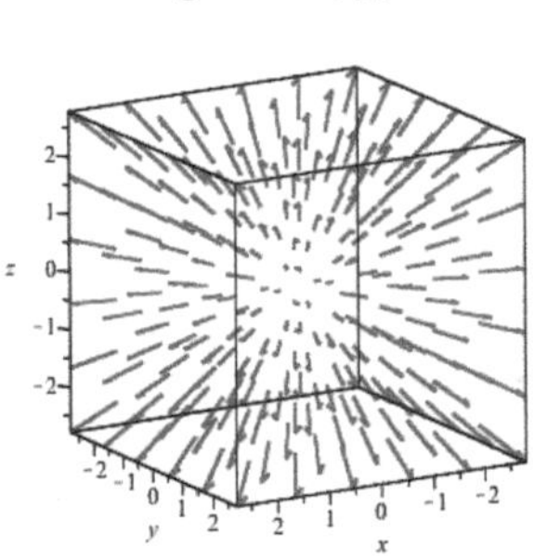

Example 1: Fig. 2 shows the radial field $\mathbf{F}(x,y,z) = \langle x,y,z \rangle$. Based on this figure, is div **F** positive, negative or zero at P=(1,2,0), Q=(1,0,2) and R=(0,0,0).

Solution: The vectors are increasing in magnitude as we move away from the origin so more are leaving little spheres at each of these points (in fact, at every point in 3D) so div **F** > 0 at every point in 3D .

> **Definition: Divergence F in 3D** div **F**
>
> For a vector field $\mathbf{F}(x,y,z) = M\mathbf{i} + N\mathbf{j} + P\mathbf{k}$ with continuous partial derivatives,
>
> the divergence of F at Point P is $\text{div } \mathbf{F}(P) = \frac{\partial M}{\partial x} + \frac{\partial N}{\partial y} + \frac{\partial P}{\partial z}$. div $\mathbf{F} = \nabla \bullet \mathbf{F}$

Example 2: Calculate div **F** for $\mathbf{F}(x,y,z) = \langle x,y,z \rangle$ at the points in Example 1.

Solution: $\text{div } \mathbf{F} = \frac{\partial M}{\partial x} + \frac{\partial N}{\partial y} + \frac{\partial P}{\partial z} = 1 + 1 + 1 = 3$ at every point (x,y,z) so more water is leaving the is entering every tiny sphere in 3D.

Practice 1: Calculate div **F** for $\mathbf{F}(x,y,z) = \langle x^2, y^2, z \rangle$ at the points at (1,1,1), (2, –3,4) and (–0.5, 0,3).

Divergence of the inverse-square radial field $\mathbf{F} = \langle x, y, z\rangle / \sqrt{x^2+y^2+z^2}$

Radial inverse-square vector fields are very common in applications such as electricity and magnetism so it is worth calculating the divergence of such a field even if it is a bit tedious.

If $\mathbf{r} = \langle x, y, z\rangle$ then $|\mathbf{r}| = \sqrt{x^2+y^2+z^2}$ and the direction of r is $\frac{\mathbf{r}}{|\mathbf{r}|}$. If a field **F** has the same direction a **r** (a radial field) and follows an inverse-square law with magnitude $\frac{1}{|\mathbf{r}|^2}$ then $\mathbf{F} = \left(\frac{1}{|\mathbf{r}|^2}\right)\left(\frac{\mathbf{r}}{|\mathbf{r}|}\right) = \frac{\mathbf{r}}{|\mathbf{r}|^3}$.

$\frac{\partial |\mathbf{r}|^3}{\partial x} = \frac{\partial}{\partial x}\left(x^2+y^2+z^2\right)^{3/2} = 3x\cdot\left(x^2+y^2+z^2\right)^{1/2} = 3x|\mathbf{r}|$ and similarly for y and z, $\frac{\partial |\mathbf{r}|^3}{\partial y} = 3y|\mathbf{r}|$ and $\frac{\partial |\mathbf{r}|^3}{\partial z} = 3z|\mathbf{r}|$. But to calculate the divergence of $\frac{\mathbf{r}}{|\mathbf{r}|^3}$ we need the quotient rule for each partial derivative. $\frac{\partial}{\partial x}\frac{x}{|\mathbf{r}|^3} = \frac{|\mathbf{r}|^3\left(\frac{\partial x}{\partial x}\right) - x\cdot\left(\frac{\partial |\mathbf{r}|^3}{\partial x}\right)}{\left(|\mathbf{r}|^3\right)^2} = \frac{|\mathbf{r}|^3\left(\frac{\partial x}{\partial x}\right) - x\cdot 3x|\mathbf{r}|}{\left(|\mathbf{r}|^3\right)^2} = \frac{1}{|\mathbf{r}|^3} - \frac{3x^2}{|\mathbf{r}|^5}$. Similarly

$\frac{\partial}{\partial x}\frac{y}{|\mathbf{r}|^3} = \frac{1}{|\mathbf{r}|^3} - \frac{3y^2}{|\mathbf{r}|^5}$ and $\frac{\partial}{\partial x}\frac{z}{|\mathbf{r}|^3} = \frac{1}{|\mathbf{r}|^3} - \frac{3z^2}{|\mathbf{r}|^5}$ so $\text{div }\mathbf{F} = \text{div }\frac{\mathbf{r}}{|\mathbf{r}|^3} = \frac{3}{|\mathbf{r}|^3} - \frac{3(x^2+y^2+z^2)}{|\mathbf{r}|^5}$

$= \frac{3}{|\mathbf{r}|^3} - \frac{3|\mathbf{r}|^2}{|\mathbf{r}|^5} = 0$. The divergence of thc inverse-square vector field **F** is 0 everywhere except at the origin where the field is not defined.

Curl: curl **F** in 3D

In section 15.2 the curl of a vector field F at a point P was introduced and described as a measure of the counterclockwise rotation of a small paddle wheel or circle at P caused by the vector field. (Fig. 3) We could also view this as vectors in the xy-plane causing rotation about an axis in the z direction. In 3D, instead of a small circle, imagine a small sphere that is fixed at its center and rotates about that center point (Fig. 4). The 3D curl is a vector with two important properties (these will proved in section 15.9 using Stoke's Theorem):

* the magnitude of the curl gives the rate of the fluid's rotation, and
* the direction of the curl is normal to the plane of greatest circulation and points in the direction so that the circulation at the point has a right hand orientation (Fig. 5).

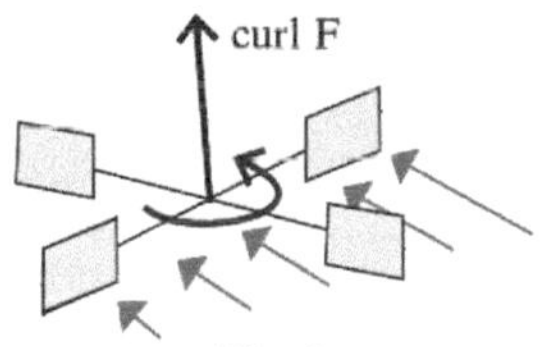

Fig. 3

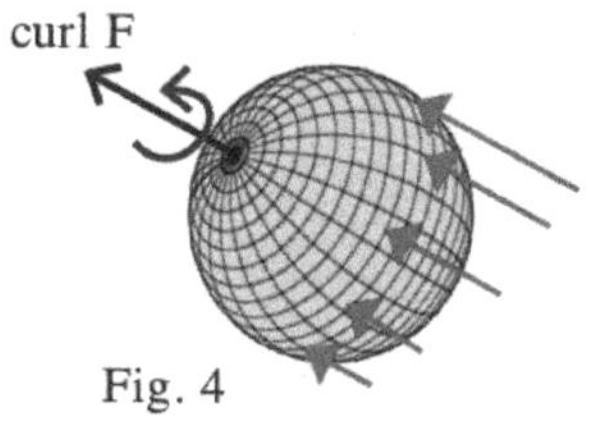

Fig. 4

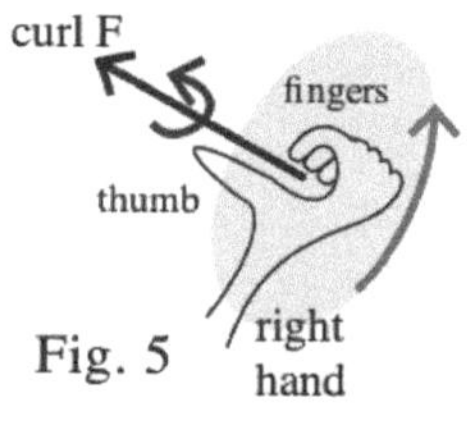

Fig. 5

If we have a small paddle wheel at point P and tilt it in different directions, then the wheel will spin fastest with a right-hand orientation when the axis points in the direction of the curl vector.

Definition: Curl F in 3D

For a vector field $\mathbf{F}(x,y,z) = M\mathbf{i} + N\mathbf{j} + P\mathbf{k}$ with continuous partial derivatives, then the curl of F at Point P is $\text{curl F} = \left(\frac{\partial P}{\partial y} - \frac{\partial N}{\partial z}\right)\mathbf{i} + \left(\frac{\partial M}{\partial z} - \frac{\partial P}{\partial x}\right)\mathbf{j} + \left(\frac{\partial N}{\partial x} - \frac{\partial M}{\partial y}\right)\mathbf{k}$. curl $\mathbf{F} = \nabla \mathbf{x} \mathbf{F}$

Note: The k component of curl, $\frac{\partial N}{\partial x} - \frac{\partial M}{\partial y}$, is just the 2D curl **F** from section 15.2.

Example 3: Calculate curl **F** for $\mathbf{F}(x,y,z) = \langle y-z,\ z-2x,\ x+3z\rangle$.

Solution:

$$\text{curl } \mathbf{F} = \nabla \mathbf{x} \mathbf{F} = \begin{vmatrix} \mathbf{i} & \mathbf{j} & \mathbf{k} \\ \frac{\partial}{\partial x} & \frac{\partial}{\partial y} & \frac{\partial}{\partial z} \\ y-z & z-2x & x+3z \end{vmatrix} = \langle 0-1,\ -(1+1),\ -2-1\rangle = \langle -1,\ -2,\ -3\rangle \ .$$

The curl of this field is the same at every point in 3D, and the little paddle wheel will spin fastest if the axis of the wheel is oriented in the direction of $\langle -1,\ -2,\ -3\rangle$.

Practice 2: Calculate curl **F** for $\mathbf{F}(x,y,z) = \langle 2y+z,\ x^2,\ 3z\rangle$.

Example 4: (a) Calculate the gradient vector field **F** of $f(x,y,z) = x^3z + 3xy^2 + 4z$.

(b) Calculate curl **F** .

Solution: (a) $\mathbf{F} = \nabla f = \langle 3x^2z + 3y^2,\ 6xy.\ x^3 + 4\ \rangle$.

(b)

$$\text{curl } \mathbf{F} = \nabla \mathbf{x} \mathbf{F} = \begin{vmatrix} \mathbf{i} & \mathbf{j} & \mathbf{k} \\ \frac{\partial}{\partial x} & \frac{\partial}{\partial y} & \frac{\partial}{\partial z} \\ 3x^2z+3y^2 & 6xy & x^3+4 \end{vmatrix} = \langle 0-0,\ -(3x^2-3x^2),\ 6y-6y\rangle = \langle 0,\ 0,\ 0\rangle \ .$$

The result in the previous example was not a lucky accident – the curl of every gradient field is 0 everywhere.

Theorem: If **F** is a conservative field ($\mathbf{F} = \nabla f$),

then curl $\mathbf{F} = \langle 0,\ 0,\ 0\rangle$ at every point in 3D. curl $(\nabla f) = \mathbf{0}$

If curl $\mathbf{F} \neq \mathbf{0}$ at any point, then **F** is a not a conservative field.

Proof: If **F** is a gradient field then there is a potential function f so that $\mathbf{F} = \nabla f = \left\langle \frac{\partial f}{\partial x}, \frac{\partial f}{\partial y}, \frac{\partial f}{\partial z}\right\rangle$.

Then $\text{curl } \mathbf{F} = \nabla\mathbf{x}\mathbf{F} = \begin{vmatrix} \mathbf{i} & \mathbf{j} & \mathbf{k} \\ \frac{\partial}{\partial x} & \frac{\partial}{\partial y} & \frac{\partial}{\partial z} \\ \frac{\partial f}{\partial x} & \frac{\partial f}{\partial y} & \frac{\partial f}{\partial z} \end{vmatrix}$.

The $\mathbf{i}$ component is $\frac{\partial}{\partial y}\left(\frac{\partial f}{\partial z}\right) - \frac{\partial}{\partial z}\left(\frac{\partial f}{\partial y}\right) = 0$ by Clairaut's Theorem of Mixed Partial Derivatives.

You can easily verify that the $\mathbf{j}$ and $\mathbf{k}$ components of this curl are also 0.

Practice 3: Show that $\mathbf{F} = \langle -y,x,z \rangle$ is not a conservative vector field.

Unfortunately, knowing that curl F = 0 is not sufficient to guarantee that F is a conservative field. However there is a partial converse to the previous theorem.

> Theorem: If **F** is defined and has continuous partial derivatives at every point in 3D
> and curl **F** = 0
> then **F** is a conservative field.

The proof requires Stoke's Theorem and is not given here.

Curls of the radial and inverse-square radial fields

We could use the definition of curl to calculate the curls for $\mathbf{r} = \langle x, y, z \rangle$ and $\mathbf{F} = \left(\frac{1}{|\mathbf{r}|^2}\right)\left(\frac{\mathbf{r}}{|\mathbf{r}|}\right) = \frac{\mathbf{r}}{|\mathbf{r}|^3}$ but it is much easier to recognize that both **r** and **F** are gradient fields and then invoke the theorem. It is easy to check that $\mathbf{r} = \nabla\left[\frac{1}{2}(x^2 + y^2 + z^2)\right]$ and $\mathbf{F} = \nabla\sqrt{x^2 + y^2 + z^2}$ so curl $\mathbf{r} = 0$ and curl $\mathbf{F} = 0$ everywhere where each of them is defined. **r** is defined everywhere and is a conservative field. **F** is not defined at the origin and is not a conservative field in a domain that includes the origin.

> Theorem: If **F** has continuous partial second derivatives, then div (curl **F**) = 0.

Proof: The proof is straightforward and the last step depends on Clairaut's Theorem.

$$\text{div (curl } \mathbf{F}) = \text{div}\left\{\left(\frac{\partial P}{\partial y} - \frac{\partial N}{\partial z}\right)\mathbf{i} + \left(\frac{\partial M}{\partial z} - \frac{\partial P}{\partial x}\right)\mathbf{j} + \left(\frac{\partial N}{\partial x} - \frac{\partial M}{\partial y}\right)\mathbf{k}\right\}$$

$$= \frac{\partial}{\partial x}\left(\frac{\partial P}{\partial y} - \frac{\partial N}{\partial z}\right) + \frac{\partial}{\partial y}\left(\frac{\partial M}{\partial z} - \frac{\partial P}{\partial x}\right) + \frac{\partial}{\partial z}\left(\frac{\partial N}{\partial x} - \frac{\partial M}{\partial y}\right)$$

$$= \frac{\partial^2 P}{\partial x \partial y} - \frac{\partial^2 N}{\partial x \partial z} + \frac{\partial^2 M}{\partial y \partial z} - \frac{\partial^2 P}{\partial y \partial x} + \frac{\partial^2 N}{\partial z \partial x} - \frac{\partial^2 M}{\partial z \partial y} = 0$$

Problems

In problems 1 to 6, calculate the divergence and curl of each vector field.

1. $\mathbf{F} = \langle x^2y,\ xyz,\ xz^3 \rangle$
2. $\mathbf{F} = \langle yz,\ xz,\ xy + 2 \rangle$
3. $\mathbf{F} = \langle x \cdot e^z,\ z \cdot e^y,\ y \cdot e^x \rangle$
4. $\mathbf{F} = \langle y,\ z,\ x^3 \rangle$
5. $\mathbf{F} = \langle x + y,\ y + z,\ z + x \rangle$
6. $\mathbf{F} = \langle x^2 + y^2,\ y^2 + z^2,\ z^2 + x^2 \rangle$

In problems 7 to 10, F is a vector field and f is a scalar function in 3D. Determine whether the given calculation is meaningful. If it is not, explain why. If it is, determine whether the result is a scalar or a vector.

7. div f , div F , div (curl F) , gradient (curl F), curl F
8. curl f , curl (div F) , gradient (div F) , div (curl (gradient f))
9. gradient f , curl (curl F) , curl (div (gradient f))
10. gradient F , curl (gradient F) , div (div F)

In problems 11 to 16, determine if the vector field F is conservative. If F is conservative, find a function f so the F=gradient f.

11. $\mathbf{F} = \langle x^2y,\ xyz,\ xz^3 \rangle$
12. $\mathbf{F} = \langle yz,\ xz,\ xy + 2 \rangle$
13. $\mathbf{F} = \langle y,\ z,\ x^3 \rangle$
14. $\mathbf{F} = \langle y^2,\ 2xy,\ 3z^2 \rangle$
15. $\mathbf{F} = \langle \sin(y \cdot z),\ x \cdot z \cdot \cos(y \cdot z),\ x \cdot y \cdot \cos(y \cdot z) + 2 \rangle$
16. $\mathbf{F} = \langle y^3,\ 3xy^2 + 5z,\ 5y \rangle$
17. Suppose curl $\mathbf{F}(1,2,3) = \langle 2,\ 4,\ 1 \rangle$. If you are at the location (5, 10, 5) and look towards (1,2,3) will you see the rotation at (1,2,3) to be clockwise or counterclockwise?
18. Suppose curl $\mathbf{F}(5,1,4) = \langle -1,\ 3,\ 1 \rangle$. If you are at the location (2, 10, 7) and look towards (5,1,4) will you see the rotation at (5,1,4) to be clockwise or counterclockwise?
19. Suppose curl $\mathbf{F}(6,3,2) = \langle 2,\ -1,\ 1 \rangle$. If you are at the location (4, 4, 3) and look towards (6,3,2) will you see the rotation at (6,3,2) to be clockwise or counterclockwise?
20. Suppose curl $\mathbf{F}(7,-4,5) = \langle -1,\ -2,\ 3 \rangle$. If you are at the location (9, 0, –1) and look towards (7,–4,5) will you see the rotation at (7,–4,5) to be clockwise or counterclockwise?

Practice Answers

Practice 1: div $\mathbf{F} = 2x + 2y + 1$ so div $\mathbf{F}(1,1,1) = 5$, div $\mathbf{F}(2,-3,4) = -1$ and div $\mathbf{F}(-0.5,0,3) = 0$.

Practice 2:

$$\text{curl } \mathbf{F} = \nabla \mathbf{x} \mathbf{F} = \begin{vmatrix} \mathbf{i} & \mathbf{j} & \mathbf{k} \\ \frac{\partial}{\partial x} & \frac{\partial}{\partial y} & \frac{\partial}{\partial z} \\ 2y+z & x^2 & 3z \end{vmatrix} = \langle 0-0,\ -(0-1),\ 2x-2 \rangle = \langle 0,\ 1,\ 2x-2 \rangle \ .$$

Practice 3:

$$\mathbf{F} = \langle -y,x,z \rangle.\ \text{curl } \mathbf{F} = \begin{vmatrix} \mathbf{i} & \mathbf{j} & \mathbf{k} \\ \frac{\partial}{\partial x} & \frac{\partial}{\partial y} & \frac{\partial}{\partial z} \\ -y & x & z \end{vmatrix} = (0-0)\mathbf{i} + (0-0)\mathbf{j} + (1-(-1))\mathbf{k} = \langle 0,0,2 \rangle \neq \mathbf{0}$$

so $\mathbf{F}$ is not a conservative field.

15.8 Parametric Surfaces

In earlier work we extended the basic idea of a function of one variable, y = f(x), to parametric equations in which both x and y were functions of a single parameter t: x=x(t) and y=y(t) (in 3D, z=(t), also). Basically this mapped a 1-dimensional object (think of a piece of wire) into 2 or 3 dimensions (Fig. 1). And by treating this mapping as a vector-valued function, $\mathbf{r}(t) = \langle x(t), y(t), z(t) \rangle$, we could use the ideas and tools of vectors. In this section we will do something similar, except now we will map a 2-dimension object (think of a sheet of paper) into 3 dimensions by treating x, y, and z as functions of two parameters, u and v: x=x(u,v), y=y(u,v) and z=z(u,v). These graphs will be surfaces in 3D. By using parametric surfaces we can work with more complicated and more general shapes (Fig. 2). And by treating these surfaces as vector-valued functions, $\mathbf{r}(u,v) = \langle x(u,v), y(u,v), z(u,v) \rangle$, we can again use the ideas and tools of vectors.

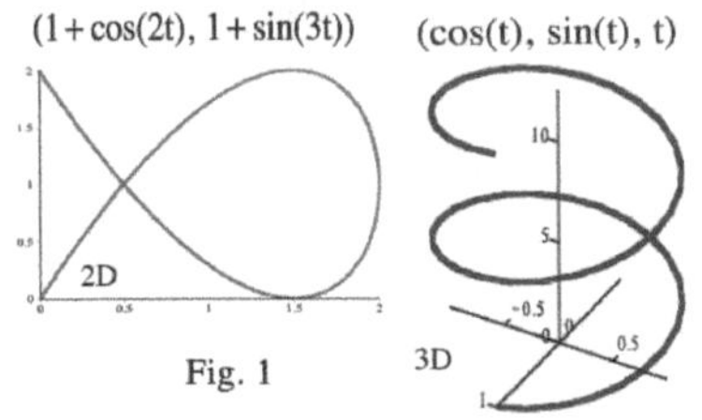

Fig. 1

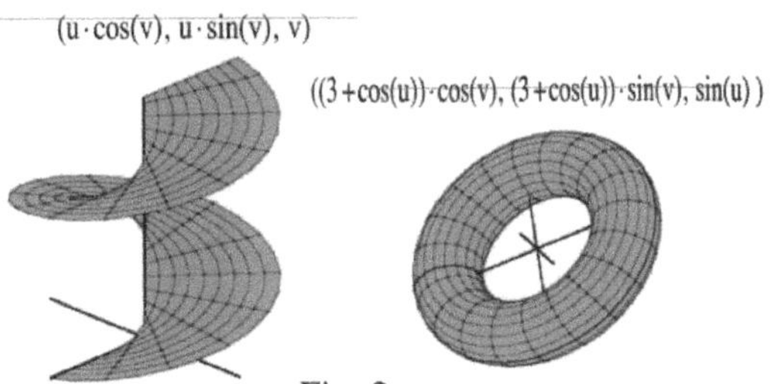

Fig. 2

Definitions: Parametric Function and Parametric Surface

Let x, y and z be functions of the parameters u and v for all (u, v) in a region D.
The vector-valued function $\mathbf{r}(u,v) = \langle x(u,v), y(u,v), z(u,v) \rangle$ is called a **parametric function** with domain D.
The set of points $S = (x(u,v), y(u,v), z(u,v))$ is called the **parametric surface** of the function **r** on domain D.

Note: As with the parametric functions of 1 variable, $\mathbf{r}(t) = \langle x(t), y(t), z(t) \rangle$, we will work with **r** as a vector $\langle x(u,v), y(u,v), z(u,v) \rangle$ but only plot the points $S = (x(u,v), y(u,v), z(u,v))$.

All of the rectangular, cylindrical and spherical coordinate functions we have used so far can be easily converted into parametric functions, and parametric functions give us even more freedom.

Example 1: (a) Convert $f(x,y) = x^2 + y^2$ with $-2 \le x \le 2$ and $-2 \le y \le 2$ into parametric form with u and v.

(b) Convert the spherical coordinate function $(3, \theta, \varphi)$, $0 \le \theta \le 2\pi$, $0 \le \varphi \le \pi/2$ (the top half of a sphere) into parametric form with u and v.

Solution: (a) Simply replace x with u and y with v and rewrite the z coordinate in terms of u and v: (u, v, u^2+v^2) with $-2\le u\le 2$ and $-2\le v\le 2$ (Fig. 3).

(b) $x=\rho\cdot\sin(\varphi)\cdot\cos(\theta)$, $y=\rho\cdot\sin(\varphi)\cdot\sin(\theta)$, $z=\rho\cdot\cos(\varphi)$ so we can replace

θ and φ with u and v and rewrite x, y and z as

$x=3\cdot\sin(v)\cdot\cos(u)$, $y=3\cdot\sin(v)\cdot\sin(u)$, $z=3\cdot\cos(v)$

with $0\le u\le 2\pi$, $0\le v\le \pi/2$. (Fig. 4)

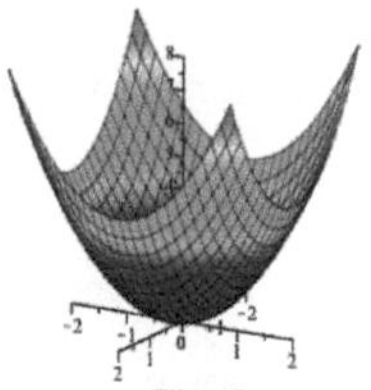
Fig. 3

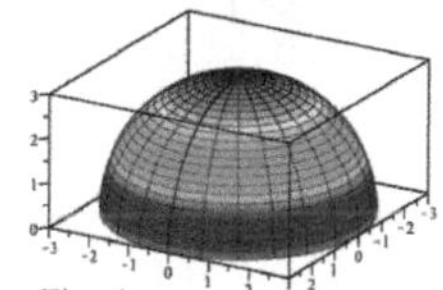
Fig. 4

Practice 1: Convert the cylindrical coordinate function $(r, \theta, 1+\sin(3\theta))$, $1\le r\le 2$, $0\le\theta\le 2\pi$ (Fig. 5) into parametric form with u and v.

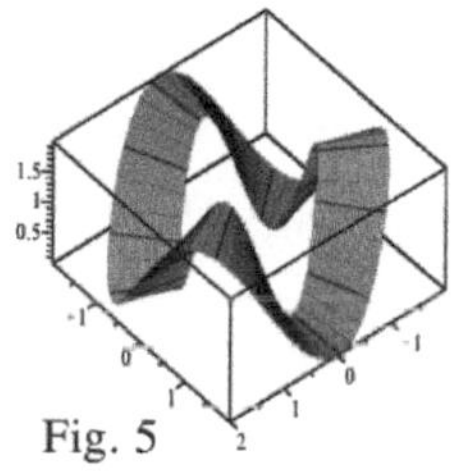
Fig. 5

One common parametric surface is the torus which can be thought of as a small circular tube around a larger circle (Fig. 6). The parametric equation for a torus with large radius R and small radius r is

$$((R-r\cdot\cos(u))\cdot\cos(v),\ (R-r\cdot\cos(u))\cdot\sin(v),\ r\cdot\sin(u))$$

The Appendix discusses the derivation of this surface as well as how to create a parametric representations of a small tubes around other curves in space.

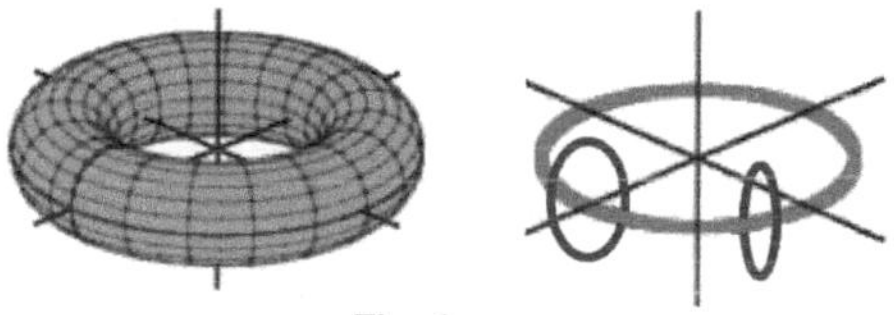
Fig. 6

Example 2: Write parametric equations for the surface generated by rotating the curve $z=\sqrt{y}$ around the y-axis for $0\le y\le 4$. (Fig. 7)

Solution: Put y=u for $0\le u\le 4$. Then the radius of the circle of revolution is $\sqrt{u}$ so $x=\sqrt{u}\cdot\cos(v)$ and $z=\sqrt{u}\cdot\sin(v)$ works.

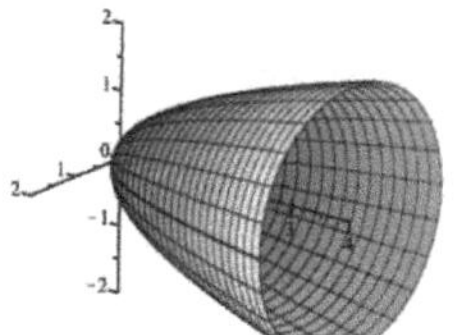
Fig. 7

Practice 2: Write parametric equations for the surface generated by rotating the curve $x=2+\sin(z)$ for $0\le z\le 5$ around the z-axis.

Surface Area of a Parametric Surface

The derivation and result for the surface area of a parametric surface **r**(u,v) are similar to the method for a surface defined by z = f(x,y) in Section 14.5. First partition the uv-domain D into small Δu by Δv rectangles (Fig. 8) with (u,v) at the lower left point of the rectangle. Call this rectangle R. Then **r**(u,v) maps this rectangle R onto a patch S on the parametric surface (Fig. 9). The bottom

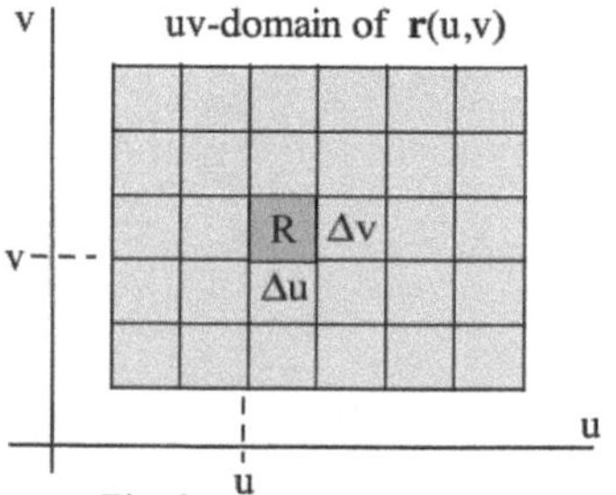

Fig. 8

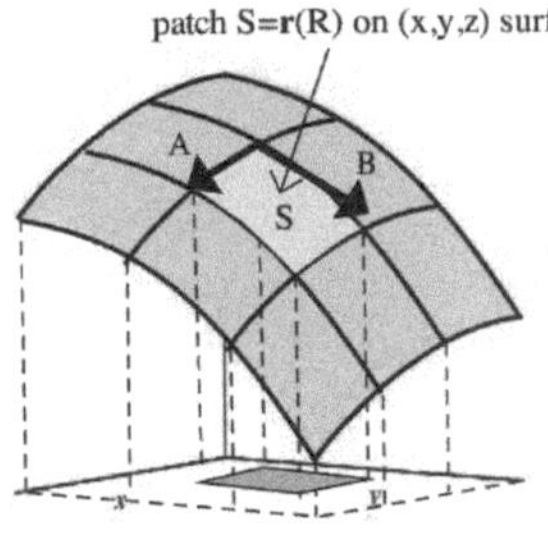

Fig. 9

corners of the R rectangle, (u,v) and (u+ Δu ,v), are mapped to **r**(u,v) and **r**(u+ Δu,v). The left edge corners of R, (u,v) and (u,v+ Δv), are mapped to **r**(u,v) and **r**(u, v+ Δv). The area of the patch S on the parametric surface is approximated by the area of the rectangle whose corners are the images under r of the corners of the R rectangle.

Let A be the vector from **r**(u,v) and **r**(u+ Δu,v), and B be the vector from **r**(u,v) and **r**(u, v+ Δv). Then (Fig. 10)

A= **r**(u+ Δu,v)– **r**(u,v) and **B**= **r**(u, v+ Δv)– **r** (u,v) so $\mathbf{A} = \frac{\mathbf{r}(u+\Delta u,v) - \mathbf{r}(u,v)}{\Delta u} \cdot \Delta u$ and

$\mathbf{B} = \frac{\mathbf{r}(u,v+\Delta v) - \mathbf{r}(u,v)}{\Delta v} \cdot \Delta v$. If Δu and Δv are small, then $\frac{\mathbf{r}(u+\Delta u,v) - \mathbf{r}(u,v)}{\Delta u} \approx \frac{\partial \mathbf{r}(u,v)}{\partial u} = \mathbf{r}_u(u,v)$

and $\frac{\mathbf{r}(u,v+\Delta v) - \mathbf{r}(u,v)}{\Delta v} \approx \frac{\partial \mathbf{r}(u,v)}{\partial v} = \mathbf{r}_v(u,v)$ so $\mathbf{A} \approx \mathbf{r}_u(u,v) \cdot \Delta u$ and $\mathbf{B} \approx \mathbf{r}_v(u,v) \cdot \Delta v$.

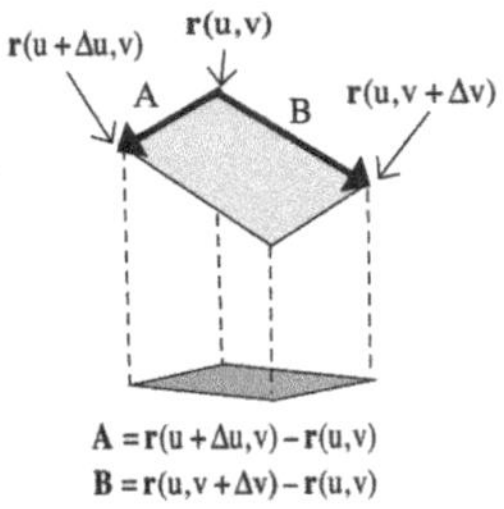

Fig. 10

Finally, the area of the patch S is approximately $|\mathbf{r}_u \mathrm{x} \mathbf{r}_v| \cdot \Delta u \cdot \Delta v$. Summing over all of the rectangles R in the uv-domain and taking limits as Δu and Δv both approach 0,

$$\sum_v \sum_u |\mathbf{r}_u \mathrm{x} \mathbf{r}_v| \cdot \Delta u \cdot \Delta v \to \iint_D |\mathbf{r}_u \mathrm{x} \mathbf{r}_v| \cdot dA \cdot$$

Surface Area of a Parametric Surface

If S is a smooth surface given by **r** (u,v) $= \langle x(u,v),\ y(u,v),\ z(u,v) \rangle$ with uv-domain D,

then {surface area of S} = $\iint_D |\mathbf{r}_u \mathrm{x} \mathbf{r}_v| \cdot dA$.

(Note: The surface S may fold over on itself, but that will not happen if the R rectangle is very small.)

If z=f(x,y), then we could parameterize the surface by x=u, y=v and z=f(u,v), and it is straightforward to derive that {Surface Area} = $\iint_R \sqrt{1+\left(\frac{\partial z}{\partial x}\right)^2+\left(\frac{\partial z}{\partial y}\right)^2} \cdot dA$ as we did in Section 14.5.

Proof: Parameterize this surface by x=u, y=v and z=f(u,v). Then $\mathbf{r}(u,v) = \langle u,v,f \rangle$ so $\mathbf{r}_u = \frac{\partial \mathbf{r}}{\partial u} = \langle 1,0,f_u \rangle$ and

$\mathbf{r}_v = \frac{\partial \mathbf{r}}{\partial v} = \langle 0,1,f_v \rangle$. $\mathbf{r}_u \mathrm{x} \mathbf{r}_v = \langle -f_u, -f_v, 1 \rangle$ so $|\mathbf{r}_u \mathrm{x} \mathbf{r}_v| = \sqrt{(f_u)^2 + (f_v)^2 + 1}$ and the surface area

is $\iint_R \sqrt{1+\left(\frac{\partial z}{\partial x}\right)^2+\left(\frac{\partial z}{\partial y}\right)^2} \cdot dA$.

Example 3: Use the parametric surface form to find the surface area of the curve $z = \sqrt{y}$ around the y-axis for $0 \le y \le 4$.

Solution: This surface was parameterized in Example 2 by y=u for $0 \le u \le 4$, $x = \sqrt{u} \cdot \cos(v)$ and $z = \sqrt{u} \cdot \sin(v)$ for $0 \le v \le 2\pi$. $\mathbf{r}_u = \frac{\partial \mathbf{r}}{\partial u} = \left\langle \frac{1}{2\sqrt{u}} \cdot \cos(v),\ 1,\ \frac{1}{2\sqrt{u}} \cdot \sin(v) \right\rangle$.

$\mathbf{r}_v = \frac{\partial \mathbf{r}}{\partial v} = \left\langle -\sqrt{u} \cdot \sin(v),\ 0,\ \sqrt{u} \cdot \cos(v) \right\rangle$, $\mathbf{r}_u \mathbf{x} \mathbf{r}_v = \left\langle \sqrt{u} \cdot \cos(v),\ \frac{1}{2},\ \sqrt{u} \cdot \sin(v) \right\rangle$ so

$|\mathbf{r}_u \mathbf{x} \mathbf{r}_v| = \sqrt{u \cdot \sin^2(v) + \frac{1}{4} + u \cdot \cos^2(v)} = \sqrt{\frac{1}{4} + u}$. Finally,

$$\text{Surface area} = \iint_D |\mathbf{r}_u \mathbf{x} \mathbf{r}_v| \cdot dA = \int_{v=0}^{2\pi} \int_{u=0}^{4} \sqrt{\frac{1}{4} + u}\ du\, dv$$

$$= \int_{v=0}^{2\pi} \frac{2}{3}\left(\frac{1}{4} + u\right)^{3/2} \Big|_{u=0}^{4} dv = \left(\frac{2}{3}\left(\frac{17}{4}\right)^{3/2} - \frac{2}{3}\left(\frac{1}{4}\right)^{3/2} \right) \cdot 2\pi = \left(\frac{1}{12}(17)^{3/2} - \frac{1}{12} \right) \cdot 2\pi$$

the same result as Practice 3 in Section 14.5.

And now we can extend surface area calculations to other systems such as cylindrical coordinates (r, θ, z).

Surface Area in Cylindrical Coordinates

If S is a smooth surface in cylindrical coordinates $(r, \theta, f(r, \theta))$ on domain D,

then $\{\text{surface area of S}\} = \iint_D \sqrt{r^2 (f_r)^2 + (f_\theta)^2 + r^2} \cdot dr \cdot d\theta$.

Proof: Parameterize this surface by $x = u \cdot \cos(v)$, $y = u \cdot \sin(v)$ (so u=r and v= θ) and $z = f(u \cdot \cos(v),\ u \cdot \sin(v))$. Then

$\mathbf{r}_u = \frac{\partial \mathbf{r}}{\partial u} = \langle \cos(v),\ \sin(v),\ f_u \rangle$ and $\mathbf{r}_v = \frac{\partial \mathbf{r}}{\partial v} = \langle -u \cdot \sin(v),\ u \cdot \cos(v),\ f_v \rangle$ so

$$\mathbf{r}_u \mathbf{x} \mathbf{r}_v = \begin{vmatrix} \mathbf{i} & \mathbf{j} & \mathbf{k} \\ \cos(v) & \sin(v) & f_u \\ -u \cdot \sin(v) & u \cdot \cos(v) & f_v \end{vmatrix} = \left\langle f_v \cdot \sin(v) - f_u \cdot u \cdot \cos(v), -(f_v \cdot \cos(v) + f_u \cdot u \cdot \sin(v)),\ u \cdot \cos^2(v) + u \cdot \sin^2(v) \right\rangle.$$

Finally, after some simplifying, $|\mathbf{r}_u \mathbf{x} \mathbf{r}_v| = \iint_D \sqrt{u^2 (f_u)^2 + (f_v)^2 + u^2} \cdot du \cdot dv$.

Example 4: Use the cylindrical coordinate integral form to calculate the surface area of the hemisphere $f(r, \theta) = \sqrt{R^2 - r^2}$ for $0 \le r \le R$ and $0 \le \theta \le 2\pi$.

Solution: $f_r = \dfrac{-r}{\sqrt{R^2 - r^2}}$ and $f_\theta = 0$ so

$$\iint_D \sqrt{r^2 (f_r)^2 + (f_\theta)^2 + r^2} \cdot dr \cdot d\theta = \int_0^{2\pi} \int_0^R \frac{Rr}{\sqrt{R^2 - r^2}}\, dr\, d\theta = \int_0^{2\pi} -R \cdot \sqrt{R^2 - r^2} \Big|_0^R d\theta = \int_0^{2\pi} R^2\, d\theta = 2\pi R^2 .$$

The surface area of the entire sphere is $4\pi R^2$.

Practice 3: Use the parametric surface form to find the surface area of $z = x^2 - y^2$ for $0 \le x^2 + y^2 \le 4$.

Problems

For Problems 1 to 10, sketch the parametric surface.

1. $x = u,\ y = v,\ z = \sqrt{v},\ 0 \le u \le 2,\ 0 \le v \le 4$.
2. $x = u,\ y = v,\ z = u^2,\ 0 \le u \le 2,\ 0 \le v \le 1$.
3. $x = 2 \cdot \cos(u),\ y = 2 \cdot \sin(u),\ z = v,\ 0 \le u \le 2\pi,\ 0 \le v \le 3$.
4. $x = \cos(u),\ y = v,\ z = \sin(u),\ 0 \le u \le 2\pi,\ 0 \le v \le 4$.
5. $x = u \cdot \cos(v),\ y = u \cdot \sin(v),\ z = u,\ 0 \le u \le 2,\ 0 \le v \le 2\pi$.
6. $x = u \cdot \cos(v),\ y = u^2,\ z = u \cdot \sin(v),\ 0 \le u \le 2,\ 0 \le v \le 2\pi$.
7. $x = u,\ y = \sqrt{u} \cdot \cos(v),\ z = \sqrt{u} \cdot \sin(v),\ 0 \le u \le 4,\ 0 \le v \le 2\pi$.
8. $x = \sqrt{u} \cdot \cos(v),\ y = u,\ z = 2 + \sqrt{u} \cdot \sin(v),\ 0 \le u \le 4,\ 0 \le v \le 2\pi$.
9. $x = u,\ y = v,\ z = \sin(v),\ 0 \le u \le 2,\ 0 \le v \le 2\pi$.
10. $x = \sin(u),\ y = v,\ z = u,\ 0 \le u \le \pi,\ 0 \le v \le 4$.

For problems 11 to 20, write the integral that represents the surface area of each surface in problems 1 to 10.

11. Write the surface area integral for the surface in problem 1.
12. Write the surface area integral for the surface in problem 2.

and so on for problems 13 to 20.

Practice Answers

Practice 1: $x = r \cdot \cos(\theta)$, $y = r \cdot \sin(\theta)$, $z = z$ Replace r and θ with u and v so we have $(u \cdot \cos(v), u \cdot \sin(v), 1 + \sin(3v))$ with $1 \le u \le 2$, $0 \le v \le 2\pi$.

Practice 2: $x = (2 + \sin(u)) \cdot \cos(v)$, $y = (2 + \sin(u)) \cdot \sin(v)$, $z = u$ works.

Practice 3: Parameterize this surface by $x = r \cdot \cos(\theta)$, $y = r \cdot \sin(\theta)$ and

$f = r^2 \cdot \cos^2(\theta) - r^2 \cdot \sin^2(\theta) = r^2 \cdot \cos(2\theta)$ for $0 \le r \le 2$, $0 \le \theta \le 2\pi$. Then

$f_r = 2r \cdot \cos(2\theta)$, and $f_\theta = -2r^2 \cdot \sin(2\theta)$ so $r^2(2r \cdot \cos(2\theta))^2 + \left(-4r^2 \cdot \sin(2\theta)\right)^2 + r^2 = 4r^4 + r^2$

Surface area $= \int_{\theta=0}^{2\pi} \int_{r=0}^{2} r \cdot \sqrt{4r^2 + 1}\, dr\, d\theta = \left[\frac{1}{12}(17)^{3/2} - \frac{1}{12} \right] \cdot 2\pi$.

Strangely, the paraboloid of Example 3 and the hyperboloid of Practice 3 have the same surface area.

Appendix: Building a Torus and Other Tubes

A torus is the collection of little circles centered on a large circle where the plane of each small circle is perpendicular to the large circle (Fig. A1). We can extend this idea to the collection of small circles centered on any curve in 2D or 3D so that the plane of each small circle is perpendicular to the curve (Fig. A2).

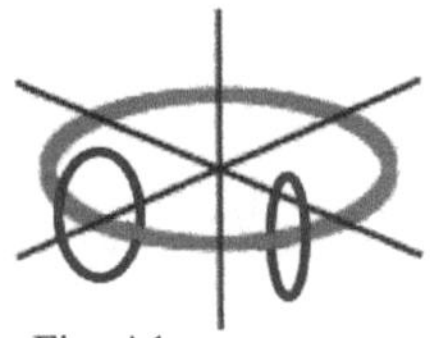

Fig. A1

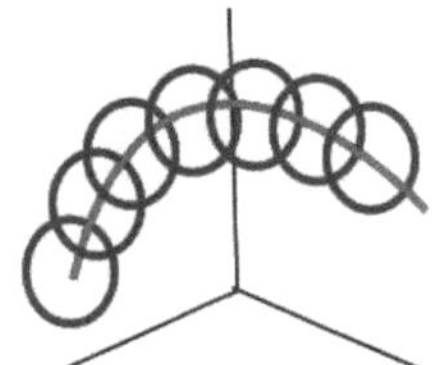

Fig. A2

In order to build these tubular surfaces (collections of small circles), we first need to know how to describe a circle in 3D at a given point, with a given radius and whose plane has a given normal vector (Fig. A3).

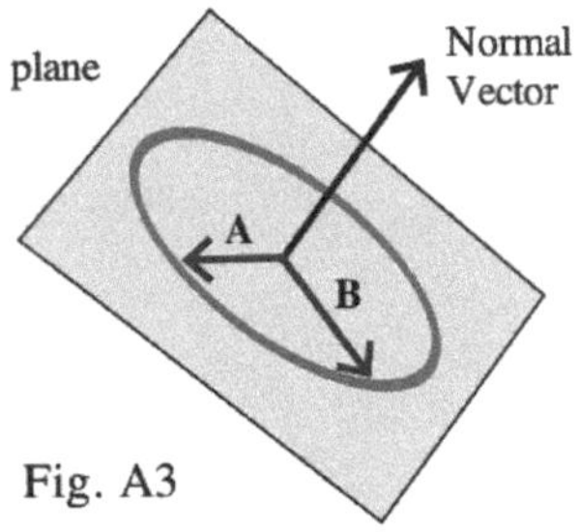

Fig. A3

Circle Algorithm: Center point at = (cx,cy,cz), radius R, and normal vector V= <vx, vy, vz>.

Then the plane of this circle is vx(x–cz)+vy(y–cy)+vz(z–cz)=0.

Pick two non-colinear unit vectors A and B perpendicular to V.

Then the equation of the circle we want is

$$\mathbf{C}(t) = \langle \text{center point} \rangle + \mathbf{A} \cdot R \cdot \cos(t) + \mathbf{B} \cdot R \cdot \sin(t).$$

Sometimes it is convenient for A and B to be perpendicular, but the only firm requirement on these unit vectors is that they are not co-linear.

Example A1: Find an equation for a circle with radius 2, center at P=(3,1,2) and normal vector $\mathbf{N} = \langle 4,2,1 \rangle$.

Solution: First we can find the equation of the plane that contains the point C and has normal vector N:

4(x–3)+2(y–1)+1(z–2)=0 so z= 2–4(x–3)–2(y-1). (Fig A4)

Next we need two unit vectors perpendicular to N:

$\mathbf{A} = \langle 1,-1,-2 \rangle / \sqrt{6}$ works as a first one.

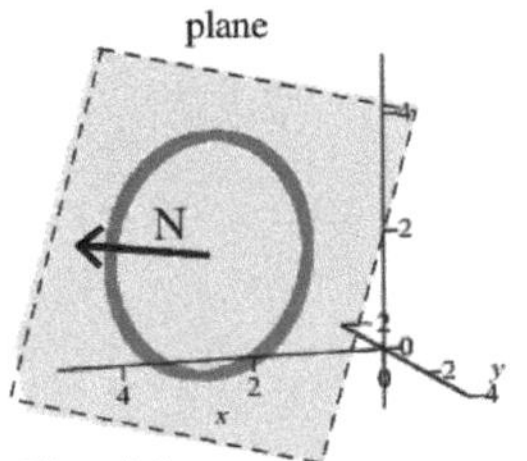

Fig. A4

For the second vector calculate $\mathbf{AxN} = \begin{vmatrix} \mathbf{i} & \mathbf{j} & \mathbf{k} \\ \frac{1}{\sqrt{6}} & \frac{-1}{\sqrt{6}} & \frac{-2}{\sqrt{6}} \\ 4 & 2 & 1 \end{vmatrix} = \frac{3}{\sqrt{6}} \langle 1,-3,2 \rangle$

and divide by its magnitude to create the second unit vector $\mathbf{B} = \langle 1,-3,2 \rangle / \sqrt{14}$ that is perpendicular to both N and A. Then the parametric equation of the circle:

$$\mathbf{C}(u) = P + \mathbf{A} \cdot 2 \cdot \cos(u) + \mathbf{B} \cdot 2 \cdot \sin(u) = \langle 3,1,2 \rangle + \langle 1,-1,-2 \rangle \cdot \frac{2}{\sqrt{6}} \cdot \cos(u) + \langle 1,-3,2 \rangle \cdot \frac{2}{\sqrt{14}} \cdot \sin(u)$$

with $0 \le u \le 2\pi$. The component functions are: $x(t) = 3 + \frac{2}{\sqrt{6}} \cdot \cos(u) + \frac{2}{\sqrt{14}} \cdot \sin(u)$,

$y(t) = 1 - \frac{2}{\sqrt{6}} \cdot \cos(u) - \frac{6}{\sqrt{14}} \cdot \sin(u)$, and $z(t) = 2 - \frac{4}{\sqrt{6}} \cdot \cos(u) + \frac{4}{\sqrt{14}} \cdot \sin(u)$.

Fig. A4 shows the vector **N**, the plane normal to **N** at P and the parametric circle.

Practice A1: Find an equation for a circle with radius 2, center at P=(2,4,3) and normal vector $\mathbf{N} = \langle -1,3,-2 \rangle$.

Once we can create a circle at a point P with those properties, then we can create a tube around a curve simply by moving the point P along the curve using the tangent vector to the curve as the vector **N**.

Tube Algorithm: To create a tube of radius r along the curve $\mathbf{F}(v) = \langle x(v),y(v),z(v) \rangle$ for $a \le v \le b$,

apply the Circle Algorithm at each point $\mathbf{F}(v)$ using $\mathbf{N} = \mathbf{F}\,'(v)$.

Example A2: Create parametric equations for a torus centered at the origin with large radius R and tube radius r . (Fig. A5)

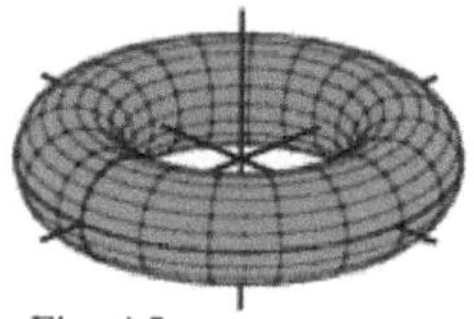
Fig. A5

Solution: For the large circle in the xy-plane, take

$\mathbf{F}(v) = \langle R \cdot \cos(v), R \cdot \sin(v), 0 \rangle$ with $0 \le v \le 2\pi$. Then the unit tangent vector is $\mathbf{N} = \mathbf{T} = \langle -\sin(v),\cos(v),0 \rangle$ and the unit vectors $\mathbf{A} = \mathbf{T}\,' = \langle -\cos(v),-\sin(v),0 \rangle$ and $\mathbf{B} = \mathbf{N}\mathbf{x}\mathbf{A} = \langle 0,0,1 \rangle$ are each perpendicular to **N** and are not co-linear. Putting this together, the parametric equation of the torus is

$C(u,v) = \mathbf{F}(v) + \mathbf{A} \cdot r \cdot \cos(u) + \mathbf{B} \cdot r \cdot \sin(u)$

$= \langle R \cdot \cos(v), R \cdot \sin(v), 0 \rangle + \langle -\cos(v), -\sin(v), 0 \rangle \cdot r \cdot \cos(u) + \langle 0,0,1 \rangle \cdot r \cdot \sin(u)$ for $0 \le u,v \le 2\pi$.

$x(u,v) = R \cdot \cos(v) - \cos(v) \cdot r \cdot \cos(u) + 0 = (R - r \cdot \cos(u)) \cdot \cos(v)$

$y(u,v) = R \cdot \sin(v) - \sin(v) \cdot r \cdot \cos(u) + 0 = (R - r \cdot \cos(u)) \cdot \sin(v)$

$z(u,v) = 0 + 0 + r \cdot \sin(u) = r \cdot \sin(u)$

Practice A2: Create parametric equations for a tube of radius r=1/2 centered on the curve $\mathbf{F}(v) = \langle 0,v,v^2 \rangle$ for $-1 \le v \le 2$.

Appendix Practice Answers

Solution A1: Thee are many correct answers depending on the unit vectors **A** and **B**, both perpendicular to **N** and not co-linear, that are chosen: $\mathbf{A} = \langle 1,1,1 \rangle/\sqrt{3}$ and $\mathbf{B} = \langle 5,-1,-4 \rangle/\sqrt{35}$ work. Then the parametric equation of the circle is

$C(u) = \mathbf{P} + \mathbf{A} \cdot 2 \cdot \cos(u) + \mathbf{B} \cdot 2 \cdot \sin(u)$

$= \langle 2,4,3 \rangle + \langle 1,1,3 \rangle \cdot \frac{2}{\sqrt{3}} \cdot \cos(u) + \langle 5,-1,-3 \rangle \cdot \frac{2}{\sqrt{35}} \cdot \sin(u)$ with $0 \le u \le 2\pi$.

Solution A2: For this curve $\mathbf{N} = \mathbf{T}(v)/|\mathbf{T}(v)| = \langle 0,1,2v\rangle/\sqrt{1+4v^2}$. Then $\mathbf{A} = \langle 1,0,0\rangle$ and $\mathbf{B} = \langle 0,2v,-1\rangle$ are both perpendicular to $\mathbf{N}$ and to each other. Putting all of this together (Fig. A6), $x(u,v) = \frac{1}{2}\cdot\cos(u)$, $y(u,v) = v + v\cdot\sin(u)$, and $z(u,v) = v^2 - \frac{1}{2}\cdot\sin(u)$ for $-1 \le v \le 2$, $0 \le u \le 2\pi$.·

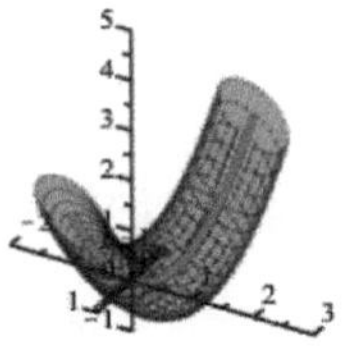

Fig. A6

15.9 Surface Integrals

Chapter 4 introduced integrals on intervals, Section 15.3 extended these ideas to integrals on paths in 2D and 3D, and Chapter 14 extended the ideas to integrals on 2D regions in the plane. This section goes one step further and considers integrals whose domains are parametric surfaces in 3D. In each previous situation the development was similar: partition, approximate on small pieces, sum, and take limits to achieve an integral. The approach here is the same. Surface integrals will be important in the coming sections on Stoke's Theorem and the Divergence Theorem and their applications.

Surface Integral for a Scalar Function

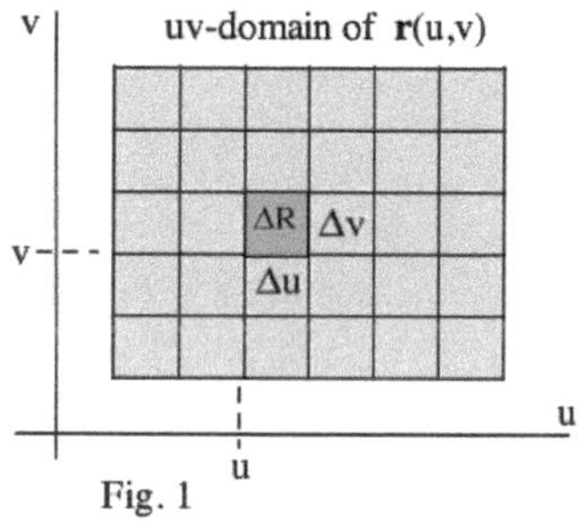

Fig. 1

If S is a smooth surface in xyz-space parameterized by $\mathbf{r}(u,v) = \langle x(u,v),\ y(u,v),\ z(u,v)\rangle$ with uv-domain R, then a partition of the uv-domain into small Δu by Δv rectangles ΔR (Fig. 1) is mapped by $\mathbf{r}$ to a partition of S into small patches in space with areas ΔS (Fig. 2), and the previous section showed that the area of each ΔS patch was $\Delta S \approx |\mathbf{r}_u x \mathbf{r}_v| \Delta u \cdot \Delta v$.

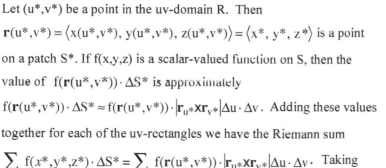

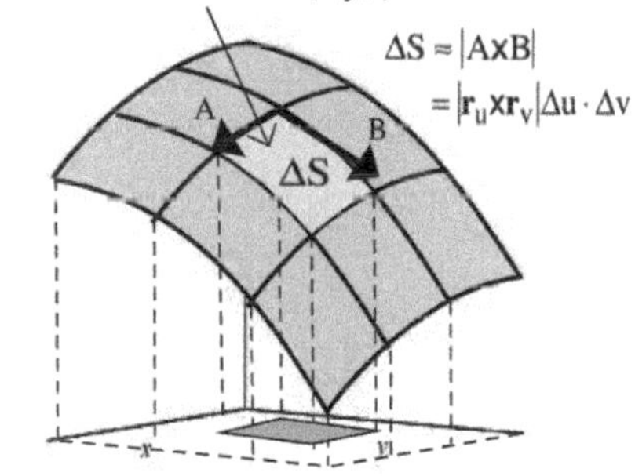

Fig. 2

Let (u*,v*) be a point in the uv-domain R. Then $\mathbf{r}(u^*,v^*) = \langle x(u^*,v^*),\ y(u^*,v^*),\ z(u^*,v^*)\rangle = \langle x^*,\ y^*,\ z^*\rangle$ is a point on a patch S*. If f(x,y,z) is a scalar-valued function on S, then the value of $f(\mathbf{r}(u^*,v^*))\cdot \Delta S^*$ is approximately $f(\mathbf{r}(u^*,v^*))\cdot \Delta S^* \approx f(\mathbf{r}(u^*,v^*))\cdot |\mathbf{r}_{u^*} x \mathbf{r}_{v^*}| \Delta u \cdot \Delta v$. Adding these values together for each of the uv-rectangles we have the Riemann sum

$$\sum_{u,v} f(x^*,y^*,z^*)\cdot \Delta S^* = \sum_{u,v} f(\mathbf{r}(u^*,v^*))\cdot |\mathbf{r}_{u^*} x \mathbf{r}_{v^*}| \Delta u \cdot \Delta v \cdot$$

Taking limits as $\Delta u,\ \Delta v \to 0$, we get

$$\iint_S f(x,y,z)\ dS = \iint_R f(\mathbf{r}(u,v))\cdot |\mathbf{r}_u x \mathbf{r}_v|\ dA .$$

> If S is a smooth surface parameterized by $\mathbf{r}(u,v)$ on domain R in the uv-domain, and f(x,y,z) is a scalar-valued function defined on S,
>
> then $\displaystyle\iint_S f(x,y,z)\ dS = \iint_R f(\mathbf{r}(u,v))\cdot |\mathbf{r}_u x \mathbf{r}_v|\ dA$.

This result enables us to evaluate many surface integrals in 3D as iterated integrals in u and v.

Note: You should notice the similarity of this result with the result for a line integral of a scalar-valued function along a curve C: $\int_C \mathbf{f}\, ds = \int_{t=a}^{b} \mathbf{f}(\mathbf{r}(t)) \cdot |\mathbf{r}'(t)|\, dt$. In this new situation the curve C is replaced with the surface S, and $|\mathbf{r}'(t)|dt$ is replaced with $|\mathbf{r}_u \mathbf{x} \mathbf{r}_v| dA$.

Note: If f(x,y,z)=1 for all (x,y,z) on S, then $\iint_S 1\ dS = \iint_R |\mathbf{r}_u \mathbf{x} \mathbf{r}_v|\ dA$ is simply the surface area of S.

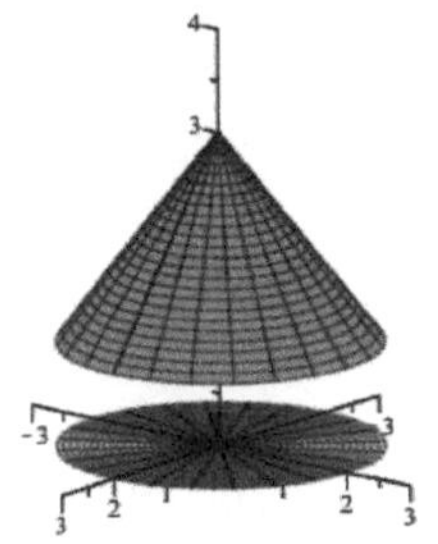
Fig. 3

Example 1: Let f(x,y,z)=1+z on the surface S parameterized by $\mathbf{r}(u,v) = \langle u \cdot \cos(v),\ u \cdot \sin(v),\ 3 - u \rangle$ with $0 \le u \le 2$ and $0 \le v \le 2\pi$. (Fig. 3) (a) Evaluate $\iint_S f(x,y,z)\ dS$.

(b) If the units of x, y and z are meters (m) and f is the surface density (g/m^2) at location (a,y,z), what are the units of $\iint_S f(x,y,z)\ dS$?

Solution: (a) $\iint_S f(x,y,z)\ dS = \iint_R f(\mathbf{r}(u,v)) \cdot |\mathbf{r}_u \mathbf{x} \mathbf{r}_v|\ dA$ so we need $f(\mathbf{r}(u,v))$, $\mathbf{r}_u$ and $\mathbf{r}_v$.

$\mathbf{r}_u = \langle \cos(v),\ \sin(v),\ -1 \rangle$, $\mathbf{r}_v = \langle -u \cdot \sin(v),\ u \cdot \cos(v),\ 0 \rangle$ and

$$\mathbf{r}_u \mathbf{x} \mathbf{r}_v = \begin{vmatrix} i & j & k \\ \cos(v) & \sin(v) & -1 \\ -u \cdot \sin(v) & u \cdot \cos(v) & 0 \end{vmatrix} = \langle -u \cdot \cos(v),\ u \cdot \sin(v),\ u \rangle \text{ so } |\mathbf{r}_u \mathbf{x} \mathbf{r}_v| = u\sqrt{2}.$$

$$\iint_S f(x,y,z)\ dS = \iint_R f(\mathbf{r}(u,v)) \cdot |\mathbf{r}_u \mathbf{x} \mathbf{r}_v|\ dA = \int_{v=0}^{2\pi} \int_{u=0}^{2} [1 + (3 - u)](u\sqrt{2})\ du\ dv$$

$$= \int_{v=0}^{2\pi} \sqrt{2}\left(-\frac{1}{3}u^3 + 2u^2 \right) \Big|_{u=0}^{2} dv = \int_{v=0}^{2\pi} \frac{16}{3}\sqrt{2}\ dv = \frac{32}{3}\sqrt{2}\pi .$$

(b) $\iint_S f(x,y,z)\ dS$ is the mass of the surface S, and the units are $(g/m^2)(m^2)$= g.

Practice 1: Evaluate $\iint_S (2 + x)\ dS$ on the surface S in Example 1.

Example 2: Let f(x,y,z)=xy on the surface S that is the part of the plane $z = 3–x–y$ that is in the first octant. Evaluate $\iint_S f(x,y,z)\ dS$.

Solution: S can be parameterized by $\mathbf{r}(u,v) = \langle u,\ v,\ 3 - u - v \rangle$ for $0 \le u \le 3$ and $0 \le v \le 3 - u$. Then $r_u = \langle 1,\ 0,\ -1 \rangle$, $r_v = \langle 0,\ 1,\ -1 \rangle$, $\mathbf{r}_u \mathbf{x} \mathbf{r}_v = \langle 1,\ 1,\ 1 \rangle$ and $|r_u x r_v| = \sqrt{3}$. $f(\mathbf{r}(u,v)) = uv$ so

$$\iint_S f(x,y,z)\ dS = \int_0^3 \int_0^{u-3} uv\sqrt{3}\ dv\ du = \int_0^{u-3} \frac{\sqrt{3}}{2} u \cdot (3 - u)^2\ dv = 3\sqrt{3}.$$

The units of the answer are (units of f)(units of S).

Practice 2: Let f(x,y,z)=x on the surface $S=\{(x,y,z):\ x^2+y^2=3,\ 0\le z\le 2\}$. Evaluate $\iint_S f(x,y,z)\ dS$.

> If the graph of f(x,y,z) with z=g(x,y) is a smooth surface S in xyz-space parameterized by $\mathbf{r}(u,v)$ on domain R in uv-space, ,
>
> then $\iint_S f(x,y,z)\ dS = \iint_R f(\mathbf{r}(u,v))\cdot\sqrt{1+(g_x)^2+(g_y)^2}\ dA$.

Proof: We can parameterize this surface by setting u=x and v=y so $\mathbf{r}(u,v)=\langle u,\ v,\ g(u,v)\rangle$.

$\mathbf{r}_u=\langle 1,\ 0,\ g_x\rangle$ and $\mathbf{r}_v=\langle 0,\ 1,\ g_y\rangle$ so $\mathbf{r}_u\mathbf{x}\mathbf{r}_v=\begin{vmatrix}\mathbf{i} & \mathbf{j} & \mathbf{k}\\ 1 & 0 & g_x\\ 0 & 1 & g_y\end{vmatrix}=\langle -g_x,\ -g_y,\ 1\rangle$ and $|\mathbf{r}_u\mathbf{x}\mathbf{r}_v|=\sqrt{1+(g_x)^2+(g_x)^2}$

Then $\iint_S f(x,y,g(x,y))\ dS=\iint_R f(\mathbf{r}(u,v))\cdot\sqrt{1+(g_x)^2+(g_x)^2}\ dA$.

Example 3: Let $f(x,y,z)=z\sqrt{x^2+y^2}$ on the heliocoid surface S parameterized by $\mathbf{r}(u,v)=\langle u\cdot\cos(v),\ u\cdot\sin(v),\ v\rangle$ with $0\le u\le 1$ and $0\le v\le 2\pi$.

Evaluate $\iint_S f(x,y,z)\ dS$. (Fig. 4 shows a heliocoid with $0\le v\le 4\pi$.)

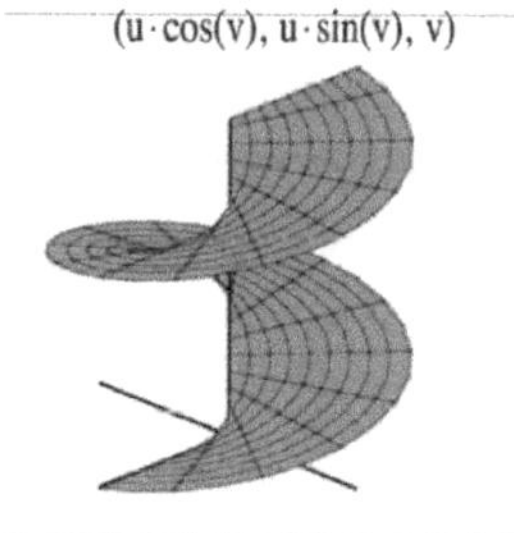

Fig. 4

Solution: $\iint_S f(x,y,z)\ dS=\iint_R f(\mathbf{r}(u,v))\cdot\sqrt{1+(g_x)^2+(g_y)^2}\ dA$ so we need $f(\mathbf{r}(u,v))$, $\mathbf{r}_u$ and $\mathbf{r}_v$. $\mathbf{r}_u=\langle\cos(v),\ \sin(v),\ 1\rangle$, $\mathbf{r}_v=\langle -u\cdot\sin(v),\ u\cdot\cos(v),\ 0\rangle$ and

$$\mathbf{r}_u\mathbf{x}\mathbf{r}_v=\begin{vmatrix} i & j & k\\ \cos(v) & \sin(v) & 1\\ -u\cdot\sin(v) & u\cdot\cos(v) & 0\end{vmatrix}=\langle -u\cdot\cos(v),\ -u\cdot\sin(v),\ u\rangle \text{ so } |\mathbf{r}_u\mathbf{x}\mathbf{r}_v|=u\sqrt{2}\ .$$

$f(x,y,z)=z\sqrt{x^2+y^2}=v\cdot u$ so

$$\iint_S f(x,y,z)\ dS=\iint_R f(\mathbf{r}(u,v))\cdot|\mathbf{r}_u\mathbf{x}\mathbf{r}_v|\ dA=\int_{v=0}^{2\pi}\int_{u=0}^{1}[v\cdot u](u\sqrt{2})\ du\ dv=\left(2\pi^2\right)\left(\frac{1}{3}\sqrt{2}\right).$$

Practice 3: Evaluate $\iint_S (1+y)\ dS$ on the surface S in Example 3.

Note: If z=g(x,y) and f(x,y,z)=1 for all (x,y), then $\iint_S f(x,y,z)\ dS$ is the surface area of S, and this area equals $\iint_R \sqrt{1+(g_x)^2+(g_x)^2}\ dA$, then same result we saw in Section 14.5 .

Oriented Surfaces and the Unit Normal Vector n

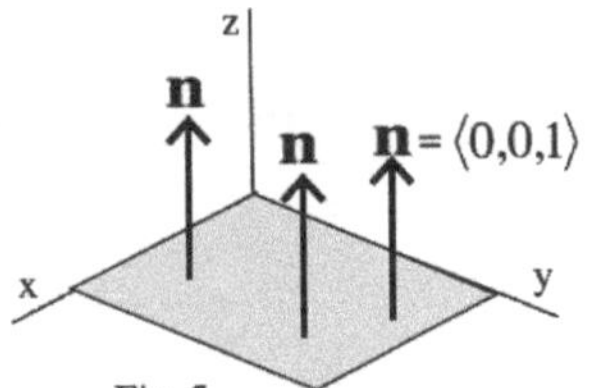

Fig. 5

Before investigating surface integrals for vector-valued functions, some vocabulary and technical issues need to be considered: oriented surfaces and an orientation for vector **n** that is normal (perpendicular) to the surface.

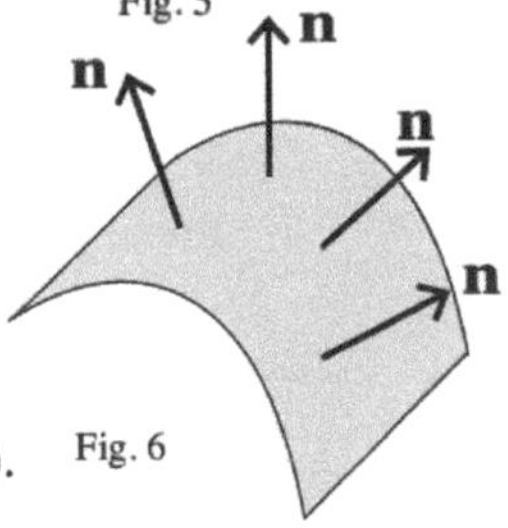

Fig. 6

A flat piece of paper in the xy-plane has a normal vector $\mathbf{n} = \langle 0,0,1 \rangle$ pointing upward at each point on the paper (Fig. 5). If we gently fold (but not crease) the paper, then the normal vector **n** will change continuously depending on its location on the paper (Fig. 6). If we follow a closed path on the paper that does not cross the paper's edge then the direction of the normal vector will change continuously and will return to the starting location pointing in the its original direction (Fig. 7). Such a surface is called oriented.

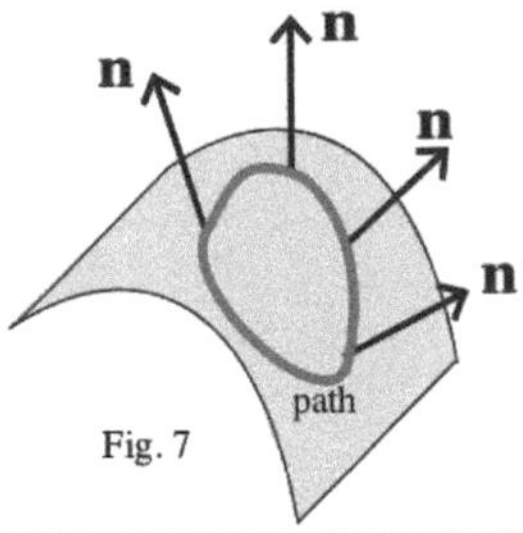

Fig. 7

A smooth surface S is **oriented** if

* S has a non–zero normal vector at each point,
* the direction of the normal vector varies continuously as we move along S (not crossing an edge),
* and, a normal vector returns to its original orientation when it returns to its original position after moving along any closed path on S (not crossing an edge).

Fortunately, most surfaces are oriented. The most famous example of a non-oriented surface is a Mobius strip (Fig. 8). If we start with a normal vector **n** at any point and travel along the middle of the strip (not crossing an edge), then we end up at the starting point again but with the normal vector now pointing in the direction –**n** (Fig. 9).

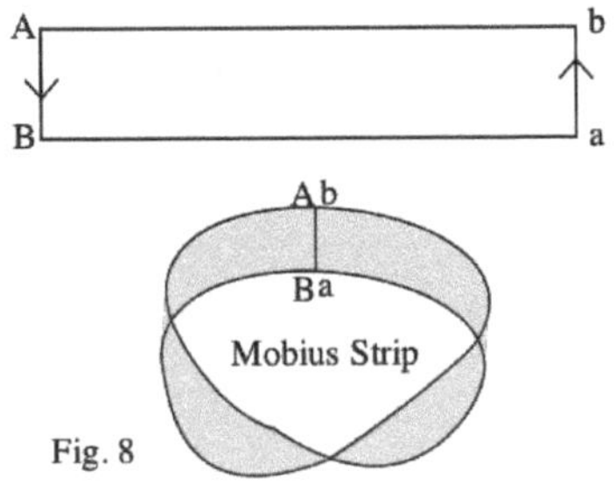

Fig. 8

The results that follow require that our surfaces be oriented.

However, at a point A on an oriented surface there are two normal vectors, and we need to select one of them for the orientation. If the surface encloses a region of space, the convention is to pick the normal vector which points outward from the enclosed region (Fig. 10).

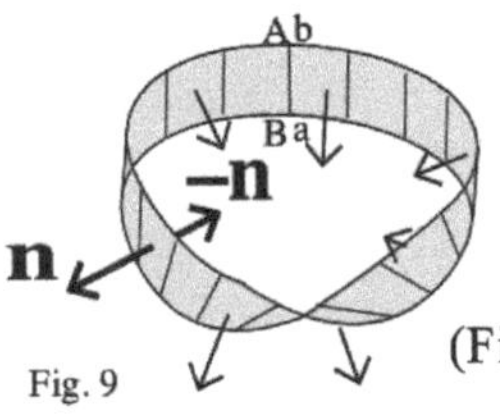

Fig. 9

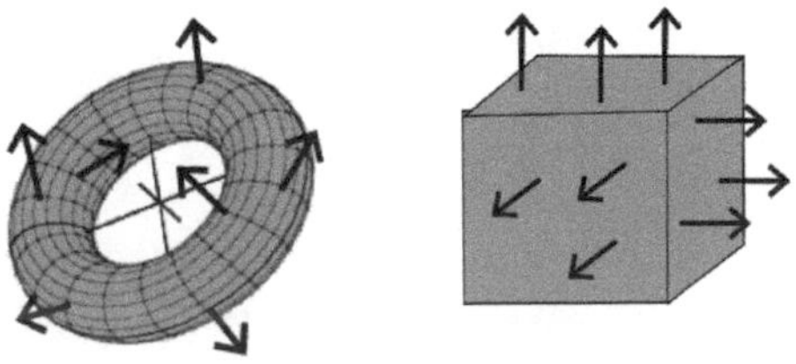
Fig. 10

Surface Integral for a Vector–Valued Function

Let ΔS be a small patch on the smooth, oriented surface S with oriented normal vector **n** (at some point of S). Then the magnitude of the vector F crossing the patch is the projection of F onto **n**. If we think of the vector field as water moving **F** at each point, then the amount of water passing through the patch ΔS in the direction of **n** is $(\mathbf{F} \bullet \mathbf{n})(\text{area of } \Delta S)$. Visually, that volume of that water (per unit of time) is the volume of the prism in Fig. 11. As before, adding the values of $(\mathbf{F} \bullet \mathbf{n})(\text{area of } \Delta S)$ for all of the patches we have the Riemann sum $\sum_{\Delta u, \Delta v} (\mathbf{F} \bullet \mathbf{n})(\text{area of } \Delta S)$.

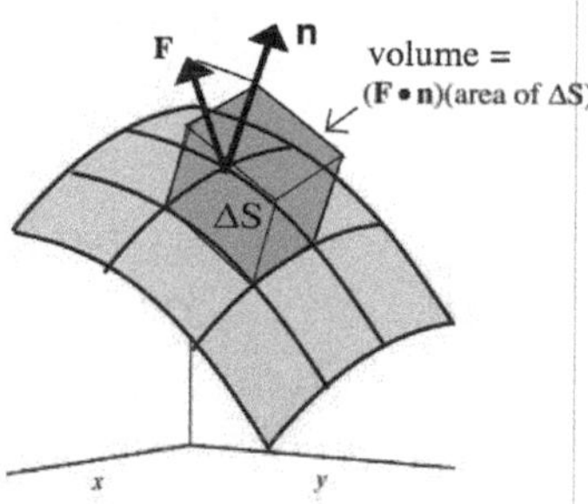

Fig. 11

Taking the limit as all of the Δu and Δv approach zero, we have the surface integral $\iint_S \mathbf{F} \cdot \mathbf{n} \, dS$. If the surface S is parameterized by the $\mathbf{r}(u,v) = (\mathbf{x}(u,v), \mathbf{y}(u,v), \mathbf{z}(u,v))$ for (u,v) in a region R, then $\Delta S = |\mathbf{r}_u \mathbf{x} \mathbf{r}_v| \cdot \Delta u \cdot \Delta v$. The vector $\mathbf{r}_u \mathbf{x} \mathbf{r}_v$ is normal to the surface S and the unit normal vector is $\mathbf{n} = \frac{\mathbf{r}_u \mathbf{x} \mathbf{r}_v}{|\mathbf{r}_u \mathbf{x} \mathbf{r}_v|}$ so $\mathbf{F} \bullet \mathbf{n} \, dS = \mathbf{F} \bullet \frac{\mathbf{r}_u \mathbf{x} \mathbf{r}_v}{|\mathbf{r}_u \mathbf{x} \mathbf{r}_v|} \cdot |\mathbf{r}_u \mathbf{x} \mathbf{r}_v| \cdot \Delta u \cdot \Delta v$ and $\mathbf{F} \bullet \mathbf{n} \, dS = \mathbf{F} \bullet (\mathbf{r}_u \mathbf{x} \mathbf{r}_v) \, dA$.

Definition: Surface Integral of **F** over S

If **F** is a continuous vector field over the oriented surface S parameterized by $\mathbf{r}(u,v)$ and having unit normal vector **n**,

then the surface integral of **F** over S is $\iint_S \mathbf{F} \cdot \mathbf{n} \, dS = \iint_{\mathbf{R}} \mathbf{F} \bullet (\mathbf{r}_u \mathbf{x} \mathbf{r}_v) \, dA$.

This integral is also called the **flux** of **F** across S.

Example 4: Suppose S is the part of the plane $3x+2y+6z=30$ with domain $0 \le x \le 4$ and $0 \le y \le 6$ (Fig. 12). This surface can be parameterized by $\mathbf{r}(u,v) = \left\langle u, v, 5 - \frac{u}{2} - \frac{v}{3} \right\rangle$ so $\mathbf{r}_u = \left\langle 1, 0, -\frac{1}{2} \right\rangle$, $\mathbf{r}_v = \left\langle 0, 1, -\frac{1}{3} \right\rangle$ and. If $\mathbf{F}(x,y,z) = \langle 0, 0, -2 \rangle$ then

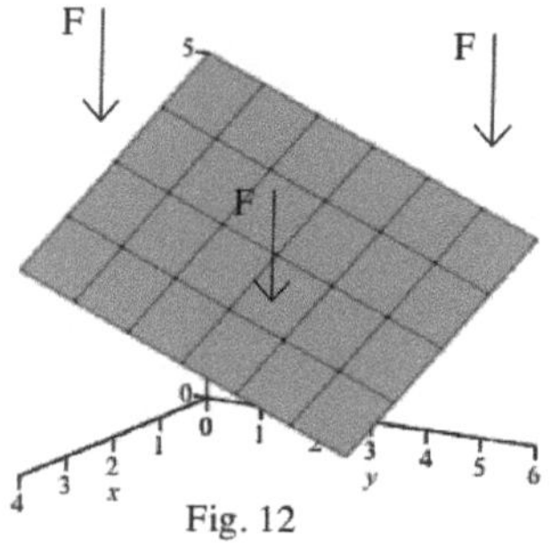

Fig. 12

$$\iint_S \mathbf{F} \cdot \mathbf{n} \, dS = \iint_{\mathbf{R}} \mathbf{F} \bullet (\mathbf{r}_u \mathbf{x} \mathbf{r}_v) \, dA = \iint_{\mathbf{R}} \langle 0, 0, -2 \rangle \bullet \left\langle \frac{1}{2}, \frac{1}{3}, 1 \right\rangle dA = \int_{u=0}^{4} \int_{v=0}^{6} -2 \, dv \, du$$

$$= (-2)(\text{area of R}) = (-2)(24) .$$

If $\mathbf{F} = \langle 0, 0, -2 \rangle$ is the velocity of water in m/s, and x and y are given in meters (m), then the units of $\iint_S \mathbf{F}\cdot\mathbf{n}\, dS$ are m^3/s : 48 m^3/s pass through the surface S. The negative sign in the answer results because the angle between $\mathbf{F}$ and $\mathbf{n}$ is greater than 90° . If we had picked the opposite normal vector, then the answer would have been +48.

Practice 4: Use the surface S from Example 3 and calculate $\iint_S \mathbf{F}\cdot\mathbf{n}\, dS$ for $\mathbf{F} = \langle 0, -3, 0 \rangle$ and for $\mathbf{F} = \langle 1, 2, 3 \rangle$.

If F is not a constant vector field as in the previous Example and Practice problems, then we need to rewrite **F**(x,y,z) as **F**(x(u,v), y(u,v), z(u,v)).

Practice 5: Suppose S is the part of the surface z=f(x,y) above the region R in the xy-plane and that $\mathbf{F} = \langle M, N, P \rangle$.

Show that $\iint_S \mathbf{F}\cdot\mathbf{n}\, dS = \iint_R \{-P\cdot f_x - Q\cdot f_y + P\}\, dA$.

When S is a sphere

Spheres occur often in applications so it is worthwhile to see the calculations for a sphere. A sphere of radius R centered at the origin is easily described in spherical coordinates as (R, θ, φ) with $0 \le \theta \le 2\pi$ and $0 \le \varphi \le \pi$. Setting $u = \theta$ and $v = \varphi$ and then converting to rectangular coordinates we have (as in 15.7 Example 1) $x(u,v) = R\cdot\sin(v)\cdot\cos(u)$, $y(u,v) = R\cdot\sin(v)\cdot\sin(u)$ and $z(u,v) = R\cdot\cos(v)$. Since $\mathbf{r}(u,v) = \langle x, y, z \rangle$, $\mathbf{r}_u = \langle -R\cdot\sin(v)\cdot\sin(u), R\cdot\sin(v)\cdot\cos(u), 0 \rangle$ and $\mathbf{r}_v = \langle R\cdot\cos(v)\cdot\cos(u), R\cdot\cos(v)\cdot\sin(u), -R\cdot\sin(v) \rangle$. Finally, $\mathbf{r}_u \mathbf{x} \mathbf{r}_v = -R^2\left\langle \sin^2(v)\cdot\cos(u), \sin^2(v)\cdot\sin(u), \sin(v)\cdot\cos(v) \right\rangle$. For an outward facing normal vector **n**, take $\mathbf{r}_u \mathbf{x} \mathbf{r}_v = R^2\left\langle \sin^2(v)\cdot\cos(u), \sin^2(v)\cdot\sin(u), \sin(v)\cdot\cos(v) \right\rangle$.

Example 5: Suppose S is a sphere of radius 1 centered at the origin and $F(x,y,z) = \langle x,y,z \rangle$ is a radial vector field.

(a) Determine the flux of **F** across S. (b) Determine the flux of **F** across S when S has radius R.

Solution: (a) $F(x,y,z) = \langle x,y,z \rangle = \langle \sin(v)\cdot\cos(u), \sin(v)\cdot\sin(u), \cos(v) \rangle$. Then

$$\iint_S \mathbf{F}\cdot\mathbf{n}\, dS = \iint_{\mathbf{R}} \mathbf{F} \bullet (\mathbf{r}_u \mathbf{x} \mathbf{r}_v)\, dA$$

$$= \iint_{\mathbf{R}} \langle \sin(v)\cdot\cos(u), \sin(v)\cdot\sin(u), \cos(v) \rangle \bullet \left\langle \sin^2(v)\cdot\cos(u), \sin^2(v)\cdot\sin(u), \sin(v)\cdot\cos(v) \right\rangle dA$$

$$= \iint_{\mathbf{R}} \sin^3(u)\cdot\cos^2(v) + \sin^3(v)\cdot\sin^2(u) + \sin(v)\cdot\cos^2(v)\, dA = \int_{u=0}^{2\pi} \int_{v=0}^{\pi} \sin(v)\, dv\, du = 4\pi \ .$$

(b) The only change in the calculation from part (a) is that now $\mathbf{r}_u \mathbf{x} \mathbf{r}_v$ has the factor R^2 so the result from part (a) needs to be multiplied by R^2 : flux = $4\pi R^2$.

Practice 6: Suppose S is the hemisphere $S = \{(x,y,z): x^2 + y^2 + z^2 = 1 \text{ and } 0 \le z\}$ and $F(x,y,z) = \langle z,x,y \rangle$. Determine the flux of F across S.

Connections with line integrals

There are nice parallels between the integrals of scalar and vector-valued functions along a curves C in 2D (section 15.3) and those on surfaces S in 3D.

scalar f on a curve C parameterized by $\mathbf{r}(t)$: $\int_C \mathbf{f}\, ds = \int_{t=a}^{b} \mathbf{f}(\mathbf{r}(t)) \cdot |\mathbf{r}'(t)|\, dt$

scalar f on a surface S parameterized by $\mathbf{r}(u,v)$: $\iint_S f(x,y,z)\, dS = \iint_R f(\mathbf{r}(u,v)) \cdot |\mathbf{r}_u \mathbf{x} \mathbf{r}_v|\, dA$

vector–valued **F** on a curve C parameterized by $\mathbf{r}(t)$: $\int_C \mathbf{F} \bullet \mathbf{T}\, ds = \int_{t=a}^{b} \mathbf{F}(\mathbf{r}(t)) \bullet \mathbf{r}'(t)\, dt$

vector–valued **F** on a surface S parameterized by $\mathbf{r}(u,v)$: $\iint_S \mathbf{F} \cdot \mathbf{n}\, dS = \iint_{\mathbf{R}} \mathbf{F} \bullet (\mathbf{r}_u \mathbf{x} \mathbf{r}_v)\, dA$

Problems

1. $f(x,y,z)=x+y+z$ and $\mathbf{r}(u,v) = \langle u + 3v, 2u - v, 3u + v \rangle$. What is the value of $f(\mathbf{r}(u,v)) \cdot |\mathbf{r}_u \mathbf{x} \mathbf{r}_v|\, \Delta A$ when u=1, v=2, $\Delta u = 0.3$ and $\Delta v = 0.1$?

2. $f(x,y,z)=x+y+z$ and $\mathbf{r}(u,v) = \langle 2u + v, 3u - v, u + 2v \rangle$. What is the value of $f(\mathbf{r}(u,v)) \cdot |\mathbf{r}_u \mathbf{x} \mathbf{r}_v|\, \Delta A$ when u=2, v=3, $\Delta u = 0.1$ and $\Delta v = 0.2$?

3. $f(x,y,z) = 2x + y^2 - z$ and $\mathbf{r}(u,v) = \langle u^2, 3u + v, v^2 \rangle$. What is the value of $f(\mathbf{r}(u,v)) \cdot |\mathbf{r}_u \mathbf{x} \mathbf{r}_v|\, \Delta A$ when u=2, v=1, $\Delta u = 0.3$ and $\Delta v = 0.2$?

4. $f(x,y,z) = 2x + y^2 - z$ and $\mathbf{r}(u,v) = \langle u^2, 3u + v, v^2 \rangle$. What is the value of $f(\mathbf{r}(u,v)) \cdot |\mathbf{r}_u \mathbf{x} \mathbf{r}_v|\, \Delta A$ when u=3, v=2, $\Delta u = 0.3$ and $\Delta v = 0.2$?

5. $f(x,y,z) = x^2 + 4y + z$ on the surface $S = \{(x,y,z): 0 \le x \le 3, 0 \le y \le 2, z = 4)$. Evaluate $\iint_S f(x,y,z)\, dS$.

6. $f(x,y,z) = x^2 + 4y + z$ on the surface $S = \{(x,y,z): 0 \le x \le 2, y = 3, 1 \le z \le 4)$. Evaluate $\iint_S f(x,y,z)\, dS$.

7. $f(x,y,z) = y^2$ on the surface $S = \{(x,y,z): x + y + z = 4 \text{ in first octant}\}$ Evaluate $\iint_S f(x,y,z)\, dS$.

8. $f(x,y,z) = xy$ on the surface $S = \{(x,y,z): z = 1 + x^2 + y^2, 0 \le x \le 2, 0 \le y \le 2\}$. Evaluate $\iint_S f(x,y,z)\, dS$.

9. $\mathbf{F}(x,y,z) = \langle x, 2y, 3z\rangle$ on the surface $S = \{(x,y,z): 0 \le x \le 3, 0 \le y \le 2, z = 4)$. Determine the flux of **F** across S. If the units of **F** are liters/second and the units of x, y and z are meters, what are the units of the flux?

10. $\mathbf{F}(x,y,z) = \langle x, -y, z\rangle$ on the surface $S = \{(x,y,z): z = x^2 + y^2, 0 \le x \le 2, 0 \le y \le 2\}$ Determine the flux of **F** across S. If the units of **F** are grams/meter^2 and the units of x, y and z are meters, what are the units of the flux?

11. $\mathbf{F}(x,y,z) = \langle x, y, z\rangle$ on the elliptical cylinder $S = \{(x,y,z): x^2 + 4y^2 = 4, 0 \le z \le 3\}$ Determine the flux of **F** across S.

12. $\mathbf{F}(x,y,z) = \langle 0, 0, K\rangle$ on the paraboloid $S = \{(x,y,z): x^2 + y^2 \le A^2, z = A^2 - x^2 - y^2\}$ Determine the flux of **F** across S.

13. Suppose **F** is the same as in Problem 12 but now S is the "stretched" paraboloid $S = \{(x,y,z): x^2 + y^2 \le A^2, z = C(A^2 - x^2 - y^2)\}$. Determine the flux of **F** across S.

Practice Answers

Practice 1: From Example 1, $|\mathbf{r}_u \mathbf{x} \mathbf{r}_v| = u\sqrt{2}$ and $x = u \cdot \cos(v)$ so

$$\iint_S (2 + x)\, dS = \int_{v=0}^{2\pi} \int_{u=0}^{2} [2 + u \cdot \cos(v)] \cdot (u\sqrt{2})\, du\, dv = \int_{v=0}^{2\pi} \left(u^2\sqrt{2} + \frac{1}{3}u^3\sqrt{2} \cdot \cos(v) \right) \Big|_{u=0}^{2} dv = 8\sqrt{2}\pi .$$

Practice 2: S can be parameterized by $\mathbf{r}(u,v) = \langle 3 \cdot \cos(u), 3 \cdot \sin(u), v\rangle$ with $0 \le u \le 2\pi$ and $0 \le v \le 2$. Then

$r_u = \langle -3 \cdot \sin(u), 3 \cdot \cos(u), 0\rangle$, $r_v = \langle 0, 0, 1\rangle$, $\mathbf{r}_u \mathbf{x} \mathbf{r}_v = \langle 3\cos(u), 3\sin(u), 0\rangle$ and $|r_u x r_v| = 3$.

$$f(\mathbf{r}(u,v)) = 3 \cdot \cos(u) \text{ so } \iint_S f(x,y,z)\, dS = \int_0^{2\pi} \int_0^{2} 3 \cdot \cos(u) \cdot 3\, dv\, du = \int_0^{2\pi} 18 \cdot \cos(u)\, dv = 0 .$$

Practice 3: From Example 2, $|\mathbf{r}_u \mathbf{x} \mathbf{r}_v| = u\sqrt{2}$ and $1 + y = 1 + u \cdot \sin(v)$ so

$$\iint_S (1 + y)\, dS = \int_{v=0}^{2\pi} \int_{u=0}^{1} [1 + u \cdot \sin(v)] \cdot (u\sqrt{2})\, du\, dv = \int_{v=0}^{2\pi} \left(\frac{1}{2}u^2\sqrt{2} - \frac{1}{3}u^3\sqrt{2} \cdot \cos(v) \right) \Big|_{u=0}^{1} dv = \sqrt{2}\pi .$$

Practice 4: For $\mathbf{F} = \langle 0, -3, 0\rangle$, $\iint_S \mathbf{F} \cdot \mathbf{n}\, dS = \iint_{\mathbf{R}} \langle 0, -3, 0\rangle \bullet \left\langle \frac{1}{2}, \frac{1}{3}, 1 \right\rangle dA = \iint_{\mathbf{R}} 1\, dA = 24$.

For $\mathbf{F} = \langle 1, 2, 3\rangle$, $\iint_S \mathbf{F} \cdot \mathbf{n}\, dS = \iint_{\mathbf{R}} \langle 1, 2, 3\rangle \bullet \left\langle \frac{1}{2}, \frac{1}{3}, 1 \right\rangle dA = \iint_{\mathbf{R}} \left(\frac{25}{6} \right) dA = 100$.

Practice 5: The surface S can be parameterized by $\mathbf{r}(u,v) = \langle u, v, f(u,v) \rangle$.

Then $\mathbf{r}_u = \langle 1, 0, f_u \rangle$, $\mathbf{r}_v = \langle 0, 1, f_v \rangle$ and $\mathbf{r}_u x \mathbf{r}_v = \langle -f_u, -f_v, 1 \rangle$ so

$$\iint_S \mathbf{F} \cdot \mathbf{n} \, dS = \iint_{\mathbf{R}} \mathbf{F} \bullet (\mathbf{r}_u x \mathbf{r}_v) \, dA = \iint_{\mathbf{R}} \langle M, N, P \rangle \bullet \langle -f_u, -f_v, 1 \rangle \, dA = \iint_R \{-P \cdot f_x - Q \cdot f_y + P\} \, dA \, .$$

Practice 6: As in the example, S can be parameterized by $x(u,v) = \sin(v) \cdot \cos(u)$, $y(u,v) = \sin(v) \cdot \sin(u)$ and $z(u,v) = \cos(v)$ with $0 \le u \le 2\pi$ and $0 \le v \le \pi/2$. Then $F(x,y,z) = \langle \cos(v), \sin(v) \cdot \cos(u), \sin(v) \cdot \sin(u) \rangle$ and

$$\iint_S \mathbf{F} \cdot \mathbf{n} \, dS = \iint_{\mathbf{R}} \mathbf{F} \bullet (\mathbf{r}_u x \mathbf{r}_v) \, dA$$

$$= \iint_{\mathbf{R}} \langle \cos(v), \sin(v) \cdot \cos(u), \sin(v) \cdot \sin(u) \rangle \bullet \left\langle \sin^2(v) \cdot \cos(u), \sin^2(v) \cdot \sin(u), \sin(v) \cdot \cos(v) \right\rangle \, dA$$

$$= \int_{v=0}^{\pi/2} \int_{u=0}^{2\pi} \cos(v) \cdot \sin^2(v) \cdot \cos(u) + \sin^3(v) \cdot \cos(u) \cdot \sin(u) + \sin^2(v) \cdot \sin(u) \cdot \cos(v) \, du \, dv \, .$$

But $\int_{u=0}^{2\pi} (\text{each term}) \, du = 0$ so the flux $=0$.

15.10 Stokes' Theorem

The circulation–curl form of Green's Theorem (section 15.5) says if $\mathbf{F} = \langle M, N \rangle$ is a 2D vector field and C is a simple, closed, piecewise smooth curve enclosing a region R then the integral of the curl of **F** on R is equal to the circulation of **F** around C (with a positive orientation): (Fig. 1)

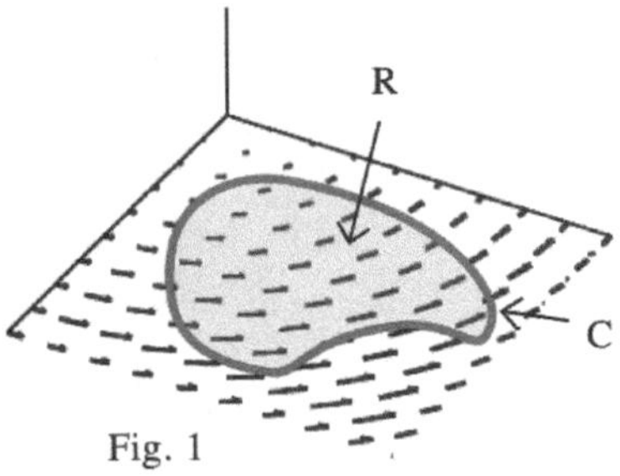

Fig. 1

$$\iint_R \text{curl } \mathbf{F} \; dA = \iint_R \left(\frac{\partial N}{\partial x} - \frac{\partial M}{\partial y} \right) dA = \oint_C \mathbf{F} \bullet \mathbf{T} \; ds = \text{circulation of } \mathbf{F} \text{ around } C .$$

Stokes' Theorem moves this result into 3D, and it has some very important consequences. If we think of Green's Theorem as applying to a flat soap film then we can think of Stokes' Theorem as giving the same result if we blow gently to create a soap bubble, a surface S in 3D. (Fig. 2)

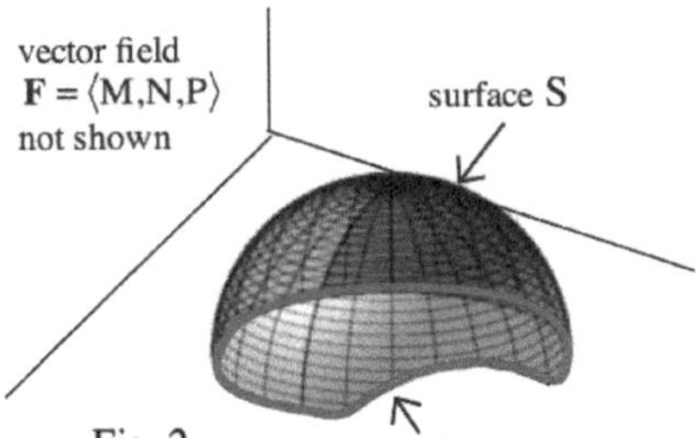

Fig. 2

Stokes' Theorem

If S is a connected, simply–connected, piecewise-smooth surface in 3D with piecewise–smooth boundary curve C, and **F** has continuous partial derivatives,

then $\iint_S \text{curl } \mathbf{F} \bullet d\mathbf{S} = \iint_S (\nabla \mathrm{x} \mathbf{F}) \bullet \mathbf{n} \; dS = \int_C \mathbf{F} \bullet d\mathbf{r} = \int_C \mathbf{F} \bullet \mathbf{T} \; ds = \text{circulation around } C$

Among the consequences of Stokes' Theorem:

* It allows us to trade 2D and 3D integrals – sometimes one of those is much easier than the other.
* It says that the integral of the curl of a vector field only depends on its values on the boundary.
* It says that the integral of the curl over a closed surface (like a sphere) is 0 since a closed surface has no boundary curve.
* It allows us to prove some of Maxwell's equations from physics (section 15.11).

The general proof of Stokes' Theorem is complicated. A proof of an easy special case is given next, and a proof for the common special case when the surface S has the form in Fig. 3 with $S = \{(x,y,z): \; z = z(x,y) \text{ is a function of x and y}\}$ is given in the Appendix.

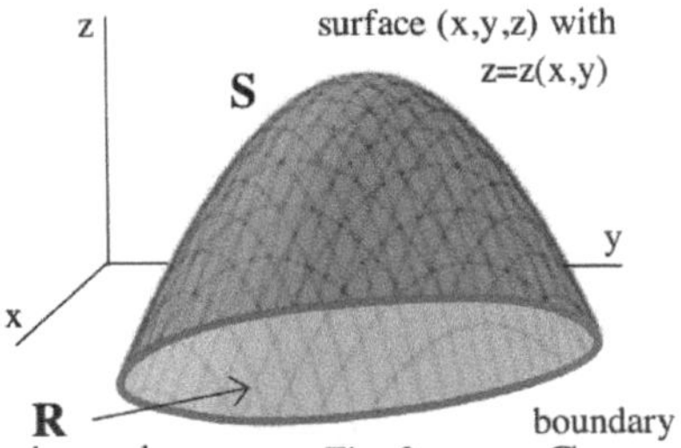

Fig. 3

Proof for an easy special case: S consists of a finite number of flat panels (Fig. 4) not necessarily in the same plane.

Then we can apply Green's Theorem to each panel and add the results together. This is very similar to the "finite Green's Theorem" we saw in section 15.5. The circulations along all of the interior edges cancel since adjacent panels have equal circulations going in opposite directions, and the only circulations remaining are those along the boundary of the region S. If S is the union of sub-regions $S_1, S_2, S_3 \ldots S_n$, then

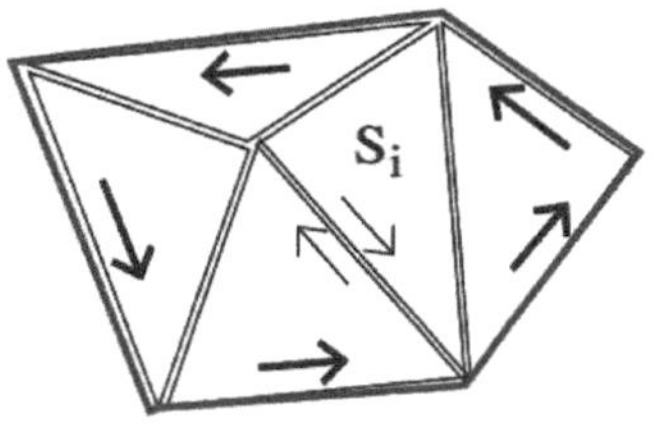

Fig. 4

$$\iint_S \text{curl } \mathbf{F} \bullet \mathbf{dS} = \sum_{i=1}^{n} \iint_{S_i} \text{curl } \mathbf{F} \bullet \mathbf{dS} = \sum_{i=1}^{n} \int_{C_i} \mathbf{F} \bullet \mathbf{T} \, dt = \int_C \mathbf{F} \bullet \mathbf{T} \, dt \, .$$

Example 1: Use Stokes' theorem to evaluate $\iint_S \text{curl } \mathbf{F} \bullet \mathbf{dS}$ where S is the hemisphere bounded by $x^2 + y^2 + z^2 = 9$ with z≥0 (Fig. 5) for the vector field $\mathbf{F} = \langle y, -x, 0 \rangle$.

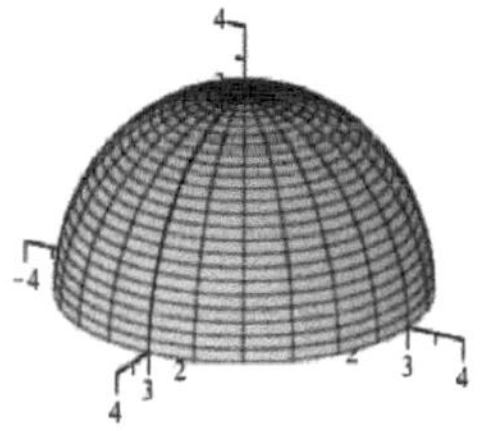
Fig. 5

Solution: By Stokes' theorem $\iint_S \text{curl } \mathbf{F} \bullet \mathbf{dS} = \int_C \mathbf{F} \bullet \mathbf{dr}$ where C is the bounding circle parameterized by $\mathbf{r}(t) = \langle 3 \cdot \cos(t), 3 \cdot \sin(t), 0 \rangle$. Then

$$\int_C \mathbf{F} \bullet \mathbf{dr} = \int_C \mathbf{F} \bullet \mathbf{T} \, dt = \int_{t=0}^{2\pi} \langle 3 \cdot \sin(t), -3 \cdot \cos(t), 0 \rangle \bullet \langle -3 \cdot \sin(t), 3 \cdot \cos(t), 0 \rangle \, dt$$

$$= \int_{t=0}^{2\pi} -9 \cdot \sin^2(t) - 9 \cdot \cos^2(t) \, dt = \int_{t=0}^{2\pi} -9 \, dt = (-9)(2\pi) = -18\pi \, .$$

It is also possible to evaluate $\iint_S \text{curl } \mathbf{F} \bullet \mathbf{dS}$ directly, but more difficult.

In this case the line integral around C was easier to evaluate than the surface integral on S.

Practice 1: Use Stokes' theorem to evaluate $\iint_S \text{curl } \mathbf{F} \bullet \mathbf{dS} = \int_C \mathbf{F} \bullet \mathbf{dr}$ where S is the paraboloid bounded by $2x^2 + 2y^2 + z = 18$ with z≥0 (Fig. 6) for the vector field $\mathbf{F} = \langle y, -x \rangle$.

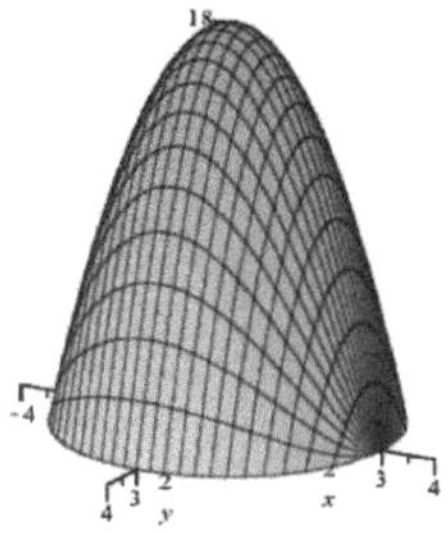
Fig. 6

Example 2: (a) Evaluate the line integral $\int_C \mathbf{F} \bullet \mathbf{dr}$ where $\mathbf{F} = \langle z, -z, x^2 - y^2 \rangle$ and C consists of the 3 line segments that bound the plane z=8–4x–2y in the first octant oriented as in Fig. 7.

(b) Evaluate $\int_C \mathbf{F} \bullet d\mathbf{r}$ for the line segment from (2, 0, 0) to (0, 4, 0).

Solution: (a) Rather than parameterizing the 3 line segments and evaluating the line integral along each of them, we can use Stoke's theorem and instead evaluate $\iint_S \text{curl } \mathbf{F} \bullet d\mathbf{S} = \iint_S (\nabla x\mathbf{F}) \bullet \mathbf{n} \ dS$.

$$\nabla x\mathbf{F} = \begin{vmatrix} \mathbf{i} & \mathbf{j} & \mathbf{k} \\ \frac{\partial}{\partial \mathbf{x}} & \frac{\partial}{\partial \mathbf{y}} & \frac{\partial}{\partial \mathbf{z}} \\ z & -z & x^2 - y^2 \end{vmatrix} = \langle 1-2y,\ 1-2x,\ 0 \rangle \text{ and } \mathbf{n} = \langle 4, 2, 1 \rangle$$

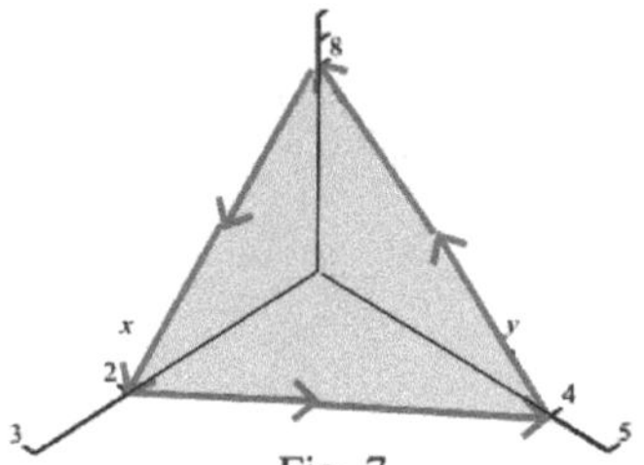
Fig. 7

is normal to the plane. Then

$$\iint_S (\nabla x\mathbf{F}) \bullet \mathbf{n} \ dS = \iint_S \langle 1-2y,\ 1-2x,\ 0 \rangle \bullet \langle 4, 2, 1 \rangle \ dS$$

$$= \int_{x=0}^{2} \int_{y=0}^{4-2x} 6 - 8y + 2x \ dy \ dx = \text{(just a standard double integral for Maple)} = -\frac{88}{3} \ .$$

(b) C is parameterized by $\mathbf{r}(t) = \langle 2 - t/2, t, 0 \rangle$ for $0 \le t \le 4$ so $\mathbf{r}\,'(t) = \langle -1/2, 1, 0 \rangle$.

$$\int_C \mathbf{F} \bullet d\mathbf{r} = \int_{t=0}^{4} \mathbf{F} \bullet \mathbf{r}\,' \ dt = \int_{t=0}^{4} \left\langle 0, 0, (2 - t/2)^2 - t^2 \right\rangle \bullet \langle -1/2, 1, 0 \rangle \ dt = \int_{t=0}^{4} 0 \ dt = 0 \ .$$

Practice 2: Calculate the circulation of $\mathbf{F} = \langle xy, xz, -2yz \rangle$ around the curve C that consists of the 3 line segments that bound the plane x+2y+2z=4 in the first octant oriented as in Fig. 8.

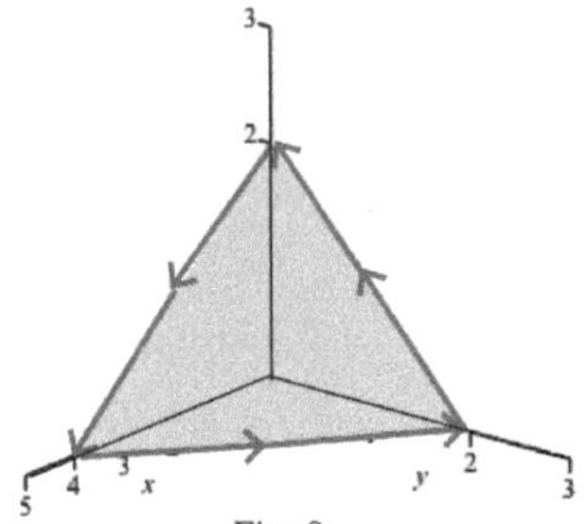
Fig. 8

Example 3: Use Stokes' theorem to evaluate $\iint_S \text{curl } \mathbf{F} \bullet d\mathbf{S}$ where S is the "cap" on the hemisphere bounded by $x^2 + y^2 + z^2 = 25$ with $z \ge 3$ (Fig. 9) for the vector field $\mathbf{F} = \langle z - y, x, -x \rangle$.

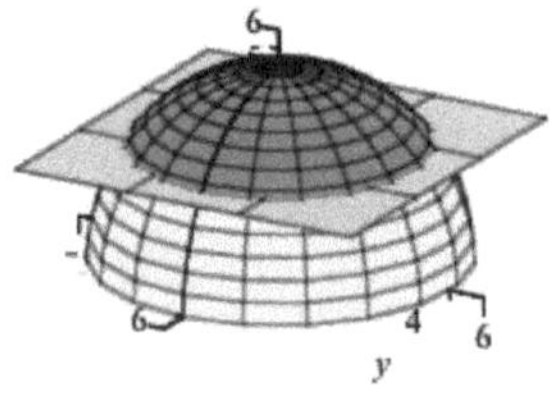
Fig. 9

Solution: $\iint_S \text{curl } \mathbf{F} \bullet d\mathbf{S} = \int_C \mathbf{F} \bullet d\mathbf{r}$ where C is the circle $\mathbf{r}(t) = \langle 4 \cdot \cos(t), 4 \cdot \sin(t), 3 \rangle$.

Then $\int_t \mathbf{F} \bullet \mathbf{r}\,' \ dr = \int_{t=0}^{2\pi} \langle 3 - 4 \cdot \sin(t), 4 \cdot \cos(t), -4 \cdot \cos(t) \rangle \bullet \langle -4 \cdot \sin(t), 4 \cdot \cos(t), 0 \rangle \ dr$

$$= \int_{t=0}^{2\pi} -12 \cdot \sin(t) + 16 \cdot \sin^2(t) + 16\cos^2(t) \ dt = 12\cos(t) + 16t \Big|_0^{2\pi} = 32\pi \ .$$

Practice 3: Use Stokes' theorem to evaluate $\iint_S \text{curl } \mathbf{F} \bullet \mathbf{dS}$ where F is the same as in Example 3, but now S is the cap with $z \geq 4$ on the same hemisphere.

Example 4: Evaluate $\iint_S \text{curl } \mathbf{F} \bullet \mathbf{dS}$ when S is the surface of the unit cube $0 \leq x \leq 1$, $0 \leq y \leq 1$, $0 \leq z \leq 1$.

Solution: Since S is a piecewise **closed** smooth surface, then S has no boundary curve C and

$$\iint_S \text{curl } \mathbf{F} \bullet \mathbf{dS} = 0 \ .$$

Practice 4: Evaluate $\iint_S \text{curl } \mathbf{F} \bullet \mathbf{dS}$ when S is the ellipsoid $x^2 + 2y^2 + 4z^2 = 16$.

If S has holes

If the surface S has a hole (Fig. 10) then the boundary of S has an additional boundary curve, and we can treat that new boundary in the same way we treated the boundary of a hole using Green's Theorem. We can create a single boundary for S by adding a path along S to the hole and then back from the hole (Fig. 11). Then the integral along this total new path will be the sum of the counterclockwise integrals around the outer boundary of S minus the sum of the counterclockwise integral around the hole. The integrals along the added paths sum to 0 since they are traveled once in each direction. Remember, for a counterclockwise orientation the region is always on our left hand side.

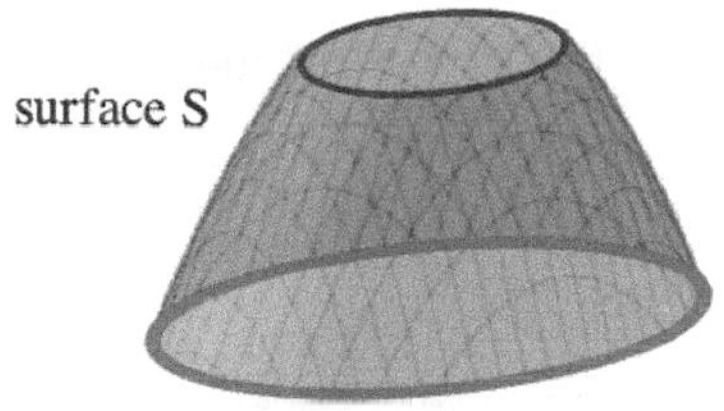

Fig. 10

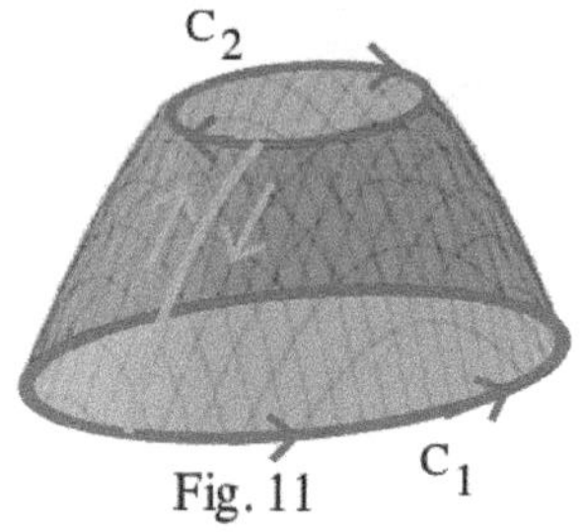

Fig. 11

Example 5: Evaluate $\iint_S \text{curl } \mathbf{F} \bullet \mathbf{dS}$ for $\mathbf{F} = \left\langle xy,\ x + z^2,\ y^3 \right\rangle$ with

$$S = \{(x,y,z):\ x^2 + y^2 + z^2 = 25,\ 0 \leq z \leq 4\}$$

Solution: This is the situation in Fig. 11. C_1 is parameterized by $\mathbf{r}_1(t) = \langle 5 \cdot \cos(t),\ 5 \cdot \sin(t),\ 0 \rangle$ and C_2 by $\mathbf{r}_2(t) = \langle 3 \cdot \cos(t),\ 3 \cdot \sin(t),\ 4 \rangle$ for $0 \leq t \leq 2\pi$ (both are counterclockwise).

$$\int_{C_1} \mathbf{F} \bullet \mathbf{dr} = \int_0^{2\pi} \left\langle 25 \cdot \sin(t) \cdot \cos(t),\ 5 \cdot \cos(t) + 0,\ 125 \cdot \sin^3(t) \right\rangle \bullet \langle -5\sin(t),\ 5\cos(t),\ 0 \rangle \, dt$$

$$= \int_0^{2\pi} -125\cdot\sin^2(t)\cdot\cos(t) + 25\cdot\cos^2(t) + 0 \, dt = 25\pi .$$

$$\int_{C_2} \mathbf{F}\bullet d\mathbf{r} = \int_0^{2\pi} \langle 9\cdot\sin(t)\cdot\cos(t),\ 3\cdot\cos(t)+4,\ 27\cdot\sin^3(t)\rangle \bullet \langle -3\cdot\sin(t),\ 3\cdot\cos(t),\ 0\rangle \, dt$$

$$= \int_0^{2\pi} -27\cdot\sin^2(t)\cdot\cos(t) + 9\cdot\cos^2(t) + 12\cdot\cos(t) + 0 \, dt = 9\pi .$$

So $\iint_S \text{curl } \mathbf{F}\bullet d\mathbf{S} = \int_{C_1} \mathbf{F}\bullet d\mathbf{r} - \int_{C_2} \mathbf{F}\bullet d\mathbf{r} = 16\pi$.

Practice 5: Evaluate $\iint_S \text{curl } \mathbf{F}\bullet d\mathbf{S}$ for $\mathbf{F} = \langle xy,\ x+z^2,\ y^3\rangle$ with $S = \{(x,y,z):\ x^2+y^2+z^2 = 25,\ 0 \le z \le 3\}$.

Meaning of the curl

In section 15.6 we claimed that the curl vector had two important properties:

- the magnitude of the curl gives the rate of the fluid's rotation, and
- the direction of the curl is normal to the plane of greatest circulation and points in the direction so that the circulation at the point has a right hand orientation (Fig. 12).

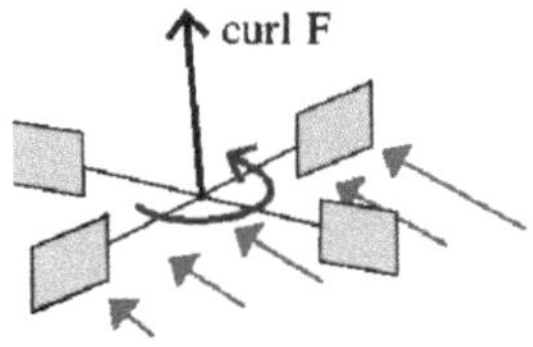

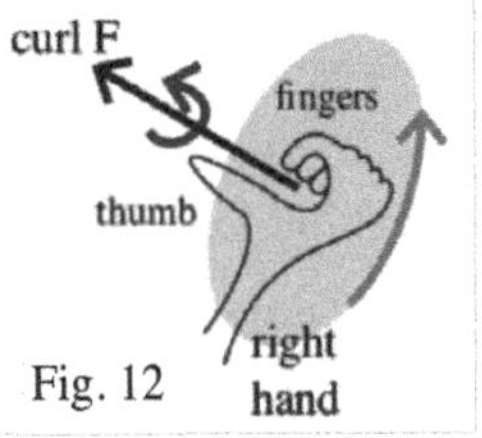

Fig. 12

If we have a small paddle wheel at point P and tilt it in different directions, then the claims say that the wheel will spin fastest with a right-hand orientation when the axis points in the direction of the curl vector.

Now we can use Stokes' Theorem to justify those claims.

Let P be a point in the vector field **F**, and let **u** be any unit vector. Suppose S is a small disk that has center at point P and radius r and that lies in the plane determined by P and u . S has a boundary circle C oriented positively so that S is always on the left side as we move along C.

Since S is small, the value of $\nabla\mathbf{x}\mathbf{F}$ is almost constant on S and so

$$\iint_S (\nabla\mathbf{x}\mathbf{F})\bullet\mathbf{u}\, dS \approx (\nabla\mathbf{x}\mathbf{F})\bullet\mathbf{u}\iint_S 1\, dS = (\nabla\mathbf{x}\mathbf{F})\bullet\mathbf{u}\,(\text{area of S}) = |\nabla\mathbf{x}\mathbf{F}|\cdot\cos(\theta)\cdot(\pi\cdot r^2)$$

But by Stokes' Theorem, $\iint_S (\nabla\mathbf{x}\mathbf{F})\bullet\mathbf{u}\, dS = \oint_C \mathbf{F}\bullet\mathbf{T}\, dt = \text{circulation of F around C}$ so

$|\nabla\mathbf{x}\mathbf{F}|\cdot\cos(\theta) = \dfrac{\text{circulation of } \mathbf{F} \text{ around C}}{\pi\cdot r^2}$ which is maximum when $\theta = 0$ and **u** has the same direction as $\nabla\mathbf{x}\mathbf{F}$.

> Together these statements say that an axis in the direction of $\nabla \mathbf{xF}$ gives the maximum circulation, and the magnitude of $\nabla \mathbf{xF}$ is the maximum rate of circulation per unit of area.

In section 15.6 we also stated the following theorem and said that a proof needed to wait until we had Stokes' Theorem.

> Theorem: If **F** is defined and has continuous partial derivatives at every point in 3D and curl **F** = 0,
> then **F** is a conservative field.

Proof: With Stokes' Theorem this is easy. If curl $\mathbf{F} = 0$ then $0 = \iint_S \text{curl } \mathbf{F}\, d\mathbf{S}$ for every simply-connected region S so by Stokes' Theorem $\int_C \mathbf{F} \bullet \mathbf{T}\ ds = \iint_S \text{curl } \mathbf{F}\, d\mathbf{S} = \mathbf{0}$ for every simple, closed, piece-wise smooth curve C. That means that **F** is path independent and conservative.

Problems

For problems 1 to 12 find the circulation of vector field **F** around the positively oriented curve C by using Stokes' Theorem and evaluating the surface integral.

1. $\mathbf{F} = \langle y, 2x, -z^2 \rangle$ and C is the ellipse $x^2 + 4y^2 = 4$.

2. $\mathbf{F} = \langle y, 2x, -z^2 \rangle$ and C is the circle $x^2 + y^2 = 9$.

3. $\mathbf{F} = \langle z, y^2, xy \rangle$ and C is the boundary of the triangle 2x+2y+2z=6 in the first octant.

4. $\mathbf{F} = \langle yz, xz, xy \rangle$ and C is the boundary of a simple closed curve in the yz=plane.

5. $\mathbf{F} = \langle z - y, x - z, x - y \rangle$ and C is the boundary of a simple closed curve in the x+y+z=5 plane.

6. $\mathbf{F} = \langle x+y^2, 3x, 2z \rangle$ and C is the boundary of the rectangle $R = \{(x,y,z):\ 0 \le x \le 3,\ 1 \le y \le 3,\ z = 0\}$ oriented in the counterclockwise direction.

7. $\mathbf{F} = \langle y^2, 3x+z, 2z+y \rangle$ and C is the boundary of the circle $R = \{(x,y,z):\ x^2+y^2 \le 4,\ z = 0\}$ oriented in the counterclockwise direction.

8. $\mathbf{F} = \langle -y^2, x+z, z^2+y \rangle$ and C is the boundary of the circle $R = \{(x,y,z):\ x^2+y^2 \le 9,\ z = 2\}$ oriented in the counterclockwise direction.

9. $\mathbf{F} = \langle y, -x, z \rangle$ and $S = \{(x,y,z):\ x^2+y^2+z^2 = 16,\ 0 \le z\}$ is a hemisphere.

10. $\mathbf{F} = \langle y, -x, z \rangle$ and $S = \{(x,y,z):\ x^2+y^2+z^2 = 16,\ 0 \le y\}$ is a hemisphere.

11. $\mathbf{F} = \langle \sin(x), \cos(y), \sin(z) \rangle$ and S is the solid tour with large radius 3 and small radius 1.

12. $\mathbf{F} = \langle xy, x^2+z^2, y^3 \rangle$ and S is the solid cube with vertices $x = \pm 1,\ y = \pm 1,\ z = \pm 1$.

In problems 13 to 18 use Stokes' Theorem to determine which of these fields are conservative.

13. $\mathbf{F} = \langle yz, xz, xy \rangle$

14. $\mathbf{F} = \langle yz+a, xz+b, xy+c \rangle$ a, b and c are constants.

15. $\mathbf{F} = \langle yza, xzb, xyc \rangle$ a, b and c are constants.

16. $\mathbf{F} = \langle y+z, x+z, x+y \rangle$

17. $\mathbf{F} = \langle x+z, y+z, x+y \rangle$

18. $\mathbf{F} = \langle yz, x-y, -x \rangle$

For problems 19 to 26 evaluate $\iint_S \text{curl } \mathbf{F} \bullet d\mathbf{S}$ for the given field **F** and surface S.

19. $\mathbf{F} = \langle z-y, x-z, y-x \rangle$ and $S = \{(x,y,z):\ x^2+y^2 = 16,\ 0 \le z \le 3\}$ is a cylinder.

20. $\mathbf{F} = \langle y, -x, z \rangle$ and $S = \{(x,y,z):\ x^2+z^2 = 9,\ 1 \le y \le 5\}$ is a cylinder.

21. $\mathbf{F} = \langle x, y^2, z^3 \rangle$ and $S = \{(x,y,z):\ x^2+y^2+z^2 = 25,\ -3 \le y\}$.

22. $\mathbf{F} = \langle \sin(x), \cos(y), \sin(z) \rangle$ and $S = \{(x,y,z):\ x^2++y^2+z^2 = 25,\ -3 \le z \le 4\}$.

23. $F = \langle -y, z, x \rangle$ and $S = \{(x,y,z):\ \frac{x^2}{4}+\frac{y^2}{9}+\frac{z^2}{25} = 1,\ 0 \le y\}$ a truncated ellipsoid.

24. $\mathrm{F} = \langle -y, z, x\rangle$ and $S = \{(x,y,z): \frac{x^2}{4} + \frac{y^2}{9} + \frac{z^2}{25} = 1,\ 0 \le z\}$ a truncated ellipsoid.

25. $\mathbf{F} = \langle y, z, x\rangle$ and $S = \{(x,y,z): x^2 + y^2 + z = 25,\ 0 \le z \le 16\}$ a truncated paraboloid.

26. $\mathbf{F} = \langle y, z, x\rangle$ and $S = \{(x,y,z): x^2 + y^2 + z = 25,\ 0 \le z \le 9\}$.

Practice Answers

Practice 1: The field F and the boundary C (a circle in the xy-plane with radius 3) are the same as in Example 1 so $\iint_S \text{curl } \mathbf{F} \bullet d\mathbf{S} = \int_C \mathbf{F} \bullet d\mathbf{r} = -18\pi$ just as in Example 1.

Practice 2: By Stokes' theorem $\int_C \mathbf{F} \bullet d\mathbf{r} = \iint_S (\nabla \mathbf{x} \mathbf{F}) \bullet \mathbf{n}\ dS$ and the second integral is easier to evaluate then doing 3 line integrals.

$$\nabla \mathbf{x} \mathbf{F} = \begin{vmatrix} \mathbf{i} & \mathbf{j} & \mathbf{k} \\ \frac{\partial}{\partial x} & \frac{\partial}{\partial y} & \frac{\partial}{\partial z} \\ xy & xz & -2yz \end{vmatrix} = \langle -2z - x,\ 0,\ z - x\rangle$$ and $\mathbf{n} = \langle 1, 2, 2\rangle$ is normal to the plane.

Then $\iint_S (\nabla \mathbf{x} \mathbf{F}) \bullet \mathbf{n}\ dS = \iint_S \langle 1-2y,\ 1-2x,\ 0\rangle \bullet \langle 4, 2, 1\rangle\ dS = \int_{x=0}^{4} \int_{y=0}^{2-x/2} -3x\ dy\ dx$

$$= \int_{x=0}^{4} -6x + \frac{3}{2}x^2\ dx = -12 .$$

Practice 3: Now $\mathbf{r}(t) = \langle 3\cdot\cos(t),\ 3\cdot\sin(t),\ 4\rangle$ so $\mathbf{r}\,'(t) = \langle -3\cdot\sin(t),\ 3\cdot\cos(t),\ 0\rangle$ and

$$\int_t \mathbf{F} \bullet \mathbf{r}\,'\ dr = \int_{t=0}^{2\pi} \langle 4 - 3\cdot\sin(t),\ 4\cdot\cos(t),\ -3\cdot\cos(t)\rangle \bullet \langle -3\cdot\sin(t),\ 3\cdot\cos(t),\ 0\rangle\ dr$$

$$= \int_{t=0}^{2\pi} -12\cdot\sin(t) + 9\cdot\sin^2(t) + 9\cos^2(t)\ dt = \int_{t=0}^{2\pi} -12\cdot\sin(t) + 9\ dt = 18\pi .$$

Practice 4: S is a smooth **closed** surface so $\iint_S \text{curl } \mathbf{F} \bullet d\mathbf{S} = 0$.

Practice 5: C_1 and $\mathbf{r}_1$ are the same as in Example 5 but now C_2 is parameterized by

$\mathbf{r}_2(t) = \langle 4\cdot\cos(t), 4\cdot\sin(t), 3\rangle$ for $0 \le t \le 2\pi$ (counterclockwise).

$$\int_{C_2} \mathbf{F}\bullet\mathbf{dr} = \int_0^{2\pi} \left\langle 16\cdot\sin(t)\cdot\cos(t), 4\cdot\cos(t)+3, 27\cdot\sin^3(t)\right\rangle \bullet \langle -4\cdot\sin(t), 4\cdot\cos(t), 0\rangle \, dt$$

$$= \int_0^{2\pi} -64\cdot\sin^2(t)\cdot\cos(t) + 16\cdot\cos^2(t) + 12\cdot\cos(t) + 0 \, dt = 16\pi .$$

Then $\iint_S \text{curl } \mathbf{F}\bullet\mathbf{dS} = \int_{C_1} \mathbf{F}\bullet\mathbf{dr} - \int_{C_2} \mathbf{F}\bullet\mathbf{dr} = 25\pi - 16\pi = 9\pi$.

Appendix: Proof of Stokes' Theorem for a common special case

Stoke's Theorem: $\iint_S \text{curl } \mathbf{F} \bullet d\mathbf{S} = \iint_S (\nabla \mathbf{x} \mathbf{F}) \bullet \mathbf{n} \ dS = \int_C \mathbf{F} \bullet d\mathbf{r} = \int_C \mathbf{F} \bullet \mathbf{T} \ dt$ = circulation around C

Common special case: S is a surface of the form (x, y, z(x,y)) (Fig. A1) This requires a lot of calculations and very careful attention to details. Let $\mathbf{F} = \langle P, Q, R \rangle$ and $S = \{(x,y,z): z = z(x,y)$ is a function of x and y$\}$. We will evaluate $\int_C \mathbf{F} \bullet d\mathbf{r}$ and $\iint_S (\nabla \mathbf{x} \mathbf{F}) \bullet \mathbf{n} \ dS$ separately and show that they are equal.

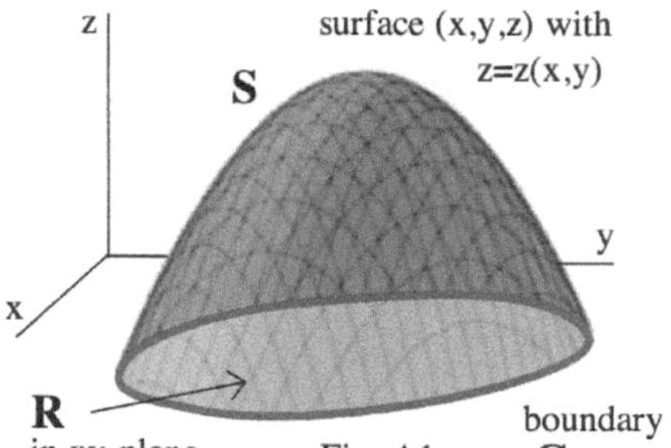

Fig. A1

$\iint_S (\nabla \mathbf{x} \mathbf{F}) \bullet \mathbf{n} \ dS$: Thinking of S as a parameterized surface (but using x and y instead of u and v) for x and y in the xy-region R (Fig. A2), then $\iint_S (\nabla \mathbf{x} \mathbf{F}) \bullet \mathbf{n} \ dS = \iint_R (\nabla \mathbf{x} \mathbf{F}) \bullet (\mathbf{r}_x \mathbf{x} \mathbf{r}_y) \ dA$. $\mathbf{r}_x = \langle 1, 0, z_x \rangle$, $\mathbf{r}_y = \langle 0, 1, z_y \rangle$ and $\mathbf{r}_x \mathbf{x} \mathbf{r}_y = \langle -z_y, -z_y, 1 \rangle$. Then

$$\iint_S (\nabla \mathbf{x} \mathbf{F}) \bullet \mathbf{n} \ dS = \iint_R \langle R_y - Q_z, P_z - R_x, Q_x - P_y \rangle \bullet \langle -z_y, -z_y, 1 \rangle \ dA \ .$$

$\int_C \mathbf{F} \bullet d\mathbf{r}$: This one is more complicated and requires the Chain Rule for functions of several variables.

$$\int_C \mathbf{F} \bullet d\mathbf{r} = \int_C \mathbf{F} \bullet \mathbf{r}' \ dt = \int_C \langle P, Q, R \rangle \bullet \langle dx, dy, dz \rangle \ dt = \int_C P \cdot dx + Q \cdot dy + R \cdot dz \ .$$

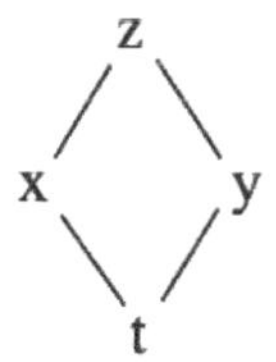

Fig. A2

But z=z(x,y) so (Fig. A2) $dz = z_x \cdot dx + z_y \cdot dy$, and

$$\int_C \mathbf{F} \bullet d\mathbf{r} = \int_C (P + R \cdot z_x) dx + (Q + R \cdot z_y) \cdot dy = \int_C M \, dx + N \, dy \cdot$$

By Green's Theorem $\int_C M \, dx + N \, dy = \iint_R N_x - M_y \ dA$ with $M = P + R \cdot z_x$ and $N = Q + R \cdot z_y$.

$$M_y = \frac{\partial}{\partial y}(P + R \cdot z_x) = P_y + P_z \cdot z_y + z_x \cdot R_y + R \cdot \frac{\partial}{\partial y} z_x$$

$$= P_y + P_z \cdot z_y + R \cdot z_{xy} + z_x \cdot (R_y + R_z \cdot z_y) \ .$$

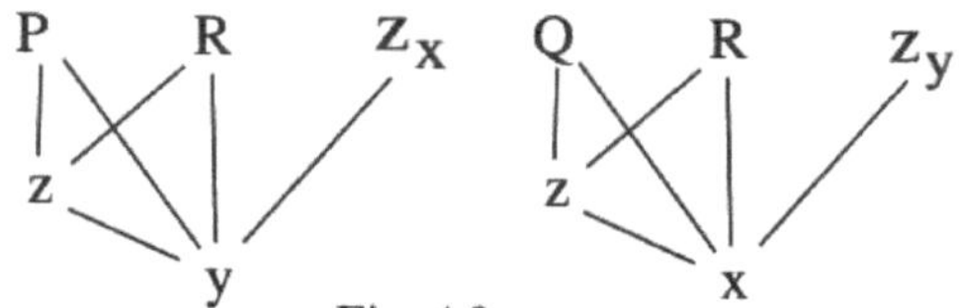

Fig. A3

Similarly, $N_x = \frac{\partial}{\partial x}(Q + R \cdot z_y) = Q_x + Q_z \cdot z_x + R \cdot z_{xy} + z_y \cdot (R_x + R_z \cdot z_x)$.

Then, after substituting and simplifying,

$$\int_C \mathbf{F} \bullet d\mathbf{r} = \iint_R N_x - M_y \, dA = \iint_R z_x(Q_z - R_y) + z_y(R_x - P_z) + (Q_x - P_y) \, dA$$

$$= \iint_R \langle R_y - Q_z, P_z - R_x, Q_x - P_y \rangle \bullet \langle -z_y, -z_y, 1 \rangle \, dA = \iint_S (\nabla \mathbf{x} \mathbf{F}) \bullet \mathbf{n} \, dS$$

so Stokes' Theorem is also true for surfaces of the form $S = \{(x,y,z): \ z = z(x,y)\}$.

Suppose we cut a hole in the surface S and attach a smooth "bump" to cover the hole (Fig. A4). Is Stokes' Theorem still true for this new surface consisting of the old S minus the hole plus the bump? You should be able to justify your answer.

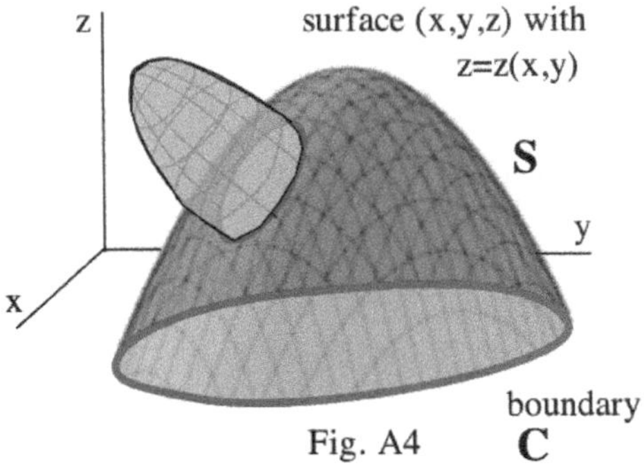

Fig. A4

15.11 Gauss/Divergence Theorem

The Gauss/Divergence Theorem is the final fundamental theorem of calculus and the final mathematical piece needed to create Maxwell's equations. Like each of the previous fundamental theorems, it relates an accumulation (integral) in some dimension to the values of a related function in a lower dimension.

The Fundamental Theorem of Calculus (section 4.5) and the Fundamental Theorem of Line Integrals (section 15.4) said that the accumulation (integral) of a function over an interval or a line is equal to a related function (the antidarivative or potential function) evaluated at the boundary (endpoints) of the interval or line.

Stoke's Theorem (15.9), like the curl-circulation form of Green's Theorem (section 15.5), said that the accumulation of a function (curl) on a 2D surface is equal to a related function (the circulation) evaluated on the boundary curve of the surface.

Finally, the Gauss/Divergence Theorem, like the divergence-flux form of Green's Theorem, says that the accumulation of the divergence in a solid 3D region is equal to a related function (flux) evaluated on the boundary surface of the region.

Gauss/Divergence Theorem

If E is a solid, closed, simple 3D region with a piecewise-smooth boundary surface S, and **n** is the outward unit normal vector to S

then for the vector field $\mathbf{F} = \langle M, N, P \rangle$ whose components have continuous partial derivatives in a open region containing E

$$\iiint_{\mathbf{E}} \text{div } \mathbf{F} \bullet dV = \iint_{S} \mathbf{F} \bullet d\mathbf{S} = \iint_{S} \mathbf{F} \bullet \mathbf{n} \, dS = \text{flux across S}$$

A proof for a common special case is given in the Appendix.

The theorem is rather obvious in the case where the 3D region E contains a finite number if sources and sinks. If the region E inside of the boundary surface S contains a number of springs (pipes adding water) and sinks (pipes removing water), then the flux of water across the boundary surface S is the (springs input)–(sinks outputs). If several pipes inside E are adding water at a total rate of 5 m^3/s and several pipes inside E are removing water at a rate of 2 m^3/s then the (outward) flux across the boundary S is 3 m^3/s. The sum of the pipes adding water (positive divergence) plus the sum of the pipes removing water (negative divergence) equals the outward flux across the boundary of the region. The Gauss/Divergence Theorem extends this finite case idea to the case where potentially every point in the region E is a source or sink.

Example 1: Suppose pipes P1 at (1,0,0) and P2 at (2,2,1) are adding water at the rates of 5 m^3/s and 3 m^3/s , respectively, and P3 at (0,2,0) and P4 at (0,0,4) are removing water at the rates of 2 m^3/s and 1 m^3/s, respectively. What is the outward flux of water across (a) the sphere S with center at (0,0,0) and radius 3.5 m, and (b) a tiny cube with center at (2,2,1)?

Solution: (a) Only P1, P2 and P3 are inside S so the flux is (5)+(3)–(2)= 6 m^3/s.

(b) Only P2 is inside this tiny cube so flux = 3 m^3/s .

Practice 1: If the same pipes as in Example 1 are adding and removing grams of water per second, what is the flux across (a) the sphere S with center at (1,1,1) and radius 5 m and (b) a tiny cube with center at (1,2,3)?

Example 2: Calculate the flux across the sphere $x^2+y^2+z^2=R^2$ for the radial vector field $\mathbf{F}=\langle x, y, z\rangle$.

Solution: E = the 3D sphere with radius R so the volume of E is $\frac{4}{3}\pi R^3$.

$$\text{div } \mathbf{F} = \nabla \bullet \mathbf{F} = \frac{\partial}{\partial x}(x) + \frac{\partial}{\partial y}(y) + \frac{\partial}{\partial z}(z) = 3 \text{ so}$$

$$\text{flux across S} = \iiint_{\mathbf{E}} \text{div } \mathbf{F} \bullet dV = 3 \cdot (\text{volume of the sphere}) = 4\pi R^3 \ .$$

Practice 2: Calculate the flux across the sphere $x^2+y^2+z^2=R^2$ for the more general radial vector field $\mathbf{F}=\langle ax, by, cz\rangle$.

Example 3: Calculate the outward flux across the boundary D of the solid unit cube $E=\{(x,y,z):\ 0\le x\le 1,\ 0\le y\le 1,\ 0\le z\le 1\}$ for the field $\mathbf{F}=\langle xy, yz, xz\rangle$.

Solution: E = the solid cube, and $\text{div } \mathbf{F} = \nabla \bullet \mathbf{F} = \frac{\partial}{\partial x}(xy) + \frac{\partial}{\partial y}(yz) + \frac{\partial}{\partial z}(xz) = y+z+x$ so

$$\text{flux across D} = \iiint_{\mathbf{E}} \text{div } \mathbf{F} \bullet dV = \int_0^1\int_0^1\int_0^1 x+y+z \, dz\, dy\, dz$$

$$= \int_0^1\int_0^1 \left(xz+yz+\frac{1}{2}z^2\right)\Big|_{z=0}^{1} dy\, dz = \int_0^1\int_0^1 \left(x+y+\frac{1}{2}\right) dy\, dz = \ldots = \frac{3}{2} \ .$$

Practice 3: Calculate the outward flux across the boundary D of the solid unit cube $E=\{(x,y,z):\ 0\le x\le 1,\ 0\le y\le 1,\ 0\le z\le 1\}$ for the field $\mathbf{F}=\langle xyz, xyz, xyz\rangle$.

Flux for the inverse–square vector field $\mathbf{F} = \dfrac{\langle x, y, z \rangle}{\sqrt{x^2+y^2+z^2}} = \dfrac{\mathbf{r}}{|\mathbf{r}|^3}$

We cannot apply the Divergence Theorem to this field on a solid that includes the origin since the field is not defined at the origin, but we can apply it to a region between two spheres (or other smooth surfaces) so that the region doe not include the origin. Let $D = \{(x,y,z) : 0 < A^2 < x^2+y^2+z^2 < B^2\}$ be the region outside of a sphere centered at the origin with radius A but inside a sphere centered at the origin with radius B. The boundary of D consists of the two spheres $S_1 = \{(x,y,z) : A^2 = x^2+y^2+z^2\}$ and $S_2 = \{(x,y,z) : x^2+y^2+z^2 = B^2\}$ but the unit normal vectors, pointing outward from the region D, have opposite directions (Fig. 1). Then $\iiint_D \text{div}\,\mathbf{F}\,dV = \iint_S \mathbf{F}\bullet\mathbf{n}\,dS = \iint_{S_1} \mathbf{F}\bullet\mathbf{n}\,dS - \iint_{S_2} \mathbf{F}\bullet\mathbf{n}\,dS$. But in Section 15.6 we determined that $\text{div}\,\mathbf{F} = 0$ for this field so $\iint_{S_1} \mathbf{F}\bullet\mathbf{n}\,dS = \iint_{S_2} \mathbf{F}\bullet\mathbf{n}\,dS$ meaning that the outward flux (away from the origin) across S_2 equals the inward flux (toward the origin) across S_1. We can evaluate $\iint_{S_1} \mathbf{F}\bullet\mathbf{n}\,dS$ by noting that on S_1 we have $|\mathbf{r}| = A$ and

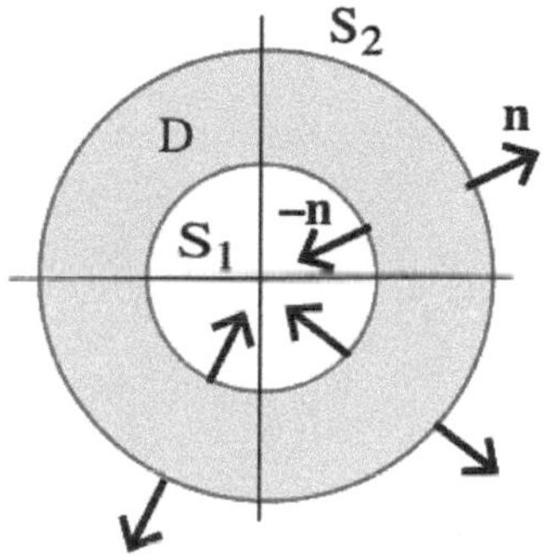

Fig. 1

$$\text{flux} = \iint_{S_1} \mathbf{F}\bullet\mathbf{n}\,dS = \iint_{S_1} \left(\frac{\mathbf{r}}{|\mathbf{r}|^3}\right)\bullet\left(\frac{\mathbf{r}}{|\mathbf{r}|}\right) dS = \iint_{S_1} \left(\frac{\mathbf{r}\bullet\mathbf{r}}{|\mathbf{r}|^4}\right) dS = \iint_{S_1} \frac{1}{|\mathbf{r}|^2}\,dS = \iint_{S_1} \frac{1}{A^2}\,dS$$

$$= \frac{1}{A^2}(\text{surface area of } S_1) = \frac{1}{A^2}(4\pi A^2) = 4\pi \,.$$

The resulting flux is the same 4π for a any region between two smooth surfaces that surround the origin.

Example 4: Determine the flux for the solid region D outside the sphere $S = \{(x,y,z) : x^2+y^2+z^2 = 1\}$ and inside the ellipsoid $E = \left\{(x,y,z) : \dfrac{x^2}{4} + \dfrac{y^2}{9} + \dfrac{z^2}{16} = 1\right\}$ for $\mathbf{F} = \dfrac{\langle x, y, z \rangle}{\sqrt{x^2+y^2+z^2}} = \dfrac{\mathbf{r}}{|\mathbf{r}|^3}$.

Solution: Based on the previous discussion we can immediately conclude that flux across D is 4π.

Practice 4: Determine the flux for the solid region D between a sphere centered at (2,3,4) with radius 4 and another sphere centered at (2,3,4) with radius 1 for $\mathbf{F} = \dfrac{\langle x, y, z \rangle}{\sqrt{x^2+y^2+z^2}} = \dfrac{\mathbf{r}}{|\mathbf{r}|^3}$.

Units of flux

Suppose **v** is a velocity vector field in meters per second (m/s) of a material that has constant density δ given in grams per cubic meter (g/m^3). Then the field $\mathbf{F} = \delta \cdot \mathbf{v}$ has units $\left(\frac{\text{g}}{\text{m}^3}\right)\left(\frac{\text{m}}{\text{s}}\right) = \left(\frac{g}{\text{m}^2 \cdot s}\right)$ which measures the amount of material per a square meter flowing past a point each second. This field **F** is sometimes called the **flux density**. But in mathematics flux is defined as a surface integral, and then the units of flux become

$$\text{flux} = \iint_S \mathbf{F} \bullet \mathbf{n}\, dS = \{\text{amount per square meter per second}\}\{\text{area of S in square meters}\}$$

$$= \left\{\frac{g}{\text{m}^2 \cdot \text{s}}\right\}\left(\text{m}^2\right) = \frac{\mathbf{g}}{\mathbf{s}} = \text{the amount (mass) per second flowing across the surface S.}$$

"In the case of fluxes, we have to take the integral, over a surface, of the flux through every element of the surface. The result of this operation is called the surface integral of the flux. It represents the quantity which passes through the surface."

—James Clerk Maxwell

In other fields the flux density **F** has other units as does the surface integral:

Heat flux density has units $\frac{\text{J}}{\text{m}^2 \cdot \text{s}}$ so the units of the surface integral (our flux) are $\frac{\text{J}}{\text{s}}$ (Joules per second).

Magnetic flux density has units $\mathbf{B} = \frac{\text{Wb}}{\text{m}^2}$ (Weber per square meter =Tesla) so the units of magnetic flux are

$$\Phi_\mathbf{B} = \iint_S \mathbf{B} \bullet d\mathbf{S} = \text{Weber.}$$

Gauss's Law for magnetism states that the total magnetic flux through a closed surface is 0: $\Phi_\mathbf{B} = \iint_S \mathbf{B} \bullet d\mathbf{S} = \mathbf{0}$ for any closed surface S (since every magnetic north pole is attached to a magnetic south pole).

Electric flux density has units $\mathbf{E} = \frac{\mathbf{F}}{\text{q}}$ (force/charge= Newtons/coulomb=volts/meter) the magnetic flux is

$$\Phi_\mathbf{E} = \iint_S \mathbf{E} \bullet d\mathbf{S} = \frac{\text{Q}}{4\pi\varepsilon_0}(4\pi) = \frac{\text{Q}}{\varepsilon_0}.$$

Gauss' Law for electric fields states that the total electric flux through a closed surface S is $\Phi_\mathbf{E} = \iint_S \mathbf{E} \bullet d\mathbf{S} = \frac{\text{Q}}{4\pi\varepsilon_0}(4\pi) = \frac{\text{Q}}{\varepsilon_0}$ where Q is the total electric charge inside S and ε_0 is a (very small) constant called the electric constant or permittivity of free space. The total electric flux need not equal 0 for a closed surface S since the divergence at a point charge inside S may not equal 0 and the flux is a constant times the total of the charges inside S.

Problems

For problems 1 to 6 pipe P1=(1,2,3) is inputting 7 m^3/s of water, P2=(–2,1,1) is removing 3 m^3/s of water, P3=(2,2,0) is inputting 2 m^3/s of water, P4=(1,1,0) is inputting 6 m^3/s of water, P5=(2,2,2) is removing 4 m^3/s of water, and P6=(3,1,3) is inputting 6 m^3/s of water,

1. What is the net flux across the sphere with center at the origin and (a) radius 2 and (b) radius 4?
2. What is the net flux across the sphere with center at the origin and (a) radius 3 and (b) radius 5?
3. What is the net flux across the sphere with center at (2,2,0) and (a) radius 1 and (b) radius 3?
4 What is the net flux across the sphere with center at (2,2,0) and (a) radius 4 and (b) radius 5?
5. What is the net flux across the boundary of the region outside the sphere with center at the origin and radius 3 and inside the sphere with center at the origin and radius 5?
6. What is the net flux across the boundary of the region outside the sphere with center at the origin and radius 1 and inside the sphere with center at the origin and radius 3?

In problems 7 to 14 find the outward flux of the field F across the boundary surface of the given solid.

7. $\mathbf{F} = \langle x,y,z \rangle$ across the solid sphere $E = \{(x,y,z) : x^2 + y^2 + z^2 \le 4\}$.
8. $\mathbf{F} = \langle -y,x,z \rangle$ across the solid sphere $E = \{(x,y,z) : x^2 + y^2 + z^2 \le 4\}$.
9. $\mathbf{F} = \langle x,\ 2y,\ -z \rangle$ across the solid sphere $E = \{(x,y,z) : x^2 + y^2 + z^2 \le 9\}$.
10. $\mathbf{F} = \langle -x,\ 2y,\ -z \rangle$ across the solid sphere $E = \{(x,y,z) : x^2 + y^2 + z^2 \le 9\}$.
11. $\mathbf{F} = \langle x,\ 2y,\ z \rangle$ across the solid box $E = \{(x,y,z) : 0 \le x \le 2,\ 0 \le y \le 3,\ 0 \le z \le 4\}$.
12. $\mathbf{F} = \langle 2x,\ 3y,\ 4z \rangle$ across the solid box $E = \{(x,y,z) : 1 \le x \le 2,\ 1 \le y \le 3,\ 1 \le z \le 4\}$.
13. $\mathbf{F} = \langle x^2,\ y^2,\ z^2 \rangle$ across the solid box $E = \{(x,y,z) : 0 \le x \le 2,\ 0 \le y \le 3,\ 0 \le z \le 4\}$.
14. $\mathbf{F} = \langle x^2,\ -y^2,\ z^2 \rangle$ across the solid box $E = \{(x,y,z) : 1 \le x \le 2,\ 1 \le y \le 3,\ 1 \le z \le 4\}$.

15. If div **F** = 0 for the region between two concentric spheres, what is the relationship between the fluxes across the surfaces of the spheres?
16. E is a convex solid with volume V. Find the flux of F across the boundary of E (a) if div **F** = 0 at every point, and (b) if div **F** = 3 at every point.
17. $\mathbf{F} = \langle 1,\ -2,\ 3 \rangle$. Without doing any calculations determine whether the flux across each sphere is positive, negative or zero. (a) Sphere with radius 2 and center at the origin. (b) Sphere with radius 1 and center at (1, 2, 3) (c) Sphere with radius 1 and center at (3, –2, 1).

18. $\mathbf{F} = \langle a, b, c \rangle$. Without doing any calculations determine whether the flux across each sphere is positive, negative or zero. (a) Sphere with radius 2 and center at the origin. (b) Sphere with radius 1 and center at (1, 2, 3) (c) Sphere with radius 1 and center at (3, –2, 1).

19. $\mathbf{F} = \langle 0, y, 0 \rangle$. Without doing any calculations determine whether the flux across each sphere is positive, negative or zero. (a) Sphere with radius 2 and center at the origin. (b) Sphere with radius 1 and center at (1, 2, 3) (c) Sphere with radius 1 and center at (3, –2, 1).

20. $\mathbf{F} = \langle 0, 1, z \rangle$. Without doing any calculations determine whether the flux across each sphere is positive, negative or zero. (a) Sphere with radius 2 and center at the origin. (b) Sphere with radius 1 and center at (1, 2, 3) (c) Sphere with radius 1 and center at (3, –2, 1).

21. $\mathbf{F} = \langle x^2, y^2, z^2 \rangle$. Without doing any calculations determine whether the flux across each sphere is positive, negative or zero. (a) Sphere with radius 2 and center at the origin. (b) Sphere with radius 1 and center at (1, 2, 3) (c) Sphere with radius 1 and center at (3, –2, 1).

22. Suppose **v** is a velocity vector field in meters per second (m/s) and δ is a density given in cows per cubic meter (c/m^3), and F is the vector field $\mathbf{F} = \delta \cdot \mathbf{v}$. What are the units of flux for this field across the boundary of a solid region?

Practice Answers

Practice 1: (a) All of the pipes are inside this sphere so flux = (5)+(3)–(2)–(1) = 5 g/s.

(b) None of the pipes are inside this tiny cube so flux = 0 g/s.

Practice 2: $\text{div } \mathbf{F} = \nabla \bullet \mathbf{F} = \frac{\partial}{\partial x}(ax) + \frac{\partial}{\partial y}(by) + \frac{\partial}{\partial z}(cz) = a + b + c$ so

$$\text{flux across S} = \iiint_{\mathbf{E}} \text{div } \mathbf{F} \bullet dV = (a+b+c)\cdot(\text{volume of the sphere}) = (a+b+c)\cdot\frac{4}{3}\pi R^3 .$$

Practice 3: E = the solid cube, and $\text{div } \mathbf{F} = \nabla \bullet \mathbf{F} = yz + xz + xy$ so

$$\text{flux across D} = \iiint_{\mathbf{E}} \text{div } \mathbf{F} \bullet dV = \int_0^1 \int_0^1 \int_0^1 (yz + xz + xy)\, dz\, dy\, dz = \ldots = \frac{3}{4} .$$

Practice 4: The smaller sphere is inside the larger one and the larger one does not contain the origin so, as was shown in section 15.5, div **F** = 0 inside the larger sphere. Then by the Divergence Theorem, the flux across each sphere is 0 so the flux across the boundary of D is 0.

Appendix: Proof of the Gauss/Divergence Theorem

There are two intuitive ways to think of the Gauss/Divergence Theorem that make the result seem "obvious." These are not proofs (see later in this Appendix), but they contain the essence of the theorem.

First intuitive approach: Imagine that the interior of the solid region E is partitioned into lots of cells and that the divergence of each cell is the total outward flow from that cell. If two cells share a boundary surface then the outward flow from one (a positive divergence) is the inward flow into the other (a negative divergence). So adding the divergences for all of the cells (a triple sum in x, y and z), all of the inside divergences sum to 0, and the final result is just the sum of those divergences on the surface S of the solid region E: $\underset{x,y,z}{\sum\sum\sum} \text{div}\,\mathbf{F} = \underset{\text{surface}}{\sum\sum} \mathbf{F}\bullet\mathbf{n}$

Second intuitive approach: Partition the solid region E into thin slices parallel to the xy-plane (Fig. A1). Then each slice will contain a 2D region R with boundary S. Applying Green's Theorem to this slice we have $\iint_R \text{div}\,\mathbf{F}\;dA = \oint_C \mathbf{F}\bullet\mathbf{n}\;ds$. And by adding these results together

$$\sum_{\Delta z}\left(\iint_R \text{div}\,\mathbf{F}\;dA\right)dz = \sum_{\Delta z}\left(\oint_C \mathbf{F}\bullet\mathbf{n}\;ds\right)dz$$ we expect the Divergence Theorem

$$\iiint_E \text{div}\,\mathbf{F}\;dV = \iint_S \mathbf{F}\bullet\mathbf{n}\;dS\,.$$

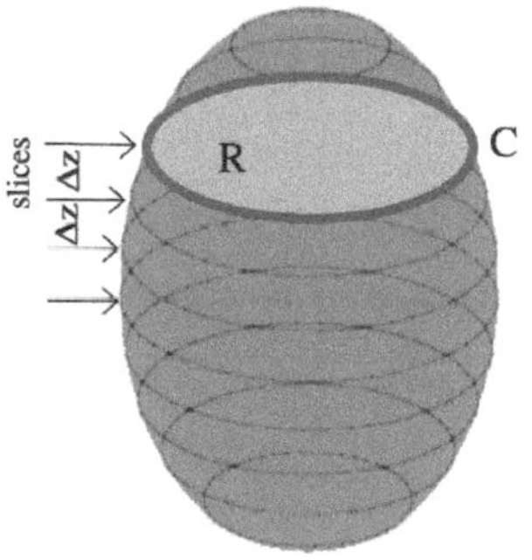

Fig. A1

Neither of these is a proof, but they might give you a better understanding of the "why" of the Divergence Theorem.

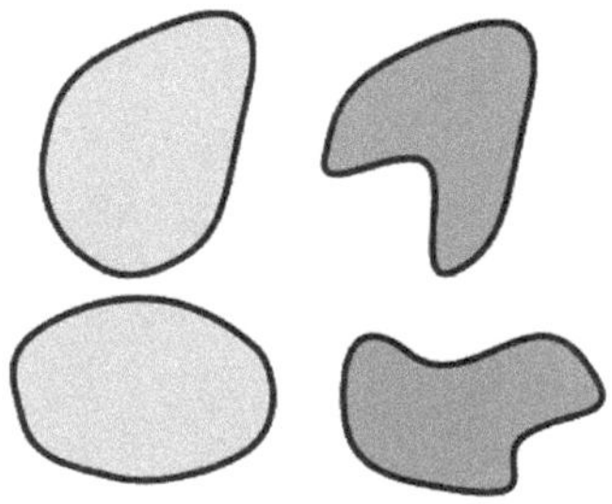

Fig. A2

Proof for a common special case: E is a convex region

A region E is called **convex** if a straight line connecting any two points in E lies in E. Fig. A2 shows some convex and non-convex 2D regions.

Assume that E is a convex region with a piecewise-smooth boundary S, that $\mathbf{F} = \langle M,N,P\rangle$ has continuous partial derivatives, and that $\mathbf{n} = \langle \mathbf{n_1}, \mathbf{n_2}, \mathbf{n_3}\rangle$ is the unit, outward pointing normal vector to S.

If E is convex, then the projection of E onto the xy, xz and yz-planes is a 2D convex region. Let D be the projection of E onto the xy-plane (Fig. A3). Then for each (x,y)

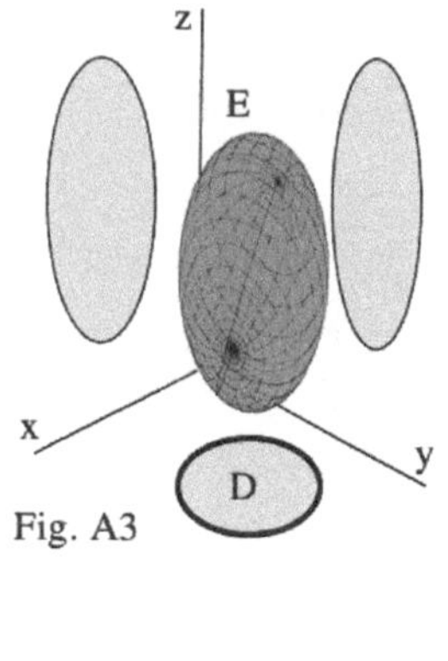

Fig. A3

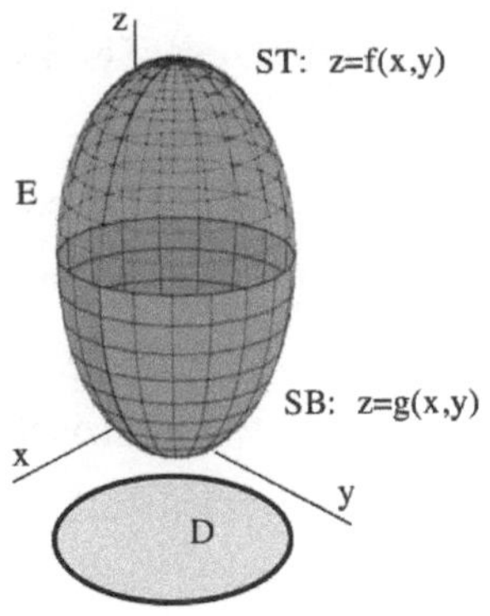

Fig. A4

in D the solid E is bounded by a top surface

ST = $\{(x,y,z): (x,y)$ is in D and $z = f(x,y)\}$ and a bottom surface

SB = $\{(x,y,z): (x,y)$ is in D and $z = g(x,y)\}$ (Fig. A4). If we can show that

$$\iiint_E \frac{\partial M}{\partial x}\, dV = \iint_S M\, dS = \iint_S \mathbf{F} \bullet \mathbf{n_1}\, dS,$$

$$\iiint_E \frac{\partial N}{\partial y}\, dV = \iint_S N\, dS = \iint_S \mathbf{F} \bullet \mathbf{n_2}\, dS, \text{ and}$$

$$\iiint_E \frac{\partial P}{\partial z}\, dV = \iint_S P\, dS = \iint_S \mathbf{F} \bullet \mathbf{n_3}\, dS \text{ then}$$

$$\iiint_E \operatorname{div} \mathbf{F}\, dV = \iiint_E \frac{\partial M}{\partial x} + \frac{\partial N}{\partial y} + \frac{\partial P}{\partial z}\, dV = \iint_S (M + N + P)\, dS = \iint_S \mathbf{F} \bullet \mathbf{n_1} + \mathbf{F} \bullet \mathbf{n_2} + \mathbf{F} \bullet \mathbf{n_3}\, dS$$

$$= \iint_S \mathbf{F} \bullet \mathbf{n}\, dS$$

which is the Divergence Theorem.

Working to show $\iiint_E \frac{\partial P}{\partial z}\, dV = \iint_S P\, dS$:

$\iiint_E \frac{\partial P}{\partial z}\, dV$: By the Fundamental Theorem of Calculus,

$$\iiint_E \frac{\partial P}{\partial z}\, dV = \iint_D \left(\int_{g(x,y)}^{f(x,y)} \frac{\partial P}{\partial z}\, dz \right) dx\, dy = \iint_D P(x,y,f(x,y)) - P(x,y,g(x,y))\, dx\, dy.$$

$\iint_S P\, dS$: The surface S consists of the two surfaces ST where z=f(x,y) and SB where z=g(x,y). On the

top surface ST, the normal **n** points up and $\mathbf{n} = \langle -f_x, -f_y, 1 \rangle$. On the bottom surface SB, the

normal vector **n** points down and $\mathbf{n} = \langle g_x, g_y, -1 \rangle$.

$$\iint_{ST} P\, dS = \iint_{ST} P(x,y,z)\, dS = \iint_D \langle 0,0,P(x,y,f(x,y)) \rangle \bullet \langle -f_x, -f_y, 1 \rangle\, dx\, dy = \iint_D P(x,y,f(x,y))\, dx\, dy .$$

$$\iint_{SB} P\, dS = \iint_{SB} P(x,y,z)\, dS = \iint_D \langle 0,0,P(x,y,g(x,y)) \rangle \bullet \langle g_x, g_y, -1 \rangle\, dx\, dy$$

$$= \iint_D -P(x,y,g(x,y))\, dx\, dy .$$

Putting these last two results together,

$$\iint_S P\, dS = \iint_{ST} P\, dS + \iint_{SB} P\, dS = \iint_D P(x,y,f(x,y))\, dx\, dy - \iint_D P(x,y,g(x,y))\, dx\, dy$$

$= \iint_D P(x,y,f(x,y)) - P(x,y,g(x,y))\, dx\, dy$ which is the same result we got for $\iiint_E \frac{\partial P}{\partial z}\, dV$.

If the ST and SB surfaces are connected by vertical walls SV, the result is still true since **n** has the form $\mathbf{n} = \langle a, b, 0\rangle$ so $\iint_{SV} P\ dS = \iint_{D} \langle 0,0,P(x,y,g(x,y))\rangle \bullet \langle a, b, 0\rangle\ dx\,dy = 0$.

Similarly $\iiint_{E} \frac{\partial M}{\partial x}\ dV = \iint_{S} M\ dS$ and $\iiint_{E} \frac{\partial N}{\partial y}\ dV = \iint_{S} N\ dS$, and then the Divergence Theorem is proven for convex solid regions.

Te proof for more general solid regions is more complicated and is not given here.

15.0 Odd Answers

1. A small eddy or whirlpool.
3. Easier from D to San Juan Island.
5. Easier from A to B.
7. Easier from E to F.

15.1 Odd Answers

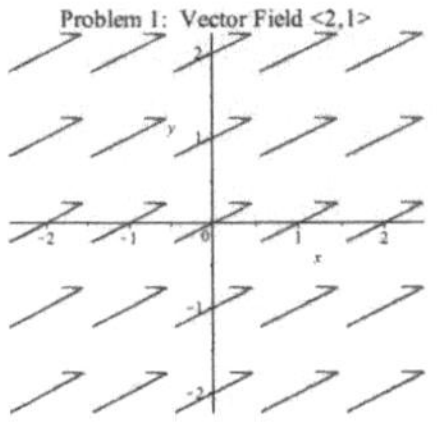

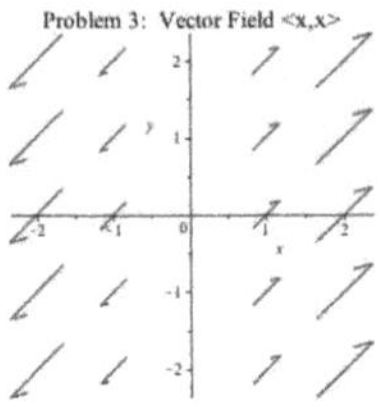

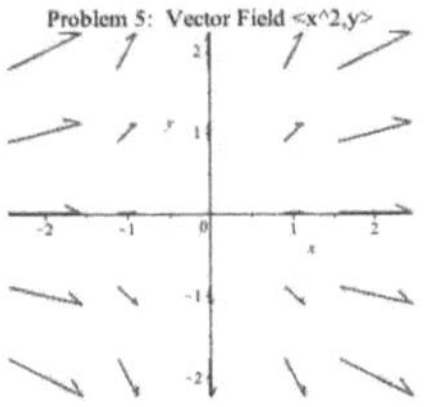

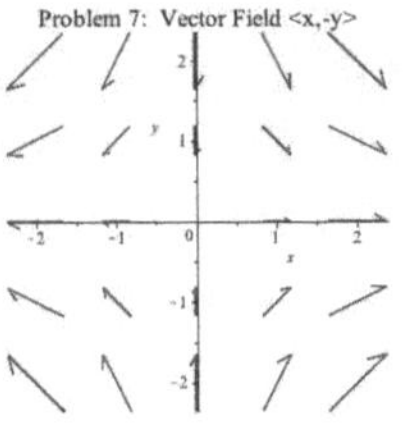

9. a-C, b-A, c-B
11. a-B, b-A, c-C
13. $\mathbf{F}(x,y) = \langle 2x, -2y \rangle$
15. $\mathbf{F}(x,y) = \langle y^2 - 2x \cdot y, 2x \cdot y - x^2 \rangle$
17. $\mathbf{F}(x,y,z) = \langle 3,2,-1 \rangle$
19. $\mathbf{F}(x,y,z) = \langle x,y,z \rangle / \sqrt{x^2 + y^2 + z^2}$
21. $\mathbf{F}(x,y,z) = \langle y \cdot z \cdot \cos(x \cdot y \cdot z), x \cdot z \cdot \cos(x \cdot y \cdot z), x \cdot y \cdot \cos(x \cdot y \cdot z) \rangle$

23.

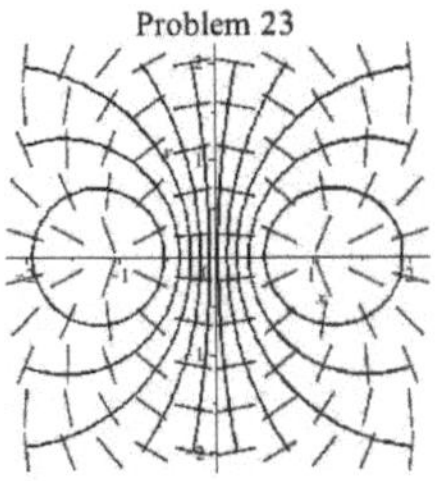

25.

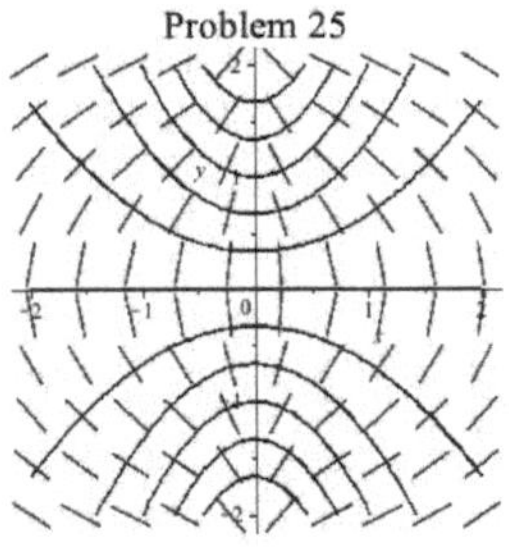

27.

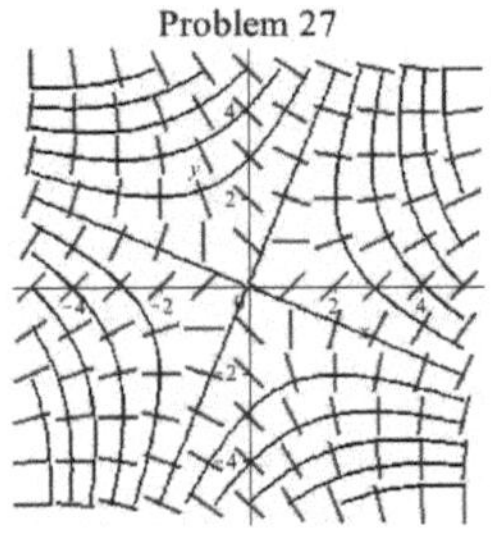

15.2 Odd Answers

1. At A div $\mathbf{F} > 0$, at B div $\mathbf{F} < 0$, at C div $\mathbf{F} = 0$.
3. div $\mathbf{F}$ = $2x + 2$. At A div $\mathbf{F} = 4$, at B div $\mathbf{F} = 6$ and at C div $\mathbf{F} = 0$.
5. div $\mathbf{F}$ = $5 + 2$. At A div $\mathbf{F} = 7$, at B div $\mathbf{F} = 7$ and at C div $\mathbf{F} = 7$.
7. See Fig. 12. More going out than coming in.
9. At A curl $\mathbf{F} > 0$, at B curl $\mathbf{F} < 0$, at C curl $\mathbf{F} = 0$.
11. curl $\mathbf{F}$ = $1 - 3$. At A curl $\mathbf{F} = -2$, at B curl $\mathbf{F} = -2$, at C curl $\mathbf{F} = -2$.
13. curl $\mathbf{F}$ = $1 - (-3)$. At A curl $\mathbf{F} = 4$, at B curl $\mathbf{F} = 4$, at C curl $\mathbf{F} = 4$.
15. See Fig. 13. Enough to rotate counterclockwise.
17. All of the partial derivatives are 0 so both the div and the curl are 0
19. No. The curl could be anything. See Fig. 14.

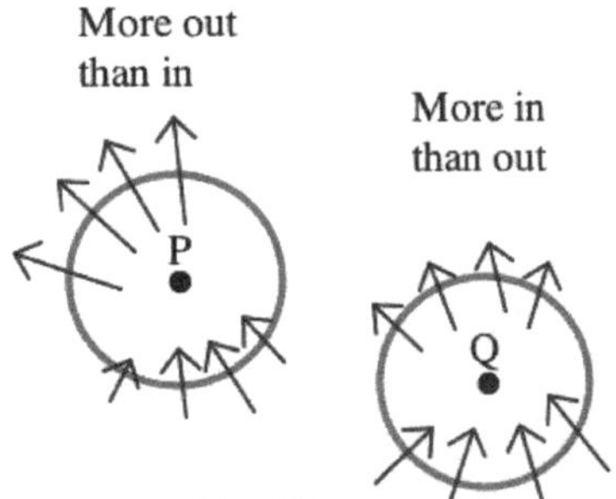

Fig. 12

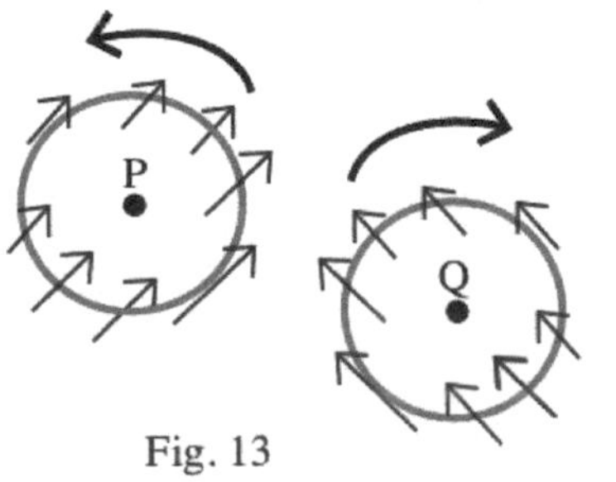

Fig. 13

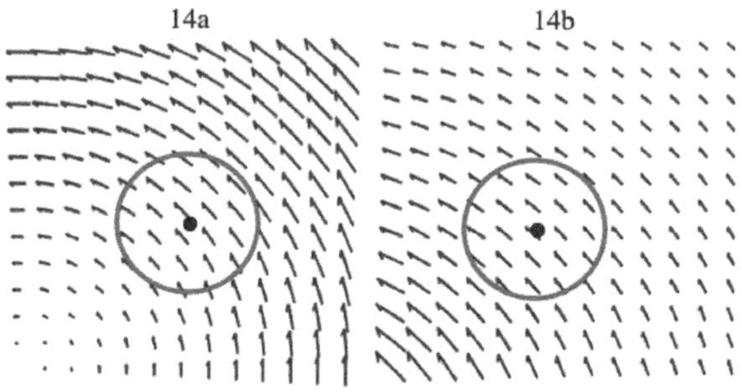

In 14a the curl is positive. In 14b the curl is negative.

21. curl $\mathbf{F} = -\frac{\partial M}{\partial y}$ might be any value.

15.3 Odd Answers

1. $\int_0^3 \left((2t+1)^2 - 4(3+t^2) + 11\right)\sqrt{(2)^2 + (2t)^2}\, dt = \int_0^3 (4t)\sqrt{4+4t^2}\; dt = \frac{8}{3}\left(t^2+1\right)^{3/2} \Big|_0^3 = \frac{8}{3}\left(10\sqrt{10}-1\right)$

3. $\int_{t=a}^{t=b} y(t)\cdot\sqrt{\left(\frac{dx}{dt}\right)^2 + \left(\frac{dz}{dt}\right)^2}\cdot dt$

4. $\int_{t=a}^{t=b} x(t)\cdot\sqrt{\left(\frac{dy}{dt}\right)^2 + \left(\frac{dz}{dt}\right)^2}\cdot dt$

5. $\text{mass} = \int_0^{\pi/2} (1 + 3\cos(t) + 6\sin(t))(3)\, dt = 27 + \frac{3}{2}\pi$

7. $\text{mass} = \int_0^2 (2+4t)(\sqrt{11})\, dt = 12\sqrt{11}$

9. $\int_0^2 (2t+9)(5)\, dt = 110$

11. $\int_0^3 (10t^2 + 10)(2t)\, dt = 495$

13. $\int_0^{\pi} (3 - \sin(t) + \cos(t))(1)\, dt = -2 + 3\pi$

15. Along A work > 0. Along B work < 0.

17. Along A flow > 0. Along B flow > 0.

19. $\text{work} = \int_0^3 (14 + 37t)\, dt = 417/2$

21. $\text{work} = \int_0^{\pi} (-\sin(t) + 2 + 3\cdot\cos(t))\, dt = 2\pi - 2$

23. $\text{flow} = \int_1^2 (3t^2 + 32t)\, dt = 55$

25. $\text{flux} = \int_0^{2\pi} (2\cdot\cos(t))(2\cdot\cos(t)) - (1 + 4\cdot\sin(t))(-\sin(t))\, dt = 8\pi$

27. All of them are possible for the flux.

For 29 to 34 consider the integral formula used for each calculation.

29. Area is unchanged.

31. Sign of work is changed.

33. Sign of circulation is changed.

15.4 Odd Answers

1. $\mathbf{F} = \langle 2x,\ 6y\rangle$

3. $\mathbf{F} = \langle 3\cos(3x+2y),\ 2\cos(3x+2y)\rangle$

5. $\mathbf{F} = \left\langle \frac{2}{2x+5y},\ \frac{5}{2x+5y} + e^y \right\rangle$

7. $f(x,y) = x^2 + y^2 \quad f(5,1) - f(1,2) = 21$

9. $M_y = 0,\ N_x = 1$ so F is not conservative. $\mathbf{r}(t) = \langle 3t,\ 2+4t\rangle$ (t from 0 to 1).

$F(r(t)) \bullet r'(t) = \langle 3t,\ 3t\rangle \bullet \langle 3,\ 4\rangle = 21t$ so $\text{work} = \int_{t=0}^{1} \mathbf{F}(\mathbf{r}(t)) \bullet \mathbf{r}'(t)\, dt = \int_{t=0}^{1} 21t\, dt = \frac{21}{2}$

11. $f(x,y,z) = xyz \quad f(4,2,1) - f(1,0,0) = 8$

13. A to B: 5   C to D: 2   E to F: –3

15. A to B: 6   C to D: 0   E to F: -2

17. $f(x,y) = x^3 + 4x + 6y$

19. $f(x,y) = \sin(xy) + x^3 y$

21. $f(x,y) = x \cdot \sin(y)$

23. $N_z \neq P_y$

15.6 Odd Answers

1. $M = x^2y$, $N = 3y$, $\frac{\partial N}{\partial x} - \frac{\partial M}{\partial y} = -x^2$ so $\iint_R -x^2 \, dA = \int_0^1 \int_0^2 -x^2 \, dx \, dy = -\frac{8}{3}$

3. $M = 3xy$, $N = 2x^2$, $\frac{\partial N}{\partial x} - \frac{\partial M}{\partial y} = 4x - 3x$ so $\iint_R x \, dA = \int_0^1 \int_0^{2x} x \, dx \, dy = \frac{2}{3}$

5. $M = ax$, $N = by$, $\frac{\partial N}{\partial x} - \frac{\partial M}{\partial y} = 0$ so $\iint_R 0 \, dA = 0$

7. $\text{circulation} = \iint_R \left(\frac{\partial N}{\partial x} - \frac{\partial M}{\partial y} \right) dA = \iint_R (-1-2) \, dA = \int_0^2 \int_0^2 -3 \, dx \, dy = -12$

$\text{flux} = \iint_R \left(\frac{\partial M}{\partial x} + \frac{\partial N}{\partial y} \right) dA = \iint_R (1+1) \, dA = \int_0^2 \int_0^2 2 \, dx \, dy = 8$

9. $\text{circulation} = \iint_R \left(\frac{\partial N}{\partial x} - \frac{\partial M}{\partial y} \right) dA = \iint_R (2x-2y) \, dA = \int_0^2 \int_0^x 2x - 2y \, dy \, dx = \frac{8}{3}$

$\text{flux} = \iint_R \left(\frac{\partial M}{\partial x} + \frac{\partial N}{\partial y} \right) dA = \iint_R (2x+2y) \, dA = \int_0^2 \int_0^x 2x + 2y \, dy \, dx = 8$

11. C_1 is from -2 to 2: $\mathbf{r}(t) = \langle t, t^2 \rangle$ so $\frac{1}{2}\int_{-2}^{2} (t)(2t) - (t^2)(1) \, dt = \frac{1}{2}\int_{-2}^{2} t^2 \, dt = \frac{8}{3}$

C_2 is from 2 to -2: $\mathbf{r}(t) = \langle t, 8 - t^2 \rangle$ so $\frac{1}{2}\int_{2}^{-2} (t)(-2t) - (8 - t^2)(1) \, dt = \frac{1}{2}\int_{-2}^{2} (-t^2 - 8) \, dt = \frac{56}{3}$

So the total area is $\frac{64}{3}$.

Note: $\int_{-2}^{2} ((8 - x^2) - (x^2) \, dx$ is much easier. Green's Theorem does not make everything easier.

13. Call D the boundary of R so $\int_D \mathbf{F} \bullet d\mathbf{r} = \int_{C_1} \mathbf{F} \bullet d\mathbf{r} - \int_{C_2} \mathbf{F} \bullet d\mathbf{r} = \int_{C_1} \mathbf{F} \bullet d\mathbf{r} - 20$ ·

But $\int_D \mathbf{F} \bullet d\mathbf{r} - \iint_R \left(\frac{\partial N}{\partial x} - \frac{\partial M}{\partial y} \right) dA = \iint_R 5 \, dA = 500$ so $\int_{C_1} \mathbf{F} \bullet d\mathbf{r} = 520$·

15. Call D the boundary of R so $\int_D \mathbf{F} \bullet d\mathbf{r} = \int_{C_1} \mathbf{F} \bullet d\mathbf{r} - \int_{C_2} \mathbf{F} \bullet d\mathbf{r} - \int_{C_3} \mathbf{F} \bullet d\mathbf{r} = \int_{C_1} \mathbf{F} \bullet d\mathbf{r} - 3\pi - 4\pi$ ∴

Area of R = $= 42 - 2\pi$.

But $\int_D \mathbf{F}\bullet d\mathbf{r} = \iint_R \left(\frac{\partial N}{\partial x}-\frac{\partial M}{\partial y}\right) dA = \iint_R 9\ dA = 9(42-2\pi) = 378-18\pi$ so $\int_{C_1} \mathbf{F}\bullet d\mathbf{r} = 378-11\pi$.

17. All of the partial derivatives are 0 so the double integrals for circulation and flux are 0.

19. $\text{circulation} = \oint_C M\,dx + N\,dy = \iint_R \left(\frac{\partial N}{\partial x}-\frac{\partial M}{\partial y}\right) dA = \iint_R (c-b)\ dA = (c-b)\cdot(\text{area of R})$

15.7 Odd Answers

1. $\text{div}\,\mathbf{F} = 3xz^2 + 2xy + xz$, $\text{curl}\,\mathbf{F} = \langle -xy, -z^3, -x^2 + yz\rangle$
3. $\text{div}\,\mathbf{F} = e^z + z\cdot e^y$, $\text{curl}\,\mathbf{F} = \langle e^x - e^y,\ xe^z - ye^x,\ 0\rangle$
5. $\text{div}\,\mathbf{F} = 3$, $\text{curl}\,\mathbf{F} = \langle -1,\ -1,\ -1\rangle$

In problems 7 to 10, X = not meaningful.

7. div f =X , div F =scalar, div (curl F)=scalar , gradient (curl F)=X, curl F=vector
9. gradient f =vector, curl (curl F)=vector , curl (div (gradient f))=X
11. Not conservative.
13. Not conservative.
15. Conservative. $f(x,y,z) = x\cdot\sin(yz) + 2z$
17. Your location = (5, 10, 5)= (1, 2, 3) + (**2**)(2, 4, 1) so you have moved in the direction of the curl. Looking back, you will see a Counterclockwise rotation.
19. Your location = (4, 4, 3)= (6, 3, 2) + (**–1**)(2, –1, 1) so you have moved in the opposite direction of the curl. Looking toward (6, 3, 2), you will see a Clockwise rotation.

15.8 Odd Answers

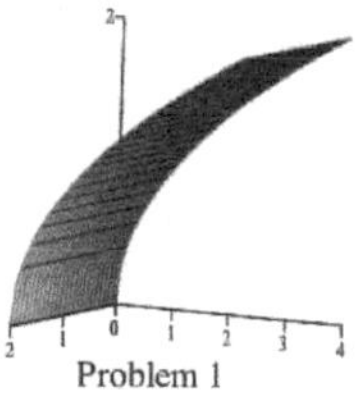
Problem 1

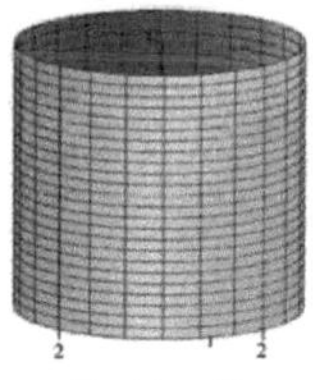
Problem 3

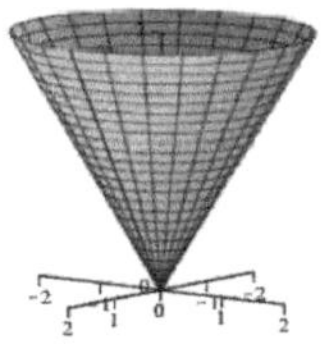
Problem 5

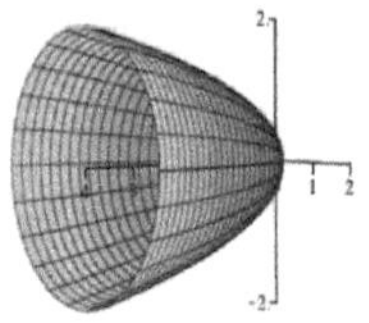
Problem 7

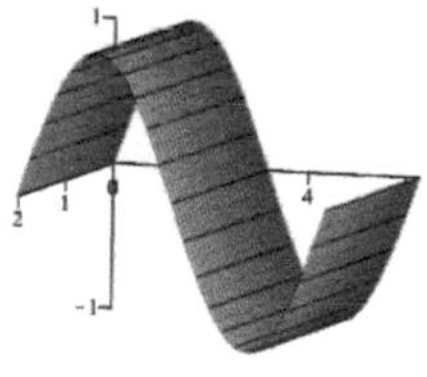
Problem 9

11. $\mathbf{r}(u,v) = \langle u,\ v,\ \sqrt{v}\rangle$, $\mathbf{r}_u = \langle 1,\ 0,\ 0\rangle$, $\mathbf{r}_v = \left\langle 0,\ 1,\ \frac{1}{2\sqrt{v}}\right\rangle$, $\mathbf{r}_u \mathbf{x} \mathbf{r}_v = \left\langle 0,\ \frac{1}{2\sqrt{v}},\ 1\right\rangle$

Surface area = $\iint_R |\mathbf{r}_u \mathbf{x} \mathbf{r}_v|\ dA = \int_{u=0}^{2}\int_{v=0}^{4} \sqrt{\frac{1}{4v}+1}\ dv\ du$.

13. $\mathbf{r}(u,v) = \langle 2\cdot\cos(u),\ 2\cdot\sin(u),\ v\rangle$, $\mathbf{r}_u = \langle -2\cdot\sin(u),\ 2\cdot\cos(u),\ 0\rangle$, $\mathbf{r}_v = \langle 0,\ 0,\ 1\rangle$,

$\mathbf{r}_u \mathbf{x} \mathbf{r}_v = \langle 2\cdot\cos(u),\ 2\cdot\sin(u),\ 0\rangle$. Surface area = $\iint_R |\mathbf{r}_u \mathbf{x} \mathbf{r}_v|\ dA = \int_{u=0}^{2\pi}\int_{v=0}^{3} 2\ dv\ du$.

15. $r(u,v) = \langle u\cdot\cos(v),\ u\cdot\sin(v),\ u\rangle$, $\mathbf{r}_u = \langle \cos(v),\ \sin(v),\ 1\rangle$, $\mathbf{r}_v = \langle -u\cdot\sin(v),\ u\cdot\cos(v),\ 0\rangle$

$r_u x r_v = \langle -u\cos(v),\ u\cdot\sin(v),\ u\rangle$. Surface area = $\iint_R |\mathbf{r}_u \mathbf{x} \mathbf{r}_v|\ dA = \int_{u=0}^{2} \int_{v=0}^{2\pi} u\sqrt{2}\ dv\ du$.

17. Surface area = $\int_{u=0}^{4} \int_{v=0}^{2\pi} \sqrt{u + \frac{1}{4}}\ dv\ du$.

19. Surface area = $\int_{u=0}^{2} \int_{v=0}^{2\pi} \sqrt{1+\cos^2(v)}\ dv\ du$.

15.9 Odd Answers

1. $r_u = \langle 1,2,3\rangle$, $r_v = \langle 3,-1,1\rangle$, $\mathbf{r}_u\mathbf{x}\mathbf{r}_v = \langle 5,8,-7\rangle$ and $|\mathbf{r}_u\mathbf{x}\mathbf{r}_v| = \sqrt{138}$. $f(\mathbf{r}(1,2)) = f(7,0,5) = 12$,

   $\Delta A = \Delta u\cdot\Delta v = 0.03$ so $f(\mathbf{r}(u,v))\cdot|\mathbf{r}_u\mathbf{x}\mathbf{r}_v|\ \Delta A = (12)(\sqrt{138})(0.03) \approx 4.23$.

3. When u=2 and v=1, then $r_u = \langle 2u,\ 3,\ 0\rangle = \langle 4,3,0\rangle$, $r_v = \langle 0,\ 1,\ 2v\rangle = \langle 0,1,2\rangle$,

   $\mathbf{r}_u\mathbf{x}\mathbf{r}_v = \langle 6v,\ -4uv,\ 2u\rangle = \langle 6,-8,4\rangle$ and $|r_u x r_v| = \sqrt{116}$. $f(\mathbf{r}(2,1)) = f(4,7,1) = 56$,

   $\Delta A = \Delta u\cdot\Delta v = 0.06$ so $f(\mathbf{r}(u,v))\cdot|\mathbf{r}_u\mathbf{x}\mathbf{r}_v|\ \Delta A = (56)(\sqrt{116})(0.06) \approx 36.19$.

5. S can be parameterized by $r(u,v) = \langle u,v,4\rangle$ with $0 \le u \le 3$ and $0 \le v \le 2$. Then $r_u = \langle 1,0,0\rangle$,

   $r_v = \langle 0,1,0\rangle$, $\mathbf{r}_u\mathbf{x}\mathbf{r}_v = \langle 0,0,1\rangle$ and $|\mathbf{r}_u\mathbf{x}\mathbf{r}_v| = 1$. $f(r(u,v)) = u^2 + 4v + 4$.

   $\iint_S f(x,y,z)\ dS = \iint_R f(\mathbf{r}(u,v))\cdot|\mathbf{r}_u\mathbf{x}\mathbf{r}_v|\ dA = \int_{v=0}^{2}\int_{u=0}^{3} u^2 + 4v + 4\ du\ dv = \int_{v=0}^{2} 9 + 12v + 12\ dv = 66$.

7. S can be parameterized by $r(u,v) = \langle u,\ v,\ 4-u-v\rangle$ with $0 \le u \le 2$ and $0 \le v \le 2$. Then $r_u = \langle 1,\ 0,\ -1\rangle$,

   $r_v = \langle 0,\ 1,\ -1\rangle$,, $\mathbf{r}_u\mathbf{x}\mathbf{r}_v = \langle 1,\ 1,\ 1\rangle$ and $|r_u x r_v| = \sqrt{3}$. $f(r(u,v)) = v^2$.

   $\iint_S f(x,y,z)\ dS = \iint_R f(\mathbf{r}(u,v))\cdot|\mathbf{r}_u\mathbf{x}\mathbf{r}_v|\ dA = \int_0^2\int_0^2 v^2\sqrt{3}\ dv\ du = \int_0^2 \frac{8}{3}\sqrt{3}\ du = \frac{16}{3}\sqrt{3}$.

9. S can be parameterized by $r(u,v) = \langle u,v,4\rangle$ with $0 \le u \le 3$ and $0 \le v \le 2$. Then $r_u = \langle 1,0,0\rangle$,
   $r_v = \langle 0,1,0\rangle$, and $\mathbf{r}_u\mathbf{x}\mathbf{r}_v = \langle 0,0,1\rangle$. $\mathbf{F}(\mathbf{r}(u,v)) = \langle u,\ 2v,\ 3\cdot 4\rangle$.

   Flux = $\iint_S \mathbf{F}\cdot\mathbf{n}\ dS = \iint_{\mathbf{R}} \mathbf{F}\bullet(\mathbf{r}_u\mathbf{x}\mathbf{r}_v)\ dA = \iint_{\mathbf{R}} \langle u,\ 2v,\ 3\cdot 4\rangle\bullet\langle 0,\ 0,\ 1\rangle\ dA = \int_0^2\int_0^3 12\ du\ dv = 72$.
   The flux units are (liters/second)(meter^2).

11. S can be parameterized by $\mathbf{r}(u,v) = \langle 2\cdot\cos(u),\ \sin(u),\ v\rangle$ with $\mathbf{0} \le u \le 2\pi$ and $0 \le v \le 3$. Then

   $\mathbf{r}_u = \langle -2\cdot\sin(u),\ \cos(u),\ 0\rangle$, $\mathbf{r}_v = \langle 0,\ 0,\ 1\rangle$ and $\mathbf{r}_u\mathbf{x}\mathbf{r}_v = \langle \cos(u),\ 2\cdot\sin(u),\ 0\rangle$. $\mathbf{F}(\mathbf{r}(u,v)) = \langle 2\cdot\cos(u),\ \sin(u),\ v\rangle$.

   Flux = $\iint_S \mathbf{F}\cdot\mathbf{n}\ dS = \iint_{\mathbf{R}} \mathbf{F}\bullet(\mathbf{r}_u\mathbf{x}\mathbf{r}_v)\ dA = \iint_{\mathbf{R}} \langle 2\cdot\cos(u),\ \sin(u),\ v\rangle\bullet\langle \cos(u),\ 2\cdot\sin(u),\ 0\rangle\ dA$

   $= \int_0^3\int_0^{2\pi} 2\ du\ dv = 12\pi$.

13. S can be parameterized by $\mathbf{r}(u,v) = \langle v\cdot\cos(u),\ v\cdot\sin(u),\ A^2 - v^2\rangle$ with $0 \le u \le 2\pi$ and $0 \le v \le A$. Then

$\mathbf{r}_u = \langle -v\cdot\sin(u),\ v\cdot\cos(u),\ 0\rangle$, $\mathbf{r}_v = \langle\cos(u),\ \sin(u),\ -2Cv\rangle$ and $\mathbf{r}_u x \mathbf{r}_v = \langle -2v^2C\cdot\cos(u),\ -2v^2C\cdot\sin(u),\ -v\rangle$.

$\mathbf{F}(\mathbf{r}(u,v)) = \langle 0,\ 0,\ K\rangle$.

$$\text{Flux} = \iint_S \mathbf{F}\cdot\mathbf{n}\ dS = \iint_{\mathbf{R}} \mathbf{F}\bullet(\mathbf{r}_u x\mathbf{r}_v)\ dA = \iint_{\mathbf{R}} \langle 0,0,K\rangle\bullet\langle -2v^2C\cdot\cos(u),\ -2v^2C\cdot\sin(u),\ -v\rangle\ dA$$

$$= \int_0^A \int_0^{2\pi} -Kv\ du\ dv = -KA^2\pi .$$

This is the same flux as in Problem 12. In fact, the flux is the same even if C=0 so that S is just the disk $x^2 + y^2 \le A^2$ in the xy-plane.

15.10 Odd Answers

1. curl $\mathbf{F} = \nabla x\mathbf{F} = \langle 0,\ 0,\ 1\rangle$, $R = \{(x,y):\ x^2 + 4y^2 \le 4\}$, area of R = $\pi(1)(2) = 2\pi$ and $\mathbf{n} = \langle 0,\ 0,\ 1\rangle$.

$$\text{circulation} = \oint_C \mathbf{F}\bullet\mathbf{T}\ ds = \iint_R (\nabla x\mathbf{F})\bullet\mathbf{n}\ dA = \iint_R \langle 0,0,1\rangle\bullet\langle 0,0,1\rangle\ dS = \text{area of R} = 2\pi .$$

3. curl $\mathbf{F} = \nabla x\mathbf{F} = \langle x,\ 1-y,\ 0\rangle$, $\mathbf{n} = \langle 2,\ 2,\ 2\rangle$, $R = \{(x,y):\ 0 \le x \le 3,\ 0 \le y \le 3 - x\}$

$$\text{circulation} = \oint_C \mathbf{F}\bullet\mathbf{T}\ ds = \iint_R (\nabla x\mathbf{F})\bullet\mathbf{n}\ dA = \iint_R \langle x,\ 1-y,\ 0\rangle\bullet\langle 2,2,2\rangle\ dS$$

$$= \int_0^3 \int_0^{3-x} 2x + 2 - 2y\ dy\ dx = 9 .$$

5. curl $\mathbf{F} = \nabla x\mathbf{F} = \langle 0,\ 0,\ 2\rangle$, and $\mathbf{n} = \langle 1,\ 1,\ 1\rangle$.

$$\text{circulation} = \oint_C \mathbf{F}\bullet\mathbf{T}\ ds = \iint_R (\nabla x\mathbf{F})\bullet\mathbf{n}\ dA = \iint_R \langle 0,0,2\rangle\bullet\langle 1,1,1\rangle\ dA = 2\cdot\iint_R 1\ dA$$

= 2(area of R) where R is the area enclosed by C.

7. curl $\mathbf{F} = \nabla x\mathbf{F} = \langle 0,\ 0,\ 3-2y\rangle$ and $\mathbf{n} = \langle 0,\ 0,\ 1\rangle$.

$$\text{circulation} = \oint_C \mathbf{F}\bullet\mathbf{T}\ ds = \iint_R (\nabla x\mathbf{F})\bullet\mathbf{n}\ dA = \iint_R \langle 0,0,3-2y\rangle\bullet\langle 0,0,1\rangle\ dA$$

$$= \int_0^2 \int_0^{2\pi} (3 - 2r\cdot\sin(\theta))\cdot r\ d\theta\, dr = \int_0^2 6\pi r\ dr = 12\pi .$$

9. The boundary of S is the circle C parameterized by $\mathbf{r}(t) = \langle 4\cdot\cos(t),\ 4\cdot\sin(t),\ 0\rangle$.

$$\iint_S \text{curl } \mathbf{F}\bullet d\mathbf{S} = \int_C \mathbf{F}\bullet d\mathbf{r} = \int_{t=0}^{2\pi} \langle 4\cdot\sin(t),\ -4\cdot\cos(t),\ 0\rangle\bullet\langle -4\cdot\sin(t),\ 4\cdot\cos(t),\ 0\rangle\ dt$$

$$= \int_{t=0}^{2\pi} -16\sin^2(t) - 16\cos^2(t)\ dt = -32\pi .$$

11. The solid torus is a closed surface so it has no boundary curve so $\iint_S \text{curl } \mathbf{F}\bullet d\mathbf{S} = 0$.

13. $\nabla \mathbf{xF} = \langle 0,0,0 \rangle$ so F is a conservative field.

15. $\nabla \mathbf{xF} = \langle cx - bx,\ ay - cy,\ bx - az \rangle$ so F is not a conservative field.

17. $\nabla \mathbf{xF} = \langle 0,0,0 \rangle$ so F is a conservative field.

19. Quick check: $\nabla \mathbf{xF} = \langle 2, 2, 2 \rangle$.

The top and bottom boundaries of S can be parameterized by $\mathbf{r}_1 = \langle 4\cdot\cos(t),\ 4\cdot\sin(t),\ 0 \rangle$ and $\mathbf{r}_2 = \langle 4\cdot\cos(t),\ 4\cdot\sin(t),\ 3 \rangle$ for $0 \le t \le 2\pi$ (both going counterclockwise).

$$\int_{C_1} \mathbf{F} \bullet \mathbf{r}_1' \ dt = \int_0^{2\pi} \langle 0 - 4\cdot\sin(t),\ 4\cdot\cos(t) - 0,\ 4\cdot\sin(t) - 4\cdot\cos(t) \rangle \bullet \langle -4\cdot\sin(t),\ 4\cdot\cos(t),\ 0 \rangle \ dt$$

$$= \int_0^{2\pi} 16\cdot\sin^2(t) + 16\cdot\cos^2(t) \ dt = 32\pi .$$

$$\int_{C_2} \mathbf{F} \bullet \mathbf{r}_2' \ dt = \int_0^{2\pi} \langle 3 - 4\cdot\sin(t),\ 4\cdot\cos(t) - 3,\ 4\cdot\sin(t) - 4\cdot\cos(t) \rangle \bullet \langle -4\cdot\sin(t),\ 4\cdot\cos(t),\ 0 \rangle \ dt$$

$$= \int_0^{2\pi} -12\cdot\sin(t) + 16\cdot\sin^2(t) + 16\cdot\cos^2(t) - 12\cdot\cos(t) \ dt = 32\pi .$$

Finally, $\iint_S \text{curl } \mathbf{F} \bullet d\mathbf{S} = \int_{C_1} F \bullet \mathbf{r}_1' \ dt - \int_{C_2} F \bullet \mathbf{r}_1' \ dt = 0$.

21. Quick check $\nabla \mathbf{xF} = \langle 0, 0, 0 \rangle$ so $\iint_S \text{curl } \mathbf{F} \bullet d\mathbf{S} = \iint_S \langle 0, 0, 0 \rangle \bullet d\mathbf{S} = 0$.

23. Quick check $\nabla \mathbf{xF} = \langle -1, -1, 1 \rangle$.

The boundary curve of S can be parameterized by $\mathbf{r} = \langle 2\cdot\cos(t),\ 0,\ -5\cdot\sin(t) \rangle$.
(goes counterclockwise looking back along the y-axis).

$$\iint_S \text{curl } \mathbf{F} \bullet d\mathbf{S} = \int_C \mathbf{F} \bullet d\mathbf{r} = \int_{t=0}^{2\pi} \langle 0,\ -5\cdot\sin(t),\ 2\cdot\cos(t) \rangle \bullet \langle -2\cdot\sin(t),\ 0,\ -5\cdot\cos(t) \rangle \ dt$$

$$= \int_{t=0}^{2\pi} -10\cdot\cos^2(t) \ dt = -10\pi .$$

25. Quick check $\nabla \mathbf{xF} = \langle -1, 0, 1 \rangle$.

When z=0, $x^2 + y^2 = 25$ is a circle of radius 5: $\mathbf{r}_1 = \langle 5\cdot\cos(t),\ 5\cdot\sin(t),\ 0 \rangle$.

$$\int_{C_1} \mathbf{F} \bullet \mathbf{r}_1' \ dt = \int_0^{2\pi} \langle 5\cdot\sin(t),\ 0,\ 5\cdot\cos(t) \rangle \bullet \langle -5\cdot\sin(\theta),\ 5\cdot\cos(t),\ 0 \rangle \ dt = \int_0^{2\pi} -25\cdot\sin^2(t) \ dt = -25\pi .$$

When z=16, $x^2 + y^2 = 9$ is a circle of radius : $\mathbf{r}_2 = \langle 3\cdot\cos(t),\ 3\cdot\sin(t),\ 16 \rangle$.

$$\int_{C_2} \mathbf{F} \bullet \mathbf{r}_2' \ dt = \int_0^{2\pi} \langle 3\cdot\sin(t),\ 16,\ 3\cdot\cos(t) \rangle \bullet \langle -3\cdot\sin(t),\ 3\cdot\cos(t),\ 0 \rangle \ dt$$

$= \int_0^{2\pi} -9\cdot\sin^2(t) + 48\cdot\cos(t) \ dt = -9\pi$. So $\iint_S \text{curl } \mathbf{F} \bullet d\mathbf{S} = \int_{C_1} F \bullet \mathbf{r}_1' \ dt - \int_{C_2} F \bullet \mathbf{r}_1' \ dt = -16\pi$.

15.11 Odd Answers

1. Distance to the origin of P1 is $\sqrt{14} \approx 3.7$, of P2 is $\sqrt{6} \approx 2.4$, of P3 is $\sqrt{8} \approx 2.8$, of P4 is $\sqrt{2} \approx 1.4$, of P5 is $\sqrt{12} \approx 3.5$, and of P6 is $\sqrt{19} \approx 4.4$.

   (a) Only P4: net flux = 6 m^3/s. (b) P1, P2, P3, P4 and P5: net flux=(7)+(–3)+(2)+(6)+(–4)= 8 m^3/s.

3. Dist to center of P1 is $\sqrt{10} \approx 3.2$, of P2 is $\sqrt{18} \approx 4.2$, of P3 is 0 , of P4 is $\sqrt{6} \approx 2.4$, of P5 is 2, and of P6 is $\sqrt{11} \approx 3.3$.

   (a) Only P3: net flux= 2 m^3/s. (b) P3, P4, P5: net flux=(2)+(–6)+(–4)= –8 m^3/s.

5. Outward flux sphere 1: 5 m^3/s. outward flux of sphere 2: 14 m^3/s. Net flux = (14)–(5)= 9 m^3/s.

7. div **F** = 3 so $\text{flux} = \iiint_E \text{div } \mathbf{F} \, dV = 3(\text{volume of E}) = 3\left(\frac{4}{3}\pi(2)^3\right) = 32\pi$.

9. div **F** = 2 so $\text{flux} = \iiint_E \text{div } \mathbf{F} \, dV = 2(\text{volume of E}) = 2\cdot\left(\frac{4}{3}\pi(3)^3\right) = 72\pi$.

11. div **F** = 4 so $\text{flux} = \iiint_E \text{div } \mathbf{F} \, dV = (4)\iiint_E 1 \, dV = 4\cdot(\text{volume of E}) = (4)\cdot(2)(3(4) = 96$.

13. div **F** = 2x + 2y + 2z so $\text{flux} = \iiint_E \text{div } \mathbf{F} \, dV = \int_0^4 \int_0^3 \int_0^2 (2x+2y+2z)\, dx\, dy\, dz = 216$.

15. $0 = \iiint_D \text{div } \mathbf{F} \, dV = \iint_S \mathbf{F}\bullet\mathbf{n} \, dS = \iint_{S_1} \mathbf{F}\bullet\mathbf{n} \, dS - \iint_{S_2} \mathbf{F}\bullet\mathbf{n} \, dS$ so the outward flux from the outer sphere equals the outward flux from the inner sphere.

17. Flux in each case is zero.

19. (a) div **F** = 0 (b) div **F** is negative (c) div **F** is negative

21. (a) div **F** is positive (b) div **F** is positive (c) div **F** is positive